Bacillus Probiotics for Sustainable Aquaculture

This book provides a comprehensive examination of the role of *Bacillus* bacteria in aquaculture, particularly focusing on finfish and shellfish. It begins with foundational chapters on the anatomy and physiology of these aquatic species, their immune systems, and the biology of *Bacillus* bacteria. The book discusses the specific interactions between *Bacillus* and the gastrointestinal tracts of fish and shellfish, examining the production and effects of *Bacillus*-derived exoenzymes and bacteriocins and their role in enhancing aquatic animal health and water quality. Additionally, the book covers the role of *Bacillus* in bioremediation, particularly in improving water quality, which is a significant concern in aquaculture. The latter sections delve into the application of *Bacillus* as probiotics and paraprobiotics in fish and shellfish culture, their integration in feed biotechnology, and a critical evaluation of their safety. The book concludes with a critical discussion on the safety and regulatory aspects of using *Bacillus* in aquaculture.

This book is intended for professionals and practitioners in the aquaculture industry and researchers, academicians, and students in the fields of microbiology, marine biology, aquaculture, and veterinary medicine.

Bacillus Probiotics for Sustainable Aquaculture

Edited by
Mehdi Soltani
Preetham Elumalai
Koushik Ghosh
Einar Ringø

CRC Press is an imprint of the
Taylor & Francis Group, an **informa** business

Cover credit: http://images.tandf.co.uk/common/jackets/agentjpg/978103282/9781032822730.jpg

First edition published 2025
by CRC Press
2385 NW Executive Center Drive, Suite 320, Boca Raton FL 33431

and by CRC Press
4 Park Square, Milton Park, Abingdon, Oxon, OX14 4RN

CRC Press is an imprint of Taylor & Francis Group, LLC

ISBN: 978-1-032-82273-0 (hbk)
ISBN: 978-1-032-82274-7 (pbk)
ISBN: 978-1-003-50381-1 (ebk)

DOI: 10.1201/9781003503811

Typeset in Times
by SPi Technologies India Pvt Ltd (Straive)

Contents

Editors

Mehdi Soltani, PhD, is a distinguished professor at the University of Tehran and an adjunct professor at Murdoch University, Australia. He earned a DVM from the University of Tehran and PhD in aquatic animal health from the University of Tasmania, Australia. Professor Soltani has an international reputation for research on aquatic animal health, with 290 published scientific papers, collaborations with researchers throughout the world, and editorship of scientific journals in fisheries and veterinary science. He has chaired government advisory committees in fisheries and aquaculture and has also worked closely with the aquaculture industry. He has developed and patented a number of fish vaccines, which are registered throughout the Middle East. He has also taught numerous undergraduate courses and supervised many higher degree students. His research interests include vaccine development for fish pathogens, immunopathogenesis of infectious agents in fish/shellfish, and development of alternative therapies such as immunostimulants, probiotics, and phytobiotics for disease control in farmed fish and shellfish.

Preetham Elumalai, PhD, is an associate professor at the Department of Marine Biology, Cochin University of Science and Technology, Kochi, Kerala, India. He earned a master's degree from the University of Madras and a PhD in biochemistry and molecular immunology from the Institute for Immunology, University of Regensburg, Germany. His research includes bioassay-guided identification of novel marine compounds, unveiling fish lectins in innate immune defense, aquatic vaccine development, evaluation of cost-effective feed additives and nutrigenomics, and effects of environmental pollutants on marine ecosystems. He has been a partner in numerous EU-, Indian-, and UK-funded projects (e.g., IVVN, BactiVac). He has written more than 70 peer-reviewed articles and has two patents in his name apart from editing five books and presenting his work at more than 60 national and international conferences. He has been awarded the prestigious INSA fellowship (2018); MASTS, Fellowship (2019); IVVN award, UK (2020); FRSB award (2021); and BactiVac award, UK (2022).

Koushik Ghosh is a professor and former head in the Department of Zoology, the University of Burdwan. He also serves as Director (Addl. Charge) of the Rural Technology Centre of the university. He received his postgraduate degree from the University of Burdwan (1997) being awarded with the University Gold Medal for obtaining 1st class 1st position in zoology. Afterwards, he received his PhD degree from the Visva Bharati University (2003). He is currently leading the Aquaculture Laboratory and conducting research fish nutrition and gut microbiota. Professor Ghosh has got extensive experience and expertise in the areas of his research, e.g., nutritional interference caused by the plant secondary metabolites, utilization of non-conventional plant resources for low-cost diet formulation, and applications of functional feed additives to promote growth and disease resistance in fish. He has published 98 research articles in reputed and peer-reviewed indexed journals. Professor Ghosh is a reviewer for numerous international journals. He is a member of the Editorial Advisory Board of the journal *Acta Ichthyologica et Piscatoria*. He has handled 6 research projects and 10 students thus far have been awarded with PhD degree under his able mentorship.

Einar Ringø is a professor emeritus at Norwegian College of Fishery Science, Faculty of Biosciences, Fisheries and Economics, UiT The Arctic University of Norway. Ringø has scientific experience in microbial ecology, lipid nutrition, and electron microscopy in various species of marine fish larvae and fry and freshwater fish species since 1982. During his academic career, Ringø has published 262 scientific publications.

Scientific Index 2023: UiT #4 (395), Norway #58 (6.736), Europe #4.539 (396.499), World # 13.814 (1.338.642)

Research.com. 2022: Microbiology: H-index 60, World ranking 1.094, National ranking 4

H-index - Google = 85; H-index – ResearchGate = 82; H-index - Web of Science = 63.

Contributors

Emmanuel Delwin Abarike
Department of Aquaculture and Fisheries Sciences
Faculty of Biosciences
University for Development Studies
Tamale, Ghana

Nermeen Abu-Elala
Faculty of Veterinary Medicine
King Salman International University (KSIU)
South Saini, Egypt
and
Department of Aquatic Animal Medicine & Management
Faculty of Veterinary Medicine
Cairo University, Egypt

Sohrab Ahmadivand
Faculty of Veterinary Medicine
Ludwig-Maximilians University Munich
Munich, Germany

Hadeer A. Amer
Faculty of Aquatic and Fisheries Sciences
Kafer El-Sheikh University
Kafer Elsheikh, Egypt

Muralidhar P. Ande
Aquaculture Division
ICAR-Central Institute of Fisheries Education
Kakinada, India

Sajna Beegum
Department of Marine Biology, Microbiology and Biochemistry
School of Marine Sciences
Cochin University of Science and Technology
Kochi, India

Saiprasad Bhusare
FNBP Division
ICAR-Central Institute of Fisheries Education
Mumbai, India

Tanmoy Gon Choudhury
Department of Aquatic Health and Environment
College of Fisheries
Central Agricultural University
Lembucherra, West Tripura, Tripura, India

Nayan Chouhan
College of Fisheries (Central Agricultural University - Imphal)
Lembucherra, Agartala, Tripura, India

S. Ferosekhan
Aquaculture Division
ICAR-Central Institute of Freshwater Aquaculture
Bhubaneswar, India

Ritam Guha
Department of Marine Biology, Microbiology and Biochemistry
School of Marine Sciences
Cochin University of Science and Technology
Kochi, India

Vivian Hlordzi
School of Biological, Earth and Environmental Sciences
and
Aquaculture and Fisheries Development Centre (AFDC)
Environmental Research Institute
University College Cork
Cork, Republic of Ireland

Fawole Femi John
Fish Nutrition and Biochemistry Unit
Department of Aquaculture and Fisheries
University of Ilorin
Ilorin, Nigeria

Nayomi John
Department of Microbiology
Mar Athanasius College (Autonomous)
Kothamangalam, Kerala, India

V. Kaleeswaran
FNBP Division
ICAR-Central Institute of Fisheries Education
Mumbai, India

Dibyendu Kamilya
Agricultural and Food Engineering Department
Indian Institute of Technology Kharagpur
Kharagpur, India

Tejaswini Kinnera
FNBP Division
ICAR-Central Institute of Fisheries Education
Mumbai, India

Felix Kofi Agbeko Kuebutornye
University of South Bohemia in České Budějovice
Faculty of Fisheries and Protection of Waters
South Bohemian Research Center of Aquaculture and Biodiversity of Hydrocenoses
Institute of Aquaculture and Protection of Waters
České Budějovice, Czech Republic

Nandha Kumar
Department of Marine Biology, Microbiology and Biochemistry
School of Marine Sciences
Cochin University of Science and Technology
Kochi, Kerala, India

Sreeja Lakshmi
King Nandhivarman College of Arts and Science
Thellar, Vandavasi, India

K. Likitha
Department of Aquaculture
Kerala University of Fisheries and Ocean Studies
Kochi, Kerala, India

Yishan Lu
College of Fisheries
Guangdong Ocean University
Huguang Yan East
Zhanjiang, Guangdong, China

L. Manjusha
FRHPHM Division
ICAR-Central Institute of Fisheries Education
Mumbai, India

Jan Mraz
University of South Bohemia in České Budějovice
Faculty of Fisheries and Protection of Waters
South Bohemian Research Center of Aquaculture and Biodiversity of Hydrocenoses
Institute of Aquaculture and Protection of Waters
České Budějovice, Czech Republic

Abdullah Bin Abdul Nazar
Kerala University of Fisheries and Ocean Studies
Panangad, Kochi, Kerala, India

Shamna Nazeemashahul
Fish Nutrition, Biochemistry and Physiology Division
ICAR-Central Institute of Fisheries Education
Mumbai, India

Lokesh Pawar
College of Fisheries (Central Agricultural University - Imphal)
Lembucherra, Agartala, Tripura, India
and
University of Algarve
Campus de Gambelas
Faro, Portugal

Naga Prasanthmadduluri
College of Fishery Science
Andhra Pradesh Fisheries University
Muttukuru, Andhra Pradesh, India

Mohammad Shafiqur Rahman
Faculty of Environmental Science
Key Laboratory of Urban Environment and Health
Institute of Urban Environment
Chinese Academy of Sciences
Xiamen, China

A.M. Babitha Rani
Aquaculture Division
ICAR-Central
Institute of Fisheries Education
Mumbai, India

Jasmine Sabu
Kerala University of Fisheries and
Ocean Studies
Panangad, Kochi, India

Upasana Sahoo
Aquaculture Division
ICAR-Central Institute of Fisheries Education
Kakinada, India

Ajay Valiyaveettil Salimkumar
Faculty of Science and Technology
Research Centre for Experimental
Marine Biology and Biotechnology
(PIE-UPV/EHU)
Areatza Pasealekua
Plentzia - Biskaia, Basque Country, Spain

Parimal Sardar
Fish Nutrition, Biochemistry and Physiology
Division
ICAR-Central Institute of Fisheries
Education
Mumbai, India

Athira Ambili Sasikumar
Inspire-SRF, CSIR NIIST
Kerala, India

Arya Singh
ICAR – Central Institute of Fisheries Education
Versova, Andheri (W)
Mumbai, Maharashtra, India

Soibam Khogen Singh
College of Fisheries (Central Agricultural
University - Imphal)
Lembucherra, Agartala, Tripura, India
Krishi Vigyan Kendra, Ukhrul
ICAR – Research Complex for NEH Region
Manipur, India

Shalini Sundi
FNBP Division
ICAR-Central Institute of Fisheries Education
Mumbai, India

Anusree Suraj
Kerala University of Fisheries and
Ocean Studies
Panangad, Kochi, India

Karthireddy Syamala
Aquaculture Division
ICAR-Central Institute of Fisheries Education
Kakinada, India

R. Vidhya
AAHE Division
ICAR-Central Institute of Brackishwater
Aquaculture
Chennai, India

V. Vidhya
FRHPHM Division
ICAR-Central Institute of Fisheries Education
Mumbai, India

Introduction

Cohn (1876), Koch (1876), and Tyndall (1877) independently discovered that certain species of bacteria spend at least part of their lives as dormant cellular structures, known as endospores. Species in the genus *Bacillus* are Gram-positive, catalase-positive, endosporing, aerobic, or facultative anaerobes, characterized by their rod-shaped morphology (between 2.5 and 10 μm). *Bacillus* is classified in the Phylum Firmicutes, Class Bacilli, and Order Bacillales. The genus comprises approximately 200 bacterial species, which are almost ubiquitous in nature: they are found in soil; in compost (*Bacillus composti* and *Bacillus thermophilus*); in extreme environments such as high pH conditions (*Bacillus firmus* OF4); high temperature (*Bacillus thermophilus*); high salt (*Bacillus halodurans*); and in aquatic environments as well as in the gastrointestinal (GI) tract of aquatic animals. They exhibit quite diverse physiological properties such as the ability to produce cellulase, phytase, tannase, chitinase, xylanase, protease, and lipase, as well as degradation of palm (*Elaeis guineensis*) biodiesel. Another favourable trait of *Bacillus* spp. is that they produce antimicrobial substances, such as peptide and lipopeptide antibiotics and bacteriocins. The sporulation capacity and the production of antimicrobial substances and enzymes provide species of *Bacillus* with the capacity to colonize different habitats and contribute to the nutrition of the host, and they have been widely used as environmental probiotics and dietary probiotics in aquaculture.

Optimal GI functionality is essential for sustainable animal production, and effective functionality of the GI tract and its gut microbiota play an important role in host health and several complex mechanisms are involved. The microbial community of the fish gut is influenced by gut anatomy, such as the presence or absence of a stomach and pyloric caeca, and the relative length of the intestine. In the absence of intestinal microbiota, normal immune development and function are impaired, reducing protection against infections and decreasing gut health. Therefore, it is crucial to increase our knowledge on beneficial gut bacteria, in the context of improved growth performance and health. As alteration of the GI tract microbiota (dysbiosis) may enhance the risk for allergies and other conditions, it is of importance to investigate the intestinal microbiota of endothermic animals as well as aquatic organisms. The presence of beneficial bacteria such as species of *Bacillus* in the GI tract of aquatic organisms merits investigations, as autochthonous bacteria, adherent to mucosa, rapidly colonize the digestive tract at early developmental larval stages. During the last 20 years, numerous studies have been published on *Bacillus* spp. in the intestine of finfish and shellfish, their potential as probiotics, pathogenicity, and their effect on the immune system of the host.

Bacillus species are spore-forming bacteria that are resistant to aggressive chemical and physical conditions, with different species exhibiting unusual physiological characteristics enabling them to survive in different environmental conditions including sediments, desert sands, hot springs, fresh waters, marine waters, Arctic soils, and the GI tract of finfish and shellfish. *Bacillus* bacteria can rapidly multiply and tolerate various environmental conditions that make them to have various beneficial effects in the aquaculture sector. Administration of *Bacillus* species in the forms of probiotics or paraprobiotic or post-biotic in feed or for bioremediation of rearing water of aquaculture provides a promising hope to sustainable aquaculture. *Bacillus* species can play a great role in removing waste products from aquaculture environments, maintaining optimum water quality, and reducing stress that leads to enhanced immunological and physiological balances, higher growth performance, and improved survival in aquatic organisms. *Bacillus* probiotics are able to increase the growth and immunological functions of aquatic animals such as fish and shellfish. *Bacillus* in the form of probiotic are able to maintain a higher density of beneficial bacteria and a lower load of pathogenic bacteria in aquaculture-rearing waters. However, more studies are required to know about the host-specific *Bacillus* in different aquatic organisms and growth condition, water quality, and diet. Details mode of actions of bacilli bacteria on physiological and immunological functions of aquatic animals and their functions as bioremediators of water quality are very interesting topics associated with aquaculture.

This book addresses the biology and life cycle of bacilli bacteria, their existence in the GI tract of finfish and shellfish, their ability to produce enzymes and antibacterial compounds, and their efficacy and potency as probiotics in aquaculture. In Chapters 1 and 2, the anatomy and physiology of the GI tract of finfish and shellfish and an overview on fish immune system are described. Chapters 3 and 4 describe the biology and physiology of spore formation for *Bacillus* probiotic production. In Chapter 5, an updated *Bacillus* bacteria isolated from the GI tract of finfish and shellfish is presented. The bacilli products including exo-enzymes and bacteriocins with an emphasis on their antimicrobial activity are described in Chapters 6 and 7. The bioremediatory effects of *Bacillus* probiotics as a tool for water treatment in aquaculture are discussed in Chapter 8. In Chapter 9, the efficacy and potency of *Bacillus* on the immune system of animals are described. The efficacy and potency of bacilli probiotics on growth, immunity, and disease resistance in finfish and shellfish are addressed in Chapters 10 and 11. *Bacillus* as paraprobiotics in aquaculture is discussed in Chapter 12. Application of gut-associated *Bacilli* in feed-biotechnology, and safety of *Bacillus* probiotics in aquaculture with an emphasis on the environmental and ecological considerations are presented in Chapters 13 and 14, respectively.

The Editors

1 Anatomy and Physiology of the Gastrointestinal Tract of Finfish and Shellfish

Shamna Nazeemashahul
ICAR-Central Institute of Fisheries Education, Mumbai, India

Fawole Femi John
University of Ilorin, Ilorin, Nigeria

S. Ferosekhan
ICAR-Central Institute of Freshwater Aquaculture, Bhubaneswar, India

Parimal Sardar
ICAR-Central Institute of Fisheries Education, Mumbai, India

V. Vidhya
ICAR-Central Institute of Fisheries Education, Mumbai, India

Saiprasad Bhusare, Tejaswini Kinnera, V. Kaleeswaran and Shalini Sundi
ICAR-Central Institute of Fisheries Education, Mumbai, India

1.1 INTRODUCTION

Fish are only one example of the diverse aquatic species that have evolved a wide range of physiological adaptations to help them be the fittest for varying aquatic environments. The gastrointestinal (GI) tract of fish is organised similar to that of other vertebrate species, with some difference in phylogeny and ontogeny, feed and nutrition, feeding behaviour, physiological conditions, and the particular activities the gut may perform. The GI tract shape varies dramatically amongst fish species. Variations in GI tract organisation ensure efficient digestion and optimal nutrient utilisation and provide an increased intestinal absorptive surface area. The changes in functional demands that arise due to growth (metamorphosis and anadromous or catadromous migrations) and daily or seasonal changes in feed availability or environmental conditions (Karila et al., 1998) are supported by the endocrine signalling pathways backed up by the enteric nervous system. The GI tract hormones and signalling molecules along with endocrine pancreatic cells help fish to rapidly and reversibly switch GI tract and other organ system characteristics to adapt the changes (Holst et al., 1996). Understanding the variations in structure and functions of different digestive organs and cells will help the nutritionist to modify the diet to enhance the feeding efficiency. Moreover, the structure and physiology of the GI tract play crucial roles in the establishment and quantitative and qualitative characteristics of its microbiota.

DOI: 10.1201/9781003503811-1

1.2 GASTROINTESTINAL TRACT

Harder (1975) classified fish GI tracts into four regions, viz., the headgut, foregut, midgut, and hindgut. The mouth and pharyngeal region form the headgut, and its role includes acquiring and mechanically processing the food. The oesophagus and stomach form the foregut, where chemical digestion of food begins. The stomach is the centre of mechanical breakdown of food which can be partial or complete, whereas the chemical digestion takes place in the midgut or intestine. Intestine is considered as the longest part of the gut. The hindgut is the last segment of the gut, which includes the rectum; however, there is no apparent physical demarcation between the midgut and the hindgut in certain situations. Fishes have an ectodermal foregut epithelium, while the midgut epithelium is endodermal.

The gut wall is made up of four concentric layers that go from the foregut to the hindgut.

- The tunica mucosa, vascularised nerves and leukocytes containing connective tissue composed of the mucosal epithelium and lamina propria.
- The submucosa is an extra connective tissue layer.
- The tunica muscularis is made up of striated or smooth muscle in circular and longitudinal layers. However, lampreys only have an oblique muscle layer.
- The tunica serosa is only found in the coelomic cavity and is made up of mesothelial cells and loose connective tissue that contains blood vessels.

1.3 DIGESTIVE TRACT AND ITS MORPHOLOGY

The GI tract is a long tube that is divided into several regions, viz., mouth, gill arch, oesophagus, stomach, intestine (mid-intestine and distal intestine), and anus (Figures 1.1, 1.2, 1.3).

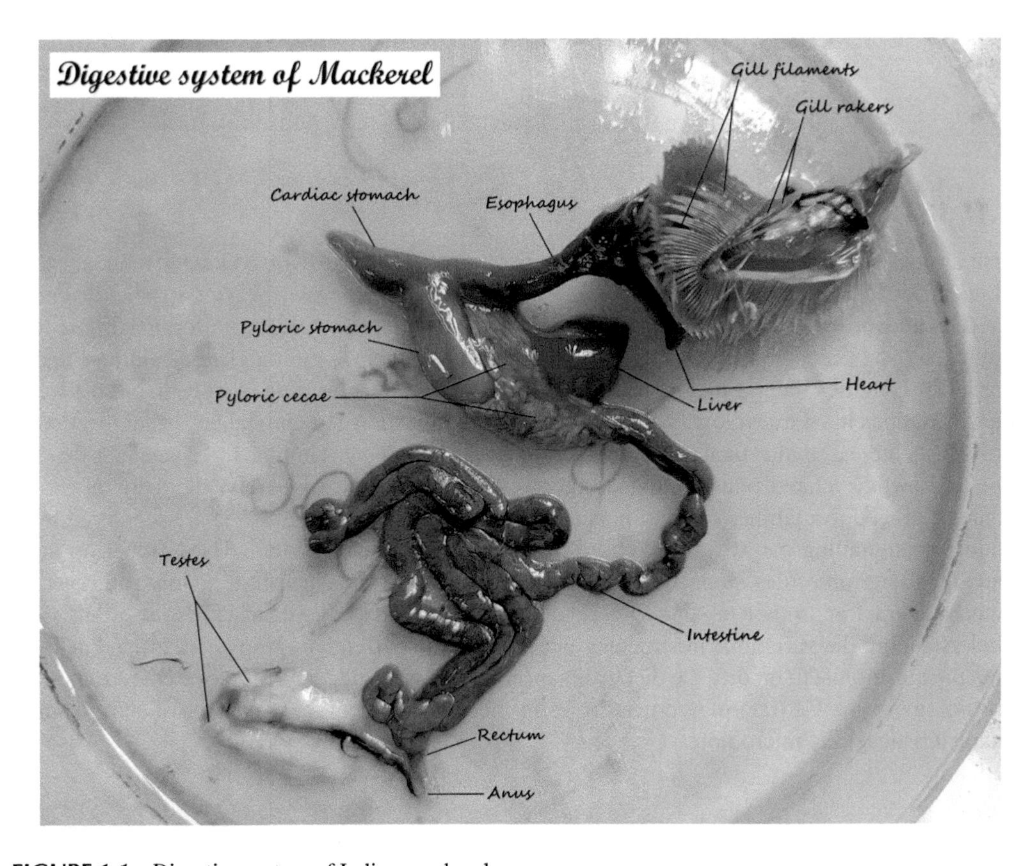

FIGURE 1.1 Digestive system of Indian mackerel.

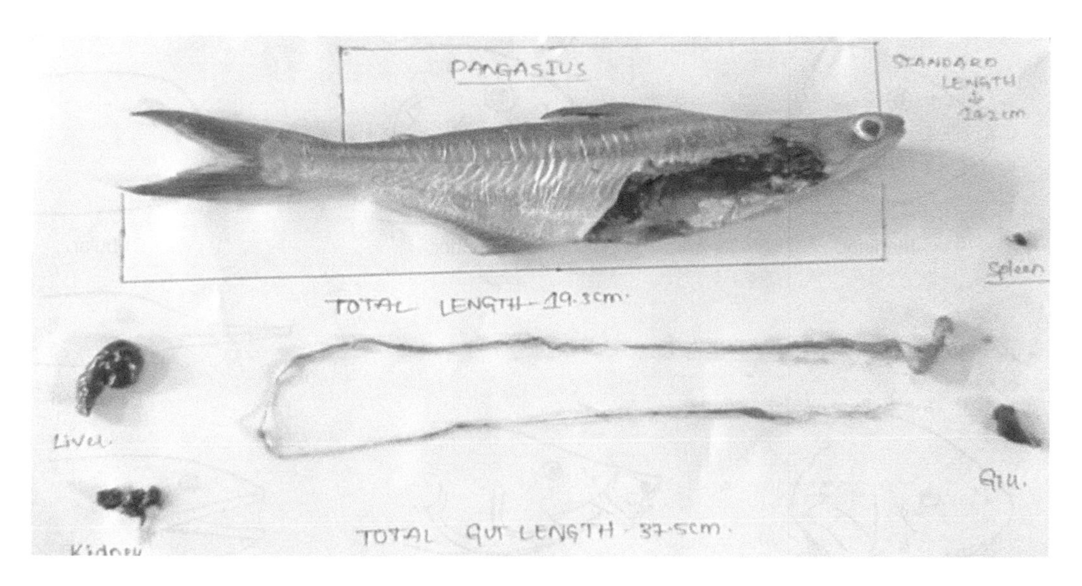

FIGURE 1.2 Digestive system of *Pangasius* sp.

FIGURE 1.3 Digestive system of tilapia fish.

1.3.1 Mouth and Mouth Parts

Mouth is the organ of respiration and feeding in fish. It is found in the anterior region of the body and is made up of four parts: the upper jaw, the lower jaw, two sets of gill arches (the operculum), and a set of teeth. All fish mouths are designed and built to suck or gobble up food from their surroundings. The form of fish mouth varies widely between species; some have huge lips that can create suction pressure to capture food items such as worms or crustaceans, while others have more pointed jaws for snatching small fishes or insects from the water column. Based on the food and method of feeding, fish have evolved different categories of mouths (Abbate et al., 2006). The four mouth shapes that are seen below are all made to aid fish in efficiently catching their next meal. Fish with protruding or terminal mouths typically eat other fishes. Superior-mouthed fish are typically ambush predators, which means they hide and wait for prey to approach them before striking. Angler fish are

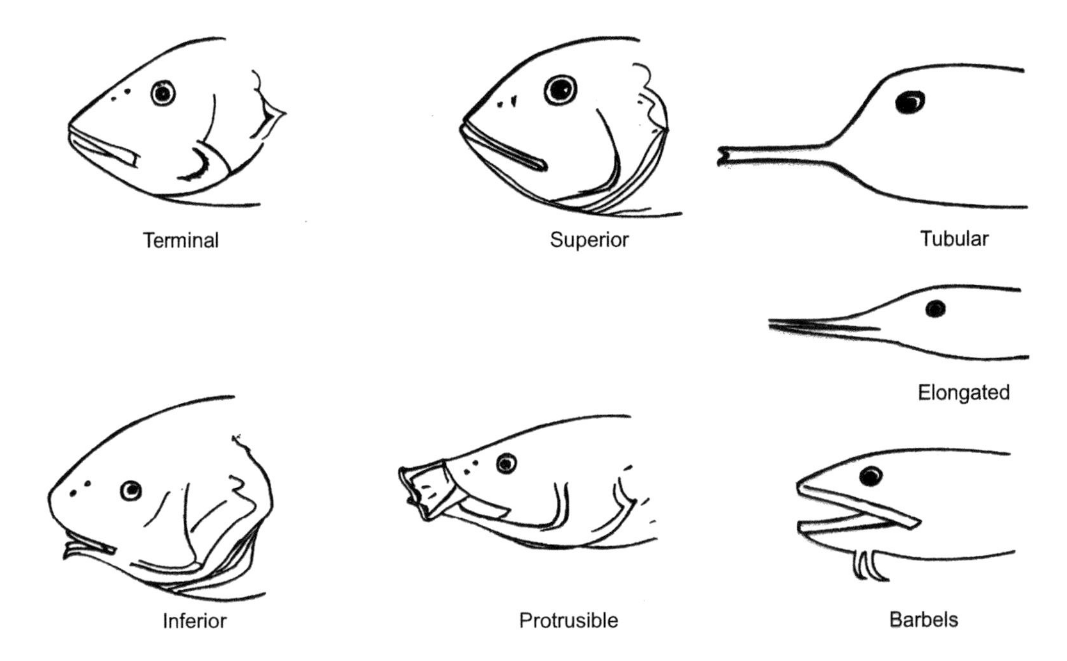

FIGURE 1.4 Sketch of types of mouth in fishes. (Adapted and modified from FU Newsletter, 2018).

typical examples of this sort of predator. The majority of fish with inferior mouth types are bottom feeders who eat things like crabs and shellfish (Figure 1.4).

Fish have evolved different types of teeth based on their food, just like other animals. For instance, most carnivorous fish (also known as carnivores) have teeth that are made to pierce, grasp, and cut their prey, but most herbivorous fish (also known as herbivores) have teeth that are better suited to shredding objects like algae (Figure 1.5). Cyprinids possess pharyngeal teeth (Figures 1.6b and 1.7).

1.3.2 Oesophagus

Fish oesophagus is a short, straight, tube with thick wall that connects the pharynx to the stomach. In catfish, it is observed that the tubular oesophagus connects the posterior end of pharynx to the anterior end of cardiac stomach (Abd El Hafez et al., 2013). Oesophagus connects to intestine in agastric fishes like carps. The cylindrical straight oesophagus connects the pharynx and intestinal bulb in grass carp (Abd El Hafez et al., 2013; Mokhtar, 2015). The oesophagus is mainly constructed morphologically for food passage and hence, the structure has mucosal layers and several spherical- or oval-shaped goblet cells. The stratified mucosal epithelium observed in teleost and elasmobranch fishes and the superficial cells of it may form micro-ridges but it has been observed that, near the posterior region, columnar epithelium is more, based on species and ambient salinity (Sullivan, 1907; Yamamoto and Hirano 1978; Meister et al., 1983). Unlike other areas of the gastrointestinal tract (GIT), stem cells are present throughout the oesophageal epithelium. Mono or binucleated polyhedral club cells, basal cells, and glands are present in the oesophagus. Glands are not normally present in the oesophagus. However, in other cases, such as the milkfish, *Chanos chanos* (Chandy 1956; Ferraris and de la Cruz 1987) and pike (Reifel and Travill 1977), the glands are obviously separate from the gastric glands. Irregular connective tissues consisting of collagenous fibres and striated fibres are present in the tunica submucosa while tunica muscularis has longitudinal skeletal fibres on the outer side and circular fibres in the inner layer.

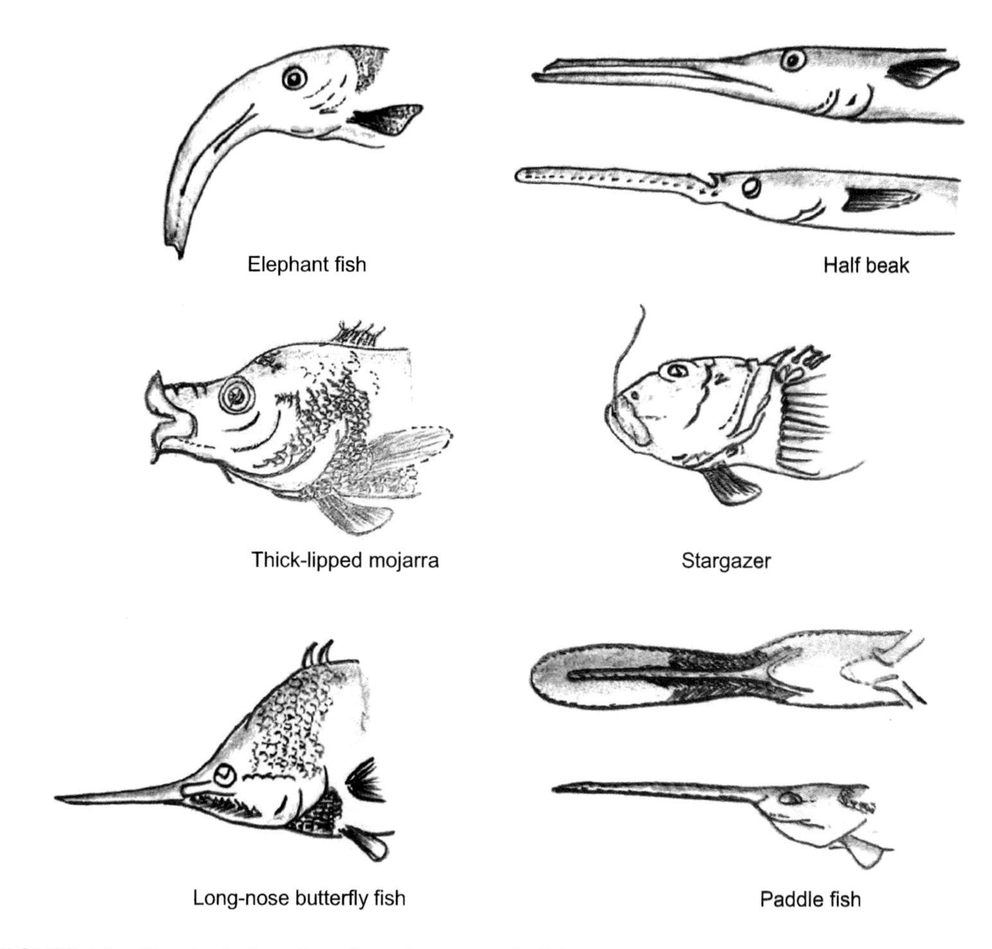

FIGURE 1.5 Sketch of adaptation of mouth structure in fish.

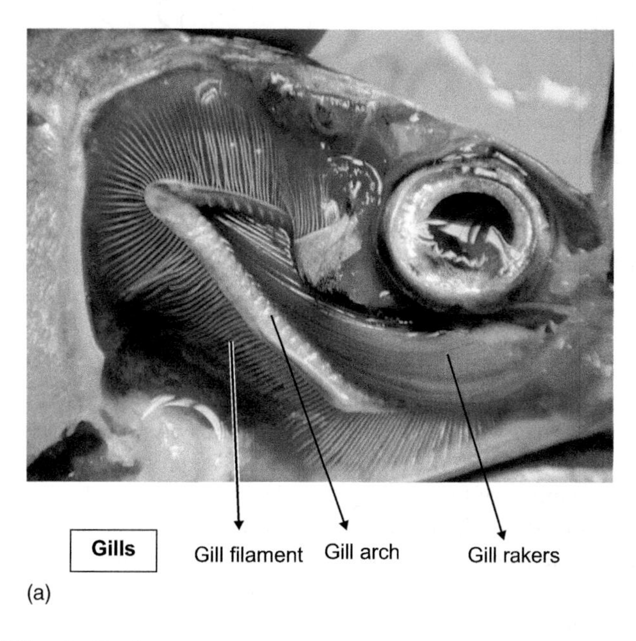

FIGURE 1.6 (a) Gill and gill rakers.

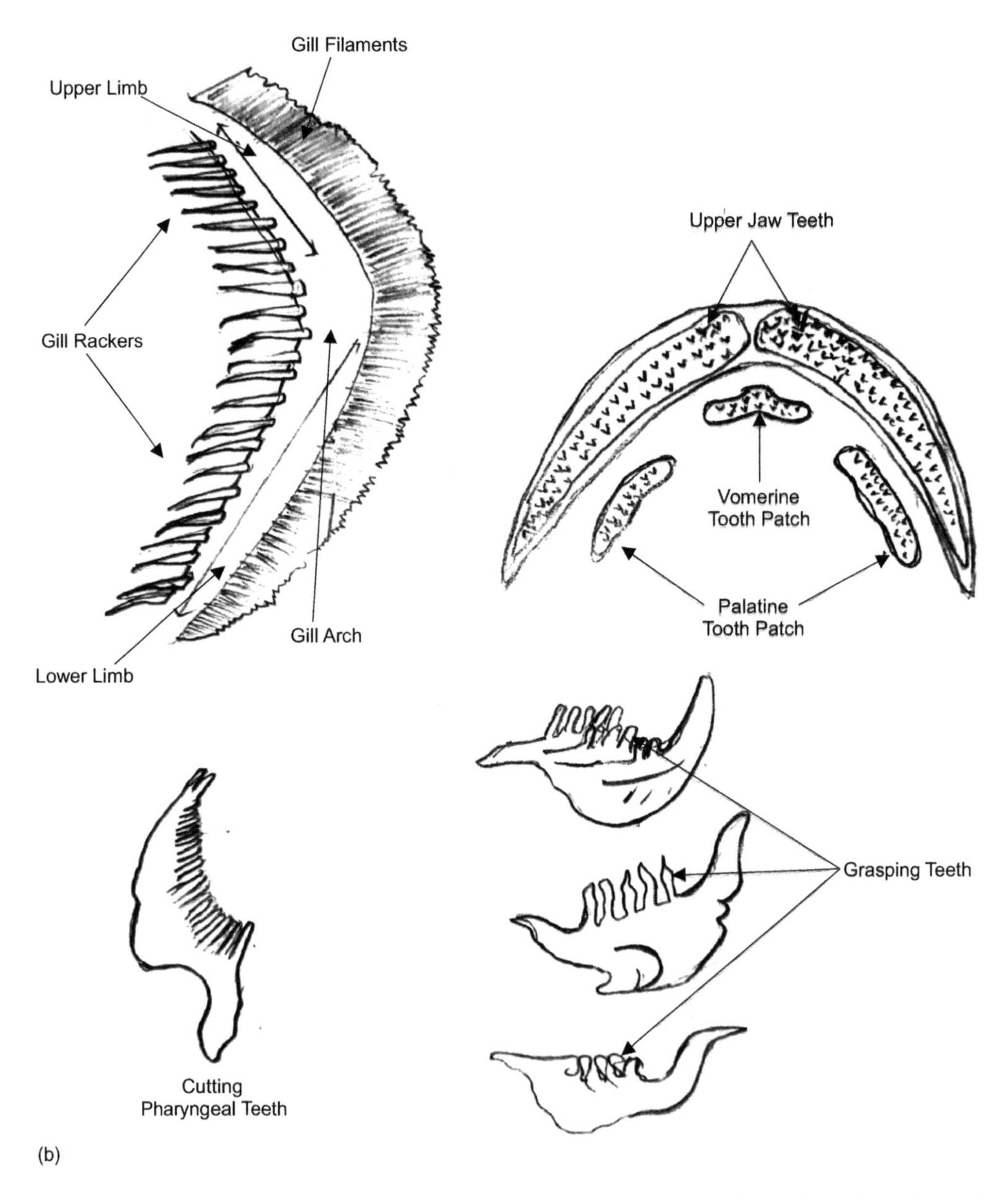

FIGURE 1.6 (b) Sketch of gill rakers, pharyngeal teeth, vomerine and palatine tooth patch, and grasping teeth in fish.

1.3.3 Stomach

The stomach is a muscular bag-like structure with an acidic environment that is located between the oesophagus and the intestine of the GI tract (Figure 1.8). The presence or lack of a stomach distinguishes the two main groupings of fish. Among the teleost fish, 85% of the fishes have stomach, and 15% of fishes are agastric (stomachless) fishes. The absence of a real stomach is the most notable aspect of the digestive systems of Cyprinidae, Cyprinodontidae, Balistidae, lampreys, hagfish, chimaeras, Labridae, Scomberesocidae, and Scaridae families. A few of the agastric fishes usually have a pesudogaster. The front section of the intestine swells in cyprinids, such as mrigal, to produce a sac-like structure known as the intestinal bulb or pseudogaster. The ingested food is transferred to

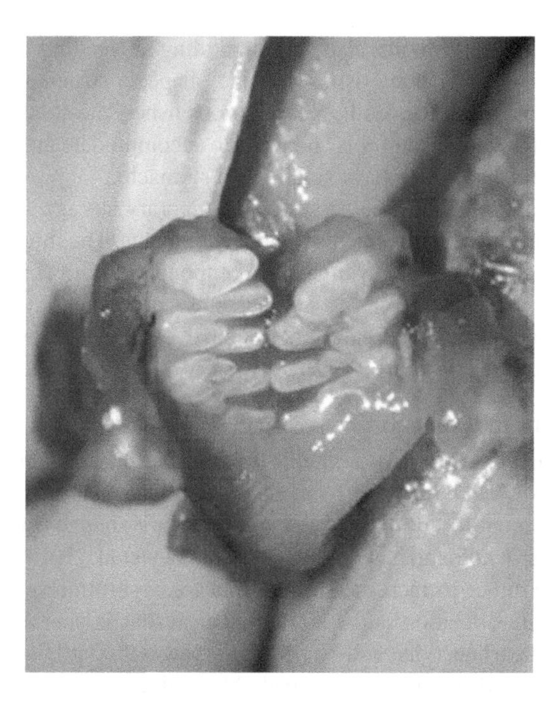

FIGURE 1.7 Pharyngeal teeth in cyprinids.

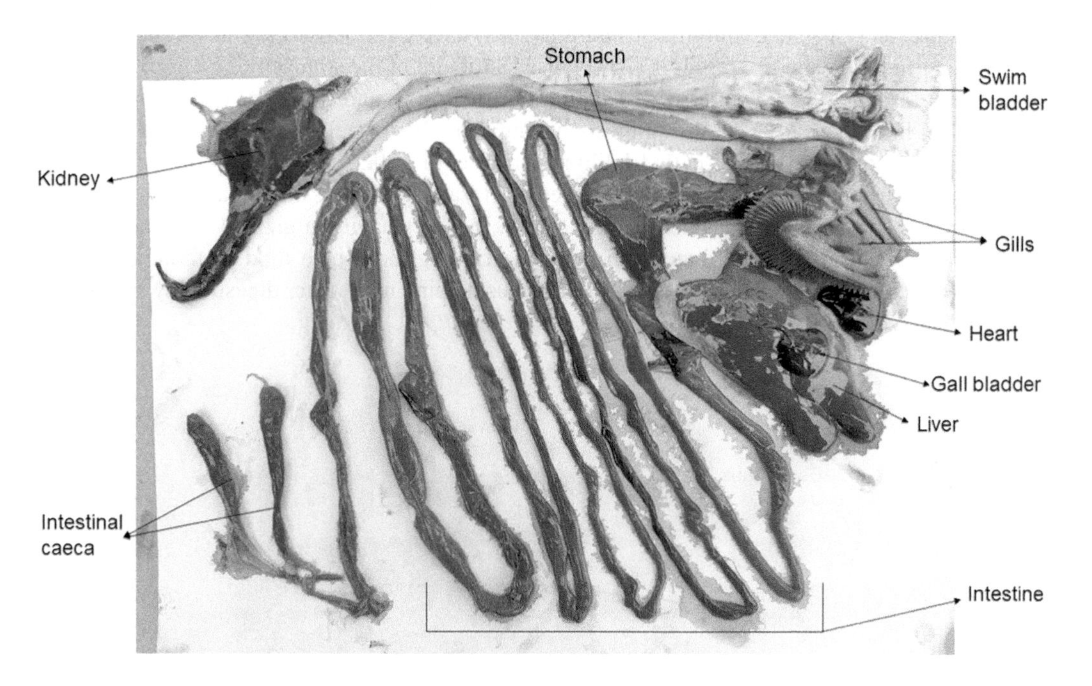

FIGURE 1.8 Stomach of *Pangasianodon hypophthalmus*.

the stomach because of the oesophageal muscular movement and the mixed and digested food bolus in stomach is then transported to the intestine. The anterior portion of the stomach is the cardiac stomach and the posterior portion of the stomach is the pyloric stomach. The size of the stomach also depends on the feeding habit, age, and size of the fish, temperature, meal size, particle size, and stress (Fänge & Grove, 1979; Bromley, 1994).

Observation of the stomach contents also helps in finding the feeding habits of fish (Beulah et al., 2021). When present, stomach characteristics like size, shape, or structure are connected to the time between meals and the nature of the food. In carnivorous fishes, the stomach is very distinctive and highly distensible with thick mucosal folds and walls to accommodate large-size prey rather than the herbivore and omnivore fishes which have small-sized stomach.

In agastric fish, the anterior intestine serves as a temporary storage site for the food consumed (Sinha, 1983) and the intestinal bulb appears to generate mucus. The mucosa is histologically similar to that of the intestine but without any digesting components (Horn et al., 2006; Manjakasy et al., 2009). Many stomachless fish compensate for their absence of a stomach by having pharyngeal teeth or gizzards to grind food. Wood-eating fishes have teeth adaptation in the form of spoon for efficiently rasping wood (Nelson et al., 1999).

1.3.3.1 Modifications in Stomach

The shape of the stomach varies from a simple straight bag-like structure (Asian seabass, lizard fish, pike, channel catfish, and halibut) to I-shaped (Carangidae, Gadidae, Scombridae, Serranidae), J-shaped (sharks), U-shaped (Mugilidae and Salmonidae), or Y-shaped (Indian mackerel, Clupeidae; Figure 1.9) stomach (Purushothaman et al., 2016; Alabssawy et al., 2019; Suyehiro, 1942). Most of these modifications are either to increase surface area for absorption or enhancing digestion. The best example for such modification is the spiral valve present in sharks. The other modifications include straight tube (Pleuronectidae and *Esox*), U-shaped (salmonids), V-shaped (Plecoglossidae, Mugilidae, Salmonidae, and Sparidae), Y-shaped (Mugilidae and Clupeidae), and I-shaped stomachs (Carangidae, Gadidae, Scombridae, and Serranidae) (Suyehiro, 1942).

Several species belonging to families Clupeoidei, Channidae, Mugilidae, Acipenseridae, Coregoninae, and Chanidae (milkfish, *Chanos chanos*) have a specialised stomach called 'gizzard'. It helps in trituration and mixing. It is a specialised structure of the stomach, present in some fishes like mullet (*Mugil* sp., *Liza* sp.), gizzard shad (*Dorosoma* sp.), and sturgeons (*Acipenser* sp.). It is a small bag-like anterior part of the stomach, having a highly muscularised thicker wall and a thick layer of keratinous substances, called koillin (Farrag et al., 2020). The lining of the stomach is toughened with connective tissues and the lumen of gizzard stomach is also small. Gizzard stomach is mostly present in omnivorous/herbivorous fishes. In fishes having a gizzard stomach, digestion begins here because the gizzard stomach helps in grinding the coarse food material into smaller particles (like mechanical digestion) and mixing, thus helping in the later digestion by stomach and intestinal enzymes.

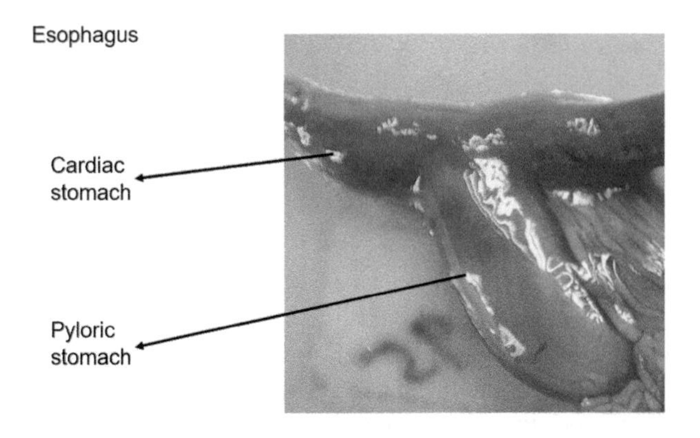

FIGURE 1.9 The Y-shaped stomach of Indian mackerel.

1.3.3.2 Histology of Stomach

Histologically, stomach also contains the mucous layers of serosa, muscularis, submucosa, and mucosa similar to intestine (Horn et al., 2006; Manjakasy et al., 2009). The inner wall of the stomach is lined with simple columnar epithelium. Moreover, histological evidences suggest that the stomach is divided into cardiac (anterior), fundic (middle), and pyloric (posterior) stomach. Mucosal folds of the stomach consist of mucous-secreting cells on the inner side of the stomach and gastric glands on the basal line of mucosal folds. Gastric glands and mucous-secreting cells are more abundant in the cardiac and fundic stomach than in the pyloric stomach which indicates the pyloric stomach is intended for storage purposes rather than digestion. Mucous production aids in the proper passage and lubrication of the food bolus in the alimentary canal.

1.3.4 Intestinal Bulb

Cyprinids (Cyprinidae), parrotfish (Scaridae), wrasses (Labridae), *Rutilus* sp., killifish (Cyprinodontidae), squaw fish (*Ptychocheilus oregonensis*), triggerfish (Balistidae), chimaera (*Holocephali*), and lungfish (*Dipnoi*) lack the true stomach and are called agastric fishes. In some of these agastric fishes, the anterior part of the proximal intestine gives a small sac-like structure called intestinal bulb. The oesophagus directly opens into the intestinal bulb and passes the food to the intestine through this intestinal bulb. The oesophageal opening has an oesophageal valve which prevents the regurgitation of digested food. The presence of pharyngeal teeth is also observed in agastric fishes to enhance the grinding of food materials (Suyehiro, 1942; Fange & Grove, 1979). There is no enzyme or acid secretion in the intestinal bulb, hence there is no digestion. It serves as a temporary storage organ for food before reaching the intestine (Sinha, 1983). Mucous cells are present to secrete mucous to facilitate the food passage into the intestine.

1.3.5 Pyloric Caeca

Several finger-like outgrowths arise from the front section of the gut in the pylorus region of a number of fish species (Figure 1.9). At the junction between the posterior part of pyloric stomach and anterior part of the proximal intestine, there are thin, blind tubules or finger-like outgrowths called pyloric caeca or intestinal caeca, and they open into the intestine lumen. They are found around the midgut and their number varies widely, a few (murrel) to several hundred (Atlantic cod). The size, state of branching, and connection of caeca to the gut of different species vary greatly. There is no clear connection between the type of the food or feeding habits and presence of the pyloric caeca, and the caeca are frequently absent or greatly diminished in herbivores and omnivores (Rust, 2002). Though pyloric caeca have been shown to increase digestive and absorptive surface area, their role in fermentation or storage is nil.

Pyloric caeca are not present in all fishes. Pyloric caeca are absent in wrasses, cyprinids, catfish, sharks, skates, and mainly agastric fishes (Barrington, 1957; Kapoor et al., 1976). Otherwise, we can say, in stomach-less fish, the pyloric caeca are always absent. They are pronouncedly found in a number of carnivores (Canan et al., 2012). However, pyloric caeca are absent in few fishes with stomach (e.g., blennies and Blenniidae). The presence of pyloric caeca cannot be considered as an indicator of the feeding behaviour of fish.

The number of pyloric caeca varies between the fish species. The number varies from one in sand eel (*Ammodytes* sp.); two in turbot (*Psetta maxima*), murrel (*Channa* sp), mullet (*Mugil cephalus*). and Gangetic leaffish (*Nandus nandus*); three in perch (*Perca fluviatilis*) and scorpionfish (*Setarches* sp.); four in *Liza abu*; five in Asian seabass (*Lates calcarifer*); nine in *L. aurata*, thirty six in brown-spotted grouper (*Epinephelus chlorostigma*); around a hundred in whiting (*Micromesistius australis*); two hundred in mackerel (*Scomber* sp.); and thousands of caeca producing a caecal mass in tuna (Farrag et al., 2020; Khayyami et al., 2015; Hassan, 2013).

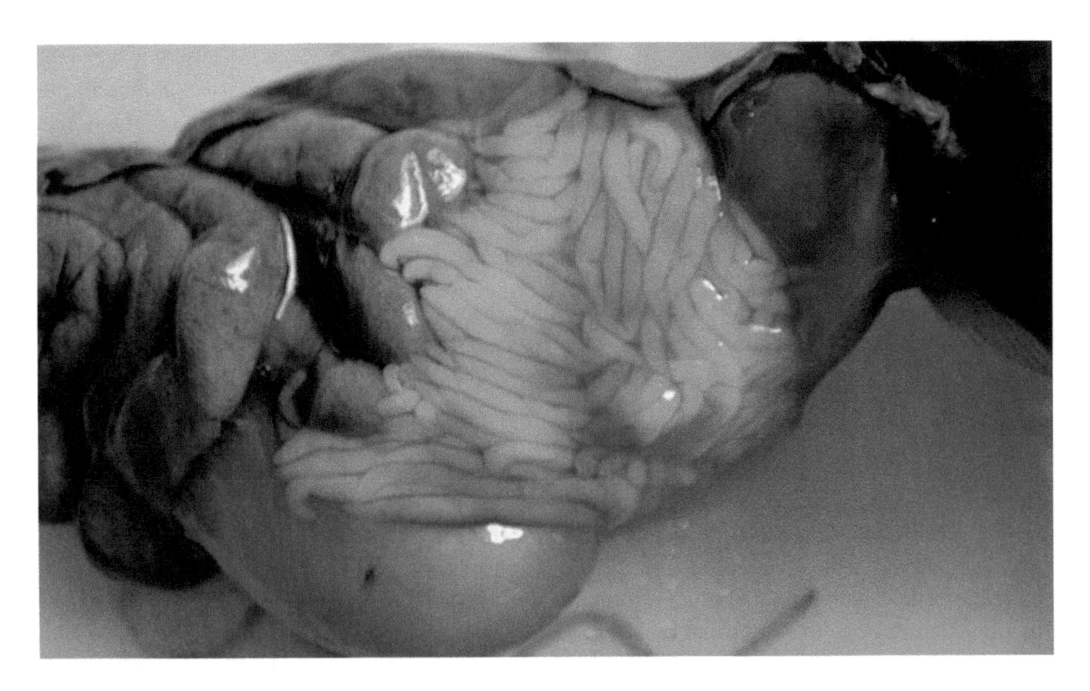

FIGURE 1.10 Pyloric caeca in *Rastrelliger kanagurta.*

Pyloric caeca develop from the anterior part of the proximal intestine (Mitra et al., 2015). The lumen of pyloric caeca opens into anterior intestine (Ray & Ringø, 2014). The pH of pyloric caeca is slightly alkaline. The number of goblet cells is also found to be lower. The wall of pyloric caeca consists of four layers, viz., serosa, muscularis, submucosa, and mucosa, from outside to the inner side of the lumen (Farrag et al., 2020; Khojasteh et al., 2009; Senarat et al., 2015; Bocina et al., 2017). The pyloric caeca are also denoted as 'intestinal caeca' by some scientists since they originate from the intestine (Hossain & Dutta, 1996).

The functions of pyloric caeca in fishes are not well established (Figure 1.10). They may be involved in the secretion of trypsin or enzymes involved in intestinal digestion, and eventually, trypsin, chymotrypsin, or alkaline proteases activity was recorded in some fishes (Caruso et al., 2009). It is also believed that they help in neutralising the acidic food bolus from the stomach before reaching the intestine (Hassan, 2013). It is strongly believed that they ultimately enhance the surface area for the post-gastric absorption of nutrients (Canan et al., 2012). In some fish species, the pyloric caeca are found to be an important region for the absorption of sugar, amino acid, and dipeptide (Buddington & Diamond, 1987). Sugiura and Ferraris (2004) found phosphorus absorption taking place in pyloric caeca through sodium–potassium cotransporter in rainbow trout. Ballesteros et al. (2013) reported the presence and expression of IgM, IgT, and interepithelial lymphocytes in the pyloric caeca of rainbow trout which is evidence for the immune function of pyloric caeca in fishes.

1.3.6 Ontogeny of Stomach and Pyloric Caeca

In precocial larvae, stomach becomes functional once the larvae start exogenous feeding and in altricial larvae, during the course of metamorphosis, the stomach becomes functional (Rønnestad et al., 2013).

Generally, in fishes, the development of the stomach and pyloric caeca starts at 1.5–2 cm in size (Pedersen & Falk-Petersen, 1992). Initially, stomach develops as a small diverticulum at the junction of the pyloric end of the oesophagus and the proximal region of intestine. Stomach and pyloric caeca become fully functional at a size greater than 2–4 cm. At this stage, stomach is structurally differentiated with gastric glands and cells to secrete acid and enzymes to begin gastric digestion (Tanaka, 1973; Govoni et al., 1986). Hence, it can be concluded that the larvae have metamorphosed into juveniles (Tanaka, 1973). In Asian seabass and turbot larvae, pepsin activity appeared at 17- and 22-days post-hatch (DPH), respectively (Walford & Lam, 1993). The size of the stomach and pyloric caeca increases, as the size of the fish increases. Thus, the pyloric caeca increase the surface area of intestine to enhance the active absorption area (Buddington & Diamond, 1987).

1.3.7 Intestine

In gastric and agastric fishes, the intestine follows the pylorus or oesophagus. The anterior intestine of some stomachless fishes may expand to form an intestinal bulb or pseudogaster and function in temporary food storage; nonetheless, gastric glands and a pylorus are absent. To fulfil a similar role, a spherically formed cecal chamber stretching from the anterior intestine following the oesophagus is observed in parrot fishes (Al-Hussaini, 1946). The fundamental purpose of the intestine is to finish the digestion processes that began in the stomach and to absorb nutrients.

The intestine is the primary digestive/absorption organ in fish (Figure 1.11). In addition to support in digestion and absorption, the intestine helps in balance of water and electrolyte, immunity, and endocrine regulation of digestion and metabolism. In different fish species, the gut varies greatly in length and organisation. Some fish species have a relative intestinal length (RIL = length of intestine/length of body) of less than one, while others have a RIL of 10–20 times their body length. RIL is highest in herbivores followed by detritivores, whereas it is lower in predatory fishes or carnivores. The intestine has a variety of looping and coiled patterns in *Cyprinids* (Figure 1.12) and *Loricariids*; in omnivorous species, the state is intermediate. In nutritional ecology, intestinal length is utilised as a morphological measure of trophic status (Horn, 1997). Aside from nutrition, other factors that determine intestine length include ontogeny, size, and age of the fish, shape of the body, feeding history, and phylogeny.

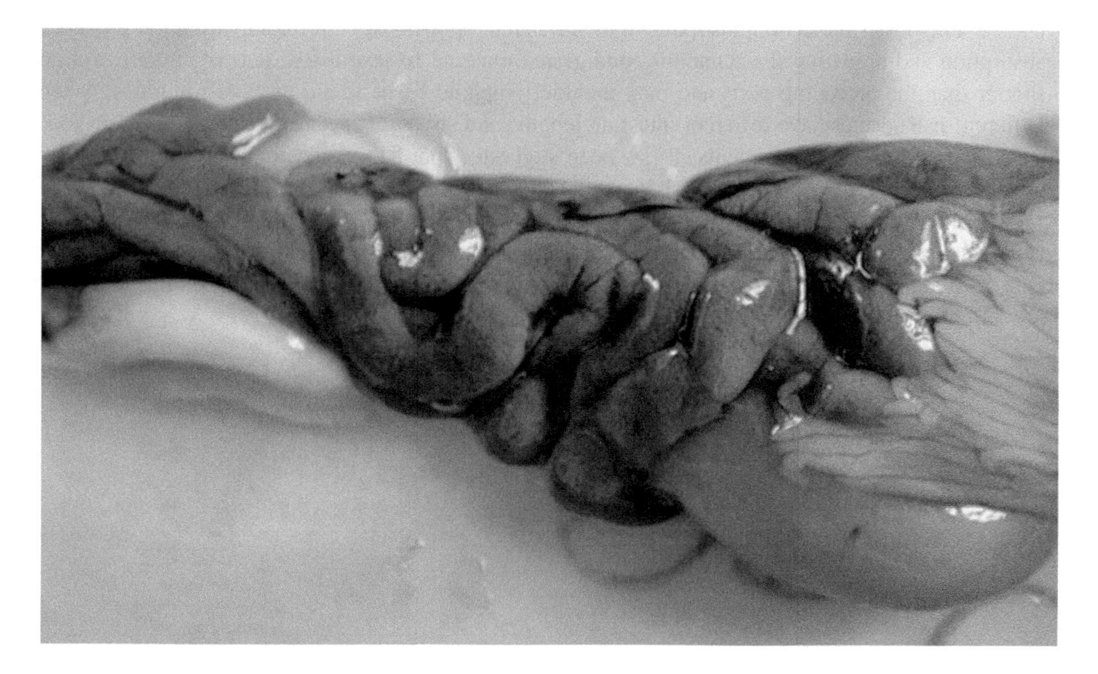

FIGURE 1.11 Intestine filled with food.

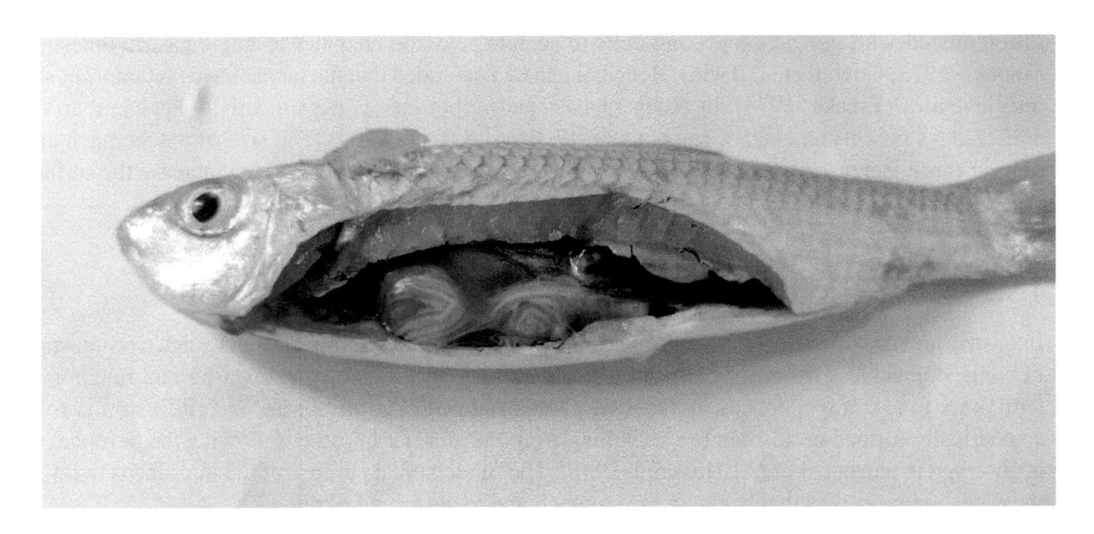

FIGURE 1.12 Coiled intestine in *Labeo rohita*.

1.3.7.1 Anatomy and Histology of the Fish Intestine

The intestine of a fish is made up of tubes that begin at the stomach and terminate at the anus. It's divided up into many sections that all do somewhat various activities. Typically, the intestine has been divided into three separate regions: the anterior, the midsection, and the posterior. The food that has been partially digested in the stomach is sent to the first portion of the small intestine, also called the duodenum, which is responsible for additional enzymatic digestion. It is distinguished by villi, the finger-like projections which enhance the surface area for the absorption of nutrients. In addition, the duodenum may include pyloric caeca, which are blind sacs that are responsible for the secretion of digesting enzymes. The middle section, also known as the jejunum, is largely responsible for the digestion and the absorption of nutrients. It has a microvilli-lined brush border, which is a highly folded inner surface. This brush border provides even more absorbent surface area, which improves nutrient uptake. The posterior section, also known as the ileum, is responsible for completing the process of absorption and preparing the remaining undigested material for expulsion. It is often narrower and shorter than the preceding parts and may include lymphoid tissue to aid the fish's immune system. Different fish species have different intestine lengths and structures because they eat different foods and live in different environments. It has been suggested that fishes' bicarbonate secretion may be greater than that of mammals since fish intestinal content has a pH range from 7 to 9 (Fard et al., 2007).

The anterior intestine mostly consists of simple columnar epithelium with brush border, mucosal cells, and microvilli structures to increase the absorptive area, whereas posterior intestine is abundant with high folding and a smaller number of villi (Figure 1.13). The rectal region has no or very less microvilli. Distal parts have abundant goblet cells.

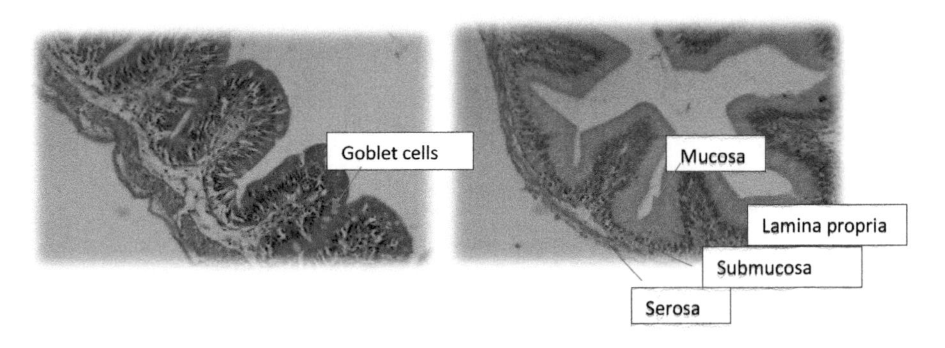

FIGURE 1.13 Histology of the intestine of *Labeo rohita* (H&E, 40X).

1.3.7.2 Functions of the Fish Intestine

The fish intestinal tract is responsible for a number of important processes that contribute to the general health of the organism.

1.3.7.2.1 Digestion

Enzymes play a crucial role in the digestive process of fish. Enzymes that digest carbs, proteins, and lipids are all secreted from the anterior part of the intestines. Enzyme secretion is facilitated by various organs, and pyloric caeca, duodenal glands, etc., play a major role in it.

1.3.7.2.2 Nutrient Absorption

The intestine of fish is equipped with a variety of specific systems, which allow for the rapid absorption of nutrients. The intestinal lining is highly folded, and it is also lined with small finger-like structures to enhance the absorptive surface area called villi and microvilli. In general, the proximal portions contribute more to nutrient absorption compared to the distal regions, as shown by a number of studies (Nordrum et al., 2000; Hernandez-Blazquez et al., 2006). Distal to the stomach, the epithelial mucosa folds and projects, increasing the intestine's surface area for nutrition absorption (de Oliveira et al., 2001). Smaller molecules like amino acids, fatty acids, glucose, and vitamins are absorbed by the fish and transported throughout the body via circulation.

1.3.7.3 Osmoregulation

Hypo-osmoregulating fish do the osmo- and iono-regulation through GI tract by engulfing saltwater. In order to absorb water from seawater throughout their journey through the digestive system, ions must undergo differential absorption across the intestinal epithelium. Intestinal anion exchange is an essential addition to the long-recognised cotransporters Na^+–Cl^- and Na^+–K^+–$2Cl^-$ in the intestines of marine teleost fish. This is because a major amount of Cl^- and water absorption in the GI tract is done by apical Cl^-/HCO^-_3 exchange (Grosell et al., 2005). It is essential for fish to keep their osmotic balance in every type of aquatic conditions. The intestinal tract actively controls the intake of water and ions. In freshwater fish, the intestine actively secretes ions to avoid excessive intake of water and to compensate for water lost through the gills. Marine fish, on the other hand, must contend with the additional issue of actively taking up ions from their surroundings and limiting water loss. The intestine modifies itself to meet these needs by developing new transport systems and ion pumps.

1.3.8 Adaptations Observed in Fish Intestine

Fish exhibit remarkable adaptations in their intestinal structure and function, allowing them to exploit various ecological niches.

1.3.8.1 Length and Surface Area

Herbivorous fish and other fish with a low nutrient intake have longer intestines. This adaptation allows for an extra transit time for the digestive process, which in turn maximises the absorption of nutrients. The longer length also provides a bigger absorption surface area.

1.3.8.2 Regional Specialisation

Some fish species have intestinal structures that are uniquely adapted to their geographic location. Carnivorous fish, for example, may have a shorter intestine with a more prominent anterior segment, allowing them to digest and absorb food more quickly. The intestine of herbivores, on the other hand, may be more elongated and abdominally enlarged in order to promote slow fermentation of plant food.

1.3.8.3 Mucus Production

The mucus that lines the fish intestine has several important functions. Mucus acts as a barrier to harmful bacteria and abrasive food particles, smooths the digestive process, and improves nutritional absorption. Mucus-producing cells are specialised cells spread throughout the mucosal lining all along the alimentary canal producing mucus. Mucus is primarily composed of water, ions, and mucins (Shephard, 1994). The mucins of both external and internal surfaces of fish are composed of many micro molecules which paly several immune-physiological functions. Mucus-producing cells are located in the mucosal lining. The number of mucus-secreting cells and, presumably, the mucus flow, differs between the mouth (buccal), oesophagus, stomach (gut), and intestines (Sklan et al., 2004).

1.3.8.4 Adaptations of the Fish Intestine for Diet and Environment

The fish intestine has developed extraordinary flexibility to accommodate the wide range of diets and environmental circumstances encountered by various fish species. Here are a few significant modifications:

The fish intestine exhibits remarkable adaptations to accommodate various diets and environmental challenges (Figure 1.14). For example:

- Herbivorous fish have **longer intestines** to allow for extended fermentation and breakdown of plant material.
- Carnivorous fish have **shorter intestines** optimised for rapid digestion and absorption of high-protein prey.
- Some fish possess **specialised structures**, such as ceca or pyloric caeca, which increase the absorptive surface area.
- Certain species, like pufferfish, have **elongated intestines** that can expand to accommodate large prey.

1.3.8.5 Specialised Digestive Enzymes

Various fish have developed specialised digestive enzymes to help them process the food they ingest. Herbivorous fish, for example, typically have the cellulase enzyme, which is useful for breaking down plant cell walls. Because of their expertise, they are able to extract nutrients from substances that most of us would consider indigestible.

1.3.8.6 Intestinal Structure Modifications

Some fish species have adapted their anatomy in order to better absorb nutrients. For example, the spiral folds or valves in the intestines of some fish enhance the absorbable surface area. On the other hand, teleosts, which lack a spiral modification, enhance intestinal surface area by either developing and multiplying pyloric caeca or by increasing the total length of the intestine, which then forms loops and is tightly enclosed inside the abdominal cavity (Wilson and Castro, 2011). These modifications improve the efficiency with which animals consuming low-nutrient-density diets absorb those nutrients.

1.3.9 Functional Features of the Digestive Tract

The digestion of food and consequent absorption of nutrients are processes that depend on the abundance and availability of adequate enzymes for digestion at the appropriate locations along the digestive tract. The presence of enzymes for digestion in fish digestive tracts has been suggested to demonstrate a natural response to meal composition. Fish have the most basic or least differentiated digestive tract of any vertebrate, which corresponds with their historical location (De Silva and Anderson, 1995). Fish have a relatively diversified digestive tube shape, demonstrating the variety

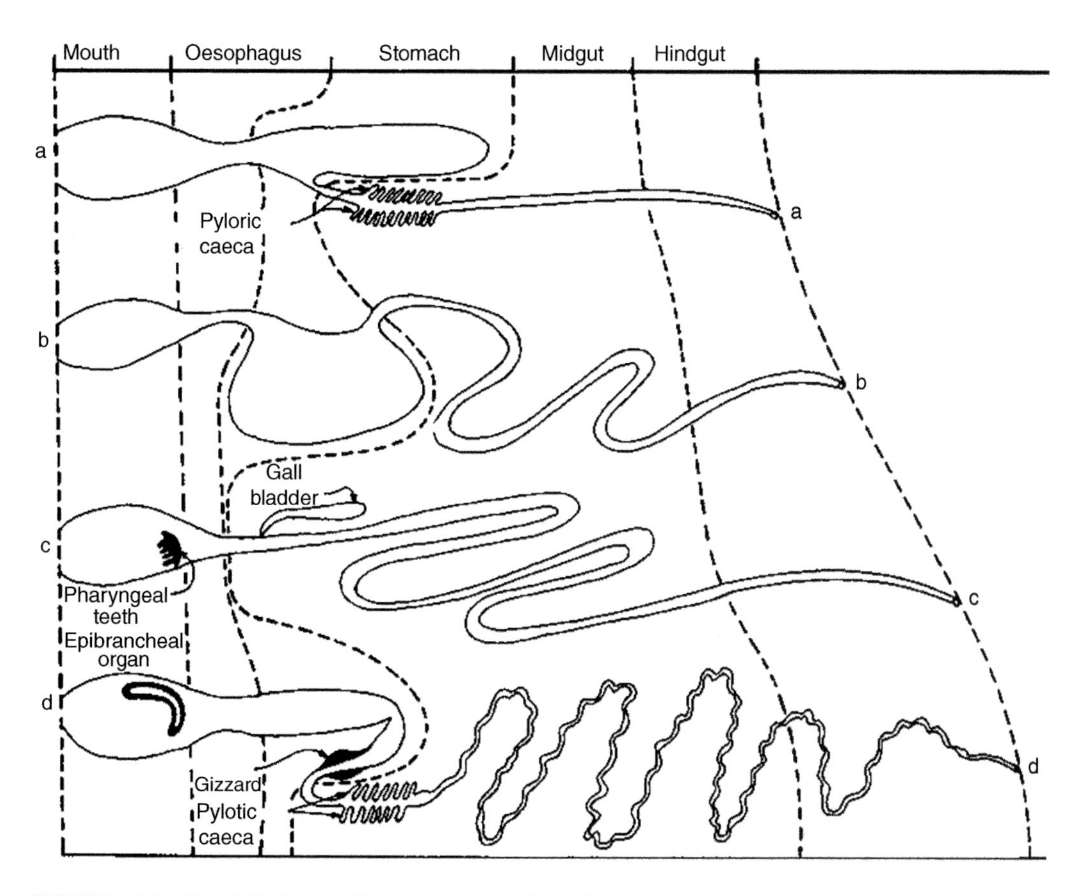

FIGURE 1.14 Four fish discussed in the text are depicted diagrammatically, with the intestine organised from shortest to longest intestine: (a) carnivore, (b) omnivore preferring animal source, (c) omnivore preferring plant sources, and (d) microphagous planktonivore. (Adapted from De Silva and Anderson,1995)

of foods and eating habits found across numerous trophic niches. Factors like the enzymatic structure of the GI tract and the adaptive qualities of these enzymes influence fish's capacity to metabolise food. Apart from the availability of GI enzymes, the efficacy of food intake and feed conversion may also be affected by the digestive system's transepithelial transport capacity (Jobling, 1995). Complete knowledge of diet, feeding patterns, and ecology niche is essential to understand any specific species' digesting technique.

The jaw was a significant vertebrate innovation that not only increased the amount of food items that could be acquired as well as allowed for a broader choice of capturing strategies (Schwenk and Rubega, 2005). Digestion enzymes are classified according to their active role in carbohydrate, fat, and protein digestion.

Type I digestion – Acid lysis is equally efficient as trituration in breaking plant cell walls in low gastric pH digestion. After acid lysis, the plant cell walls are usually intact, but the insides of the cells are liberated for digestion and absorption. Low pH also induces protein denaturation and stimulation by the endopeptidase pepsin (pHo4), which initiates meal enzymatic digestion. The stomach has thin walls (in relation to the gizzard below), but its muscles are adequate for trituration and mixing of food, providing a certain mechanical breakdown.

Type II digestion – The families Mugilidae (mullets), Kyphosidae (subfamily Girellinae, nibblers), Clupeidae (subfamily Dorosomatinae, gizzard shad), and Acipenseridae (sturgeon) all have trituration. The vast muscular pyloric portion of the stomach, along with swallowed sand grains and a thick mucus coat, is effective for triturating bacteria, cyanobacteria, algae, and detritus. Although gastric glands exist in these kinds of fish, the gastric pH is quite elevated, implying that acid peptic digestion is less relevant.

Type III digestion – The pharyngeal apparatus triturates food. In some fishes, such as carps, the stomach trituration is replaced by a strong pharyngeal mill, which sends finely ground food directly to the intestine.

Type IV digestion – Microbial fermentation is used by some herbivorous fishes to produce short-chain fats (SCFA), which act as a source of energy for the enterocytes (Horn, 1997). While hindgut chambers exist in Pomacanthidae (angelfishes) and Kyphosidae (sea chubs), fermentation by bacteria occurs in species that do not have such anatomical specialisations. Fish pyloric ceca, on the other hand, are not involved in microbial fermentation. Before the absorption phase, the food bolus may be retained and subjected to digestive factors. As a result, the duration of transit has an immediate effect on nutrient biological efficiency. GI content outflow is a complex system influenced by species, fish size, food type and amount consumed, and water temperature.

1.3.10 Role of Enzymes in Digestive Process

There are three types of vertebrate digestive enzymes (Guillaume and Choubert, 2001):

- Enzymes released as zymogens by the pancreas and stomach
- Membrane enzymes connected to microvilli
- Cellular enzymes

A few fish contain all the digestive enzymes (Table 1.1). Agastric fishes, for example, lack pepsin. Furthermore, certain regions of the fish's digestive tract have distinct enzymatic activity, whether qualitatively or quantitatively. The place of enzyme secretion can also differ between species. Some fish's stomachs, for example, may have amylohydrolytic and lipolytic activity, whereas others do not. Most of the enzymes used for digestion are found in a few fish; for example, pepsin is absent in agastric fish. Furthermore, different parts of the fish's digestive tract exhibit different levels of activity of enzymes, whether qualitatively or quantitatively. Zymogens are triggered by severing an amino acid and such peptides' hydrolysis in the pepsin molecule starts with hydrochloric acid activity, followed by autocatalytic breakdown of these peptides by pepsin. Enterokinase initially activates pancreatic zymogens, initiating a sequence of autocatalytic events comprising trypsin activity on chymotrypsinogen. In that sequence, the active enzymes harm elastase, collagenase, carboxypeptidase A and B, phospholipase, and colipase. Except for the phospholipase–colipase pair, all these enzymes use endohydrolysis to break down protein macromolecules, which produces smaller molecules that cannot be taken up by other enzymes (endo- and exopeptidases).

1.3.11 Endogenous Secretions in GIT and Luminal pH

It is well known that salivary glands are absent in fish and the digestion starts in the stomach (gastric fish) or intestine (agastric fish) region. The pH value in the GIT of fish varies and in general, the stomach bears acid values ranging between 1.6 and 4, whereas in intestinal region the pH values range from neutral to alkaline. The pH range of gastric juice varies depending upon the species and is between 1.6 and alkaline. There are individual variations within the species depending on feeding habits and digestive phase (Pegel', 1950; Deguara et al., 2003; Solovyev et al., 2018).

In most fishes, the pH of stomach varies from 1.5 to 4 and up to 5 in some marine fishes to cope with the alkaline condition of seawater. However, studies showed that strongly acidic (1.5–2) to weakly alkaline (7–8) pH are observed in the stomach of fish (Izvekova et al., 2013; Hlophe et al., 2014). The acidic condition of the stomach depends on whether the stomach is full or empty. Stretching of the stomach and partially digested protein stimulate gastrin and thus enhance acid

TABLE 1.1
Major Enzymes Reported in the GIT of Fish

Enzyme	Function	Reference
Proteases	Digestion of protein	Garcia-Carreno (1992a, 1992b)
Amylase	Assist in digestion of starch α-Amylase (random cleaving of the chain from within) β-Amylase (acts at every two glucose units)	Nelson (1944); Somogyi (1952) Hidalgo et al. (1999)
Dextrinase	Dextrinase (acts at the branched point)	Borlongan et al. (2002)
Lipase	Digestion of triglycerides into monoglycerides and free fatty acids with the support of bile acids	Winkler and Struckman (1979)
Phospholipases	Hydrolyse phospholipids	Borlongan et al. (2002); Guillaume et al. (2001)
Pepsin	Proteins hydrolysis in the stomach Peptides and some free amino acids	Anson (1938)
Trypsin and chymotrypsin	Alkaline digestion of protein, secreted from pancreas and stored in midgut and the pyloric caeca; enterokinase activates it in intestine; trypsin activates chymotrypsinogen to chymotrypsin; trypsin cleaves peptide linkages which are formed by basic amino acids (arginine, lysine, and histidine); chymotrypsin cleaves linkages with aromatic amino acids (phenylalanine, tyrosine, and tryptophan)	Preiser et al. (1975)
Aminopeptidase	Aminopeptidases are exopeptidases (act on N-terminal peptide of proteins)	Roncari and Zuber (1969)
Maltase	Convert maltose to glucose	Dahlqvist (1968)
Chitinase	Digestion of chitin	Borlongan et al. (2002)
Cellulase	The complete digestion of cellulose is mediated by two enzymes: cellulase and cellobiase; cellulase digests cellulose to disaccharide cellobiose; cellobiase converts cellobiose to glucose	Guillaume et al. (2001)
(Endopeptidase) Elastase I and II	Elastase I and II hydrolyse elastin and other structural proteins	Borlongan et al. (2002)
(Exopeptidase) Carboxypeptidase A and B	Exopeptidases cleave the C-terminal amino acid of peptides or proteins, carboxypeptidase A acts on aromatic C-terminal amino acids (phenylalanine, tyrosine, and tryptophan), carboxypeptidase B acts on peptides with basic amino acids (lysine and arginine)	Borlongan et al. (2002)

and enzyme secretion. Like other vertebrates, pepsin is the major enzyme produced in the stomach. Pepsin belongs to the category of proteases, the protein-digesting enzymes, and it requires an optimal pH ranging from 2 to 4 for its action. When the food enters the stomach, the stomach walls get distended which stimulates the glandular cells and gastric glands located in the mucosal folds, to secrete pepsinogen and hydrochloric acid (HCl), respectively. Oxynticopeptic cells produce both HCl and pepsinogen. The inactive pepsinogen, known as zymogen, is converted to the active pepsin in the presence of HCl acid. Pepsin is an endopeptidase that hydrolyses the protein by breaking the interior peptide bonds into polypeptides, peptides, and, to some extent, free amino acids (Lied & Solbakken 1984). Gastric acid also helps in the breakdown of larger nutrient compounds and releases some soluble nutrients. Pepsin is active only at acidic pH and is capable to hydrolyse complex proteins like collagen (Gildberg, 2004). Movement of stomach muscles mixes the food with pepsin and HCl, thus increasing the contact of food material with enzymes to enhance digestive activity. Mucous secretion and neutral mucopolysaccharides on the epithelial layer of the stomach protect the stomach walls against self-digestion by gastric acid and enzymes (Díaz et al., 2003; Machado et al., 2013).

Generally, the intestinal pH varies narrower in fish (6–9). The secretion of bicarbonates in the anterior region of intestine keeps the high alkalinity in this region which helps to neutralise the receiving acidic chime (Nikolopoulou et al., 2011). However, in few fishes, the pH values used to increase from anterior to posterior region (Moriarty, 1973), whereas a few species showed no variation in the intestinal pH (German & Bittong, 2009). Changes in gut microbiota and the substrates, changes in water temperature, and the time of examination can be possible reasons for changes in pH in various segments of the intestine. In marine fishes, the deposition of calcium carbonate from the drinking water raises the pH level in the intestine to higher alkaline pH (Wilson & Grosell, 2003). Alkaline proteases include enzymes which can function in a wide range of pH (Solovyev et al., 2015) and intestinal alkaline proteases functions in a pH range from 8 to 10. However, the action of alkaline proteases is observed to be species specific (Toledo-Solis et al., 2015). The activity of trypsin is found to be active in a pH range of 8–9, whereas the chymotrypsin functions from pH 8 to 10. Exception to this is observed in Prussian carp, in which the activity of trypsin and chymotrypsin was observed at pH 9–10 (Jany, 1976). The function of other enzymes like collagenase, elastase, or other proteases also occurs in higher alkaline range (Hidalgo et al., 1999). The activity of amylase and lipase also depends upon the species, pH of GI tract, the substrate, and the isoforms of the enzymes (Kapoor et al., 1976; Solovyev et al., 2018). The α-amylase showed activity at pH 4–9. Liver is an important organ which is considered as the metabolic centre in fish and hence, the absorbed nutrients will be transported to liver.

1.3.12 Passage Rate and Residence Time

GI transit determines the time taken by the food to move through the GIT and it is a complex process assisted by nerves. Gastric emptying decodes the exposure time of food material for digestion and absorption of nutrients. It is greatly influenced by physical parameters like water temperature, salinity, feeding frequency, physical characteristics, and the composition of the diet (Gerking, 2014; De et al., 2016). Increased exposure can ensure the proper enzymatic and microbial digestion and absorption of nutrients (Olsen & Ringø, 1997). It has been reported in several fish species that the decrease in gut transit time decreases the nutrient digestion and efficiency in many fish species (Jobling, 1995, Lee et al., 2000). Faster gastric evacuation time speeds up the feed intake in fish and increase in feeding frequency fastens the gastric evacuation time. Another observation was in turbots: moist feeds enhanced the gastric emptying time compared to dried pellets (Grove et al., 2001). van Nieuwenhoven et al. (2001) reported that the presence of guar gum in the diet slowed down the intestinal transit of digesta.

1.4 CRUSTACEAN DIGESTIVE SYSTEM

All crustaceans share a common feature of having bilaterally symmetrical bodies covered with a chitinous exoskeleton. This exoskeleton can vary in thickness and texture, ranging from thick and calcareous (e.g., crayfish) to delicate and transparent (e.g., water fleas). Unlike the internal skeletons of vertebrates, the exoskeleton of crustaceans does not grow with the animal and must be periodically shed during the moulting process. The moulting process is crucial for crustaceans as it allows them to grow and develop into their adult form. Adult crustaceans have segmented bodies and linked legs made of three separate segments: the head, the thorax, and the abdomen (Ceccaldi 1989; 1998). This is similar to other arthropods. The cephalothorax, a combination of the head and thorax that is characteristic of bigger crustaceans, is shielded by the carapace, a sizable portion of their exoskeleton.

1.4.1 Parts of Digestive Tract

The digestive tract is the site of nutrient digestion and absorption. The function of the digestive tract includes ingestion, nutrient transit, mechanical, chemical, and biochemical digestion and hydrolysis, absorption and transport of excreta. The lumen of the digestive tract, like that of other groups of animals, must be seen as being located outside the animal itself. Another unique feature of the crab digestive tract is that the majority of the animal's reserves are stored here. They are used during each intermoulting cycle to generate new tissues, such as following ecdysis. At the level of stomach walls, top crustaceans, particularly decapods, have a set of hard chewing pieces that are discarded and regrow after moulting. Their position and motility, on the one hand, and the muscles and nerves that contribute to movement, on the other, have distinct properties.

The decapod digestive system has three regions, viz., the foregut (oesophagus and stomach), the midgut (hepatopancreas, midgut tube, and anterior and posterior dorsal caeca), and the hindgut. The digestive tube of an adult decapod crustacean is separated into the foregut, midgut, and hindgut. The foregut typically consists of the oesophagus and a sizable portion of the stomach, which houses the masticating organs. The tubulated hepatopancreas, which secretes digestive enzymes, is found in the midgut but lacks chitin. However, the midgut tube is straight with chitin. It ends at the anus after being enlarged posteriorly into the rectum. As in most crustacean groups, the foregut, or stomadeum, and the hindgut, or protodeum, are distinguished extremely early during embryo development. The digestion system also shows the differentiation during larval development (Ceccaldi 1989; 1998). The digestive system, which was initially quite rudimentary, becomes complex after hatching. After hatching, nauplii larvae have an inoperative digestive system with a mouth but the anus will open after 2 or 3 moults.

1.4.1.1 Mouth and Mouth Parts

The maxillula, maxilla, mandibles, and maxillipeds are specialised prehensible appendages connected to the mouth. A tough labrum strengthens the front of the mouth. The oesophagus connects the mouth to the stomach and is short, straight, and vertically positioned. An anterior roll and two lateral rolls can both be seen in the cross section as an extension of the labrum. The lumen is shaped like an X. The chitin protein coat, which is of exodermal origin, covers the stomach's whole inner surface. The anterior region of the stomach is a flexible chamber with thin walls where food is chewed. Several articulated calcareous pieces provide as reinforcement and support the pyloric and posterior regions of the cardiac stomach. They are thick due to a coat of chitin.

1.4.1.2 Oesophagus

In crustaceans, oesophageal opening is positioned on the ventral side (Figure 1.15). Oesophagus is a small tube located vertically that forms a bridge between the mouth and the base of the cardiac stomach. It helps in the digestion process to transport the ingested food from mouth to cardiac stomach. The cross section of the oesophagus in crustaceans is structured like the letter X. Oesophageal lining is the extension of outer exoskeleton and is made of chitin, which is also reconstructed on each moulting cycle. Oesophageal glands are involved in the production of acidic mucopolysaccharides, which forms a component of gastric fluid in the stomach. The stomato-gastric nervous system governs the function of oesophagus and stomach (Maynard & Dando, 1974; Daur et al., 2016).

A cuticle 140 μm thick that covers the oesophageal epithelium is divided into a homogenous exocuticle and an interior endocuticle that is horizontally depressed and separated by a highly periodic acid Schiff (PAS) reaction, indicating a positive zone. According to Barker and Gibson (1978), the cuticle of the crab *Scylla serrata* is riddled by small canals (0.3 cm in diameter), which lead to glandular structures called tegumentary glands, which range in diameter from 0.07 to 0.13 mm. The oesophagus lumen is reached by the glandular structures' secretions via the canals; 0.2 mm thick longitudinal fibres and acidophilic bands make up the oesophageal musculature.

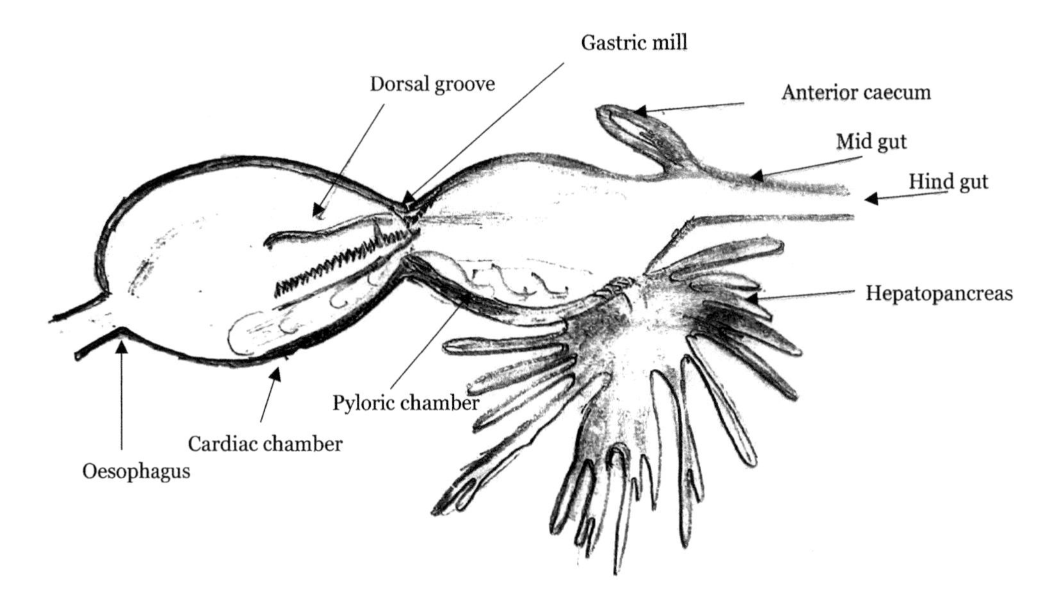

FIGURE 1.15 Sketch of different parts of GIT in crustacean.

1.4.1.3 Stomach

Stomach is a part of anterior gut and has mostly mechanical digestion of food particles. To support digestion, both the cardiac and pyloric chambers are equipped with specialised structures. Impregnated chemo and mechano-receptors along with hard tooth-like structures called ossicules assist in digestion in this region. Moreover, the stomach of crustacean has several folds and grooves which facilitate crushing of the food. The cross section of this region shows an X-shaped structure and internal lining gets renewed during the moulting process. In terms of histology, the stomach's walls resemble those of the oesophagus, but they are devoid of tegumentary glands, and the muscular components are larger and more distinct.

1.4.1.4 Midgut Gland or Hepatopancreas

Hepatopancreas is the major digestive gland in crustacean, which is a dorsally situated as a bi-lobed massive organ (Felgenhauer 1992). Hepatopancreas handles a major role in enzymatic digestion and absorption of nutrients in crustacean as it has many blind-ending tubules supported by secretory, absorptive, and storage cells (Figure 1.16). The multiple kinds of epithelial cells represent the midgut gland's diverse functions. The presence of cells with microvilli suggests an absorption role

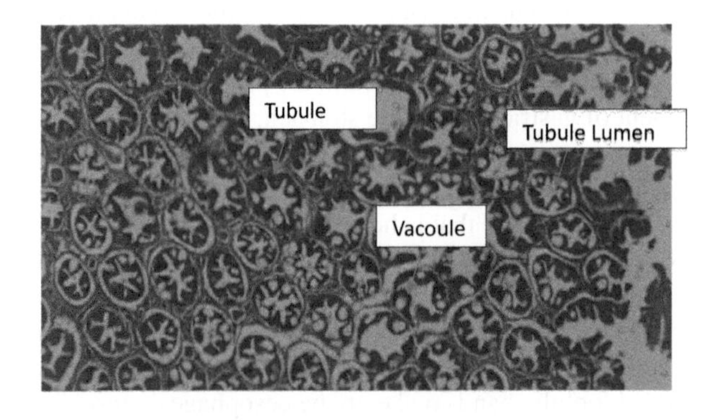

FIGURE 1.16 Histology of hepatopancreatic tissue of vannamei.

that has been well reported in various groups of decapods (Dall & Moriarty, 1983). Secretory glands, which produce vesicles that break in the GI tract, are also found. In the midgut of fewer crustaceans, such as copepods, several types of cells coexist (Yoshikoshi, 1975).

Higher crustaceans have a specialisation in hepatopancreas where distinct types of cells differentiate at the tip of the hepatopancreas tubules. According to Vogt G (2021) and Guillaume (2001), the hepatopancreas has different cells called E, M, F, and RE cells. They are the embryonic cells which give fibrillary F cells, secretory B cells and storage R cells. Additionally, B cells perform endocytosis. Although the origin of each cell type is unknown, their specialisation grows towards the base of the tubules. Brush border or microvilli are present in R cells and F cells; however, there is no definite confirmation that F cells have an equivalent absorptive role.

1.4.2 Function of Digestive Tract

The function of digestive tract includes intake, transportation, mechanical and chemical digestion, and hydrolysis, absorption, and removal of excreta. One unique feature of the crab's digestive tract is that the majority of the animal's reserves are stored here. They are used during each intermoulting cycle to generate new tissues, such as following ecdysis. At the level of stomach walls, top crustaceans, particularly decapods, have a set of hard chewing pieces that are discarded and rebuilt at each moult. Their number, location, and motility, and the muscles and nerves that contribute to movement have distinct properties. In fact, most chewing action on feed particles occurs in the stomach, and their movements are dependent on very sophisticated systems.

1.4.3 Enzymes in Digestive Tract

The crustaceans are mostly omnivorous, with a physically and functionally sophisticated digestive system. In the stomach, they feature highly efficient cuticular chewing and filtering structures that are regularly replaced by moulting. Decapods synthesise a diverse spectrum of digestive enzymes with distinct characteristics, including chitinases, cellulases, and collagenases. The said enzymes are produced in the hepatopancreas' F cells and are pre-pro-proteins coded in genome. Even digestive enzymes are unique, with particular endocrine processes that regulate their synthesis and proteases of low molecular weight. Despite the differences in the organisational schemas of the examined taxa, there are some shared features at the cellular level that should be recognised as crustacean characteristics, particularly within absorption stocking systems.

1.4.4 Role of the Microbiota in Digestion

Gut microorganisms are known to be major drivers of various metabolic processes in the host. Bacteria have also been frequently found in the digestive tract of decapods, primarily in the hindgut lumen. Gut bacteria secrete enzymes that may aid digestion in the host; however, whether these microorganisms act as symbionts with the host or simply acquire food and multiply during the gut transit (Saborowski, 2015; Martin et al., 2020) is not yet studied. The magnitude of the environmental perturbation and the physiological ability of each species influence gut contraction rates. A wide range of crustaceans have been studied to see how digesta moves from the foregut into the midgut and then through the hindgut.

Digestive system bacteria can provide food and vitamins, and even manufacture digestive enzymes. However, due to the short transit time, a direct enzymatic involvement would be very limited (Rieper, 1978). Despite the lack of particular features to aid their colonisation, crustacean hindguts are clearly excellent settings for numerous microorganisms. *Vibrio, Moraxella*, and *Flavobacterium* were identified as the most common bacterial taxa in the digestive system of *P. japonicus* by Sakata and Taruno (1987). The host taxon and its habitat had little effect on the hindgut microbiota, but the host's feeding behaviour influence the microbiota composition. Detritivores are only linked with epimural rod mats. Cocci is only found in scavengers and predators. Only detritivores had extensive colonisation, with more than 50% of the hindgut lining colonised. Harris (1993) reported that carnivorous fish had no rod bacteria.

1.5 CONCLUSION

The fish intestine is a multifunctional organ that helps with digestion, nutrition absorption, and maintaining proper body fluid balance. Fish live in a wide variety of habitats and consume many different foods, and these factors are reflected in their anatomy and physiology. The protection and sustainable management of aquatic environments will benefit from increased knowledge of fish physiology, nutrition, and ecological interactions because of ongoing research on the fish intestine. The study of anatomy and digestive physiology of crustaceans has shown major differences with vertebrates and a few unique traits. The digestive tract adaptations to food and feeding of varying food combinations, fluctuation of enzyme activities in response to these change, and further study on digestive enzymes all require much more investigation.

ACKNOWLEDGEMENTS

The authors are thankful to PhD scholars Mr. Veeramani Maruthi and Anusha E. Patel for providing the photographs and drawings of digestive system of fishes.

BIBLIOGRAPHY

Abbate, F., Germanà, G. P., De Carlos, F., Montalbano, G., Laurà, R., Levanti, M. B., and Germanà, A., 2006. The Oral Cavity of the Adult Zebrafish (Danio rerio). *Anatomia, Histologia, Embryologia*, 35(5), pp.299–304.

Abd El Hafez, E.A., Mokhtar, D.M., Abou-Elhamd, A.S. and Hassan, A.H.S., 2013. Comparative histomorphological studies on oesophagus of catfish and grass carp. *Journal of Histology*, 2013, Article ID 858674. http://doi.org/10.1155/2013/858674

Alabssawy, A.N., Khalaf-Allah, H.M. and Gafar, A.A., 2019. Anatomical and histological adaptations of digestive tract in relation to food and feeding habits of lizardfish, *Synodus variegatus* (Lacepède, 1803). *The Egyptian Journal of Aquatic Research*, 45(2), 159–165.

Al-Hussaini, A. H. (1946). The anatomy and histology of the alimentary tract of the bottomfeeder Mulloides auriflamma (Forsk.). *Journal of Morph.*,78, 121.

Anson, M.L., 1938. The estimation of pepsin, trypsin, papain, and cathepsin with hemoglobin. *The Journal of General Physiology*, *22*(1), p.79.

Ballesteros, N.A., Castro, R., Abos, B., Rodríguez Saint-Jean, S.S., Pérez-Prieto, S.I. and Tafalla, C., 2013. The pyloric caeca area is a major site for IgM+ and IgT+ B cell recruitment in response to oral vaccination in rainbow trout. *PLoS One*, 8 (6), p.e66118.

Barker, P. L. and Gibson, R. 1978. Observations of the structure of the mouthparts, histology of the alimentary tract, and digestive physiology of the mud crab *Scylla serrata* (Forskal) (Decapoda, Portunidae). *Journal of Experimental Marine Biology and Ecology*, 32, pp.177–196.

Barrington, E.J.W., 1957. The alimentary canal and digestion. *Physiology of Fishes*, 1, p.109.

Beulah, A.M., Sucharitha, K.V., Dheeraj, S. and Prameela, P., 2021. Effect of Casein edible coating on the postharvest quality of fresh Guava fruits during ambient storage conditions. *Carpathian Journal of Food Science & Technology*, 13(2), pp.5–20.

Bocina, I., Santic, Z., Restovic, I. and Topic, S., 2017. Histology of the digestive system of the garfish Belone belone (Teleostei: Belonidae). *The European Zoological Journal*, 84(1), pp.89–95.

Borlongan, I.G., Coloso, R.M. and Golez, N.V., 2002. Feeding habits and digestive physiology of fishes. In *Nutrition in Tropical Aquaculture: Essentials of fish nutrition, feeds, and feeding of tropical aquatic species*. Aquaculture Department, Southeast Asian Fisheries Development Center, pp. 77–97.

Bromley, P.J., 1994. The role of gastric evacuation experiments in quantifying the feeding rates of predatory fish. *Reviews in Fish Biology and Fisheries*, 4, pp.36–66.

Buddington, R.K. and Diamond, J.M., 1987. Pyloric ceca of fish: a" new" absorptive organ. *American Journal of Physiology-Gastrointestinal and Liver Physiology*, 252(1), pp.G65–G76.

Canan, B., Nascimento, W.S., Silva, N.B. and Chellappa, S., 2012. Morphohistology of the digestive tract of the damsel fish Stegastes fuscus (Osteichthyes: Pomacentridae). *The Scientific World Journal*, 1, p.787316.

Caruso, G., Denaro, M. and Genovese, L., 2009. Digestive enzymes in some teleost species of interest for Mediterranean aquaculture. *The Open Fish Science Journal*, 2(1), pp.74–86.

Ceccaldi, H., 1989, Anatomy and physiology of digestive tract of Crustaceans Decapods reared in aquaculture. In *Advances in Tropical Aquaculture*, Workshop at Tahiti, French Polynesia, 20 Feb-4 Mar 1989.

Ceccaldi, H.J., 1998. A synopsis of the morphology and physiology of the digestive system of some crustacean species studied in France. *Reviews in Fisheries Science*, 6(1–2), pp.13–39.

Chandy, M. 1956. On the oesophagus of the milk-fish Chanos chanos (Forskal). *Journal of the Zoological Society of India*, 8, pp.79–84.

Dahlqvist, A., 1968. Assay of intestinal disaccharidases. *Analytical Biochemistry*, 22(1), pp.99–107.

Dall, W. and Moriarty, D. J. W. 1983. Functional aspects of nutrition and digestion. In (Mantel, L.H., (Ed.)*The Biology of Crustacea, 5, Internal Anatomy and Physiological Regulation*, New York: Academic Press, pp.215–261.

Daur, N., Nadim, F. and Bucher, D., 2016. The complexity of small circuits: the stomatogastric nervous system. *Current Opinion in Neurobiology*, 41, pp.1–7.

de Oliveira, C., Taboga, S. R., Smarra, A. L. and Bonilla-Rodriguez, G. O., 2001. Microscopical aspects of accessory air breathing through a modified stomach in the armoured catfish Liposarcus anisitsi (Siluriformes, Loricariidae). *Cytobios*, 105(410), pp.153–162.

De Felice, E., Palladino, A., Tardella, F.M., Giaquinto, D., Barone, C.M.A., Crasto, A. and Scocco, P., 2021. A morphological, glycohistochemical and ultrastructural study on the stomach of adult Rainbow trout Oncorhynchus mykiss. *The European Zoological Journal*, 88(1), pp.269–278.

De Silva, S.S., Anderson, T.A., 1995. Aquaculture. In De Silva, S.S., & Anderson, T.A., (Eds.), *Fish Nutrition in Aquaculture*. Chapman and Hall, London, pp.1–14.

Deguara, S., Jauncey, K. and Agius, C., 2003. Enzyme activities and pH variations in the digestive tract of gilthead sea bream. *Journal of Fish Biology*, 62(5), pp.1033–1043.

Díaz, A.O., Garcia, A.M., Devincenti, C.V. and Goldemberg, A.L., 2003. Morphological and histochemical characterization of the mucosa of the digestive tract in Engraulis anchoita (Hubbs and Marini, 1935). *Anatomia, Histologia, Embryologia*, 32(6), pp.341–346.

Fange, R., and Grove, D., 1979. Digestion. In Hoar, W.S., Randall, D.J., & Brett, J.R. (Eds.), *Fish physiology. VIII. Bioenergetics and growth.* Academic Press, New York, pp.161–260

Fard, M. R. S., Weisheit, C. and Poynton, S. L., 2007. Intestinal pH profile in rainbow trout Oncorhynchus mykiss and microhabitat preference of the flagellate Spironucleus salmonis (Diplomonadida). *Diseases of Aquatic Organisms*, 76(3), pp.241–249.

Farrag, M.G.D., Azab, A. and Alabssawy, A., 2020. Comparative study on the histochemical structures of stomach, pyloric caeca and anterior intestine in the grey mullet, *Mugil cephalus* (Linnaeus, 1758). *Egyptian Journal of Aquatic Biology and Fisheries*, 24(7-Special issue), pp.1055–1071.

Felgenhauer, B.E., 1992. Internal anatomy of the Decapoda: an overview. *Microscopic Anatomy of Invertebrates*, 10, pp.45–75.

Ferraris, R. P., and de la Cruz, M. C. (1987). Development of the digestive tract of milkfish, Chanos chanos (Forsskal): histology and histochemistry. *Aquaculture*, 61, pp.241–257.

FU Newsletter. 2018. https://floridamuseum.ufl.edu/discover-fish/fish/anatomy/mouth-types

García-Carreño, F.L., 1992a. Protease inhibition in theory and practice. *Biotechnology Education*, 3(4), pp.145–150.

García-Carreño, F.L., 1992b. The digestive proteases of langostilla (*Pleuroncodes planipes*, Decapoda): their partial characterization, and the effect of feed on their composition. *Comparative Biochemistry and Physiology Part B: Comparative Biochemistry*, 103(3), pp.575–578.

Gerking, S.D., 2014. *Feeding ecology of fish*. Elsevier; pp.1–399.

German, D.P. and Bittong, R.A., 2009. Digestive enzyme activities and gastrointestinal fermentation in wood-eating catfishes. *Journal of Comparative Physiology* B, 179, pp.1025–1042.

Gildberg, A., 2004. Digestive enzyme activities in starved pre-slaughter farmed and wild-captured, Atlantic cod (Gadus morhua). *Aquaculture*, 238(1–4), pp.343–353.

Govoni, J.J., Boehlert, G.W. and Watanabe, Y., 1986. The physiology of digestion in fish larvae. *Environmental Biology of Fishes*, 16(1), pp.59–77.

Grosell, M., Wood, C. M., Wilson, R. W., Bury, N. R., Hogstrand, C., Rankin, C., and Jensen, F. B., 2005. Bicarbonate secretion plays a role in chloride and water absorption of the European flounder intestine. *American Journal of Physiology-Regulatory, Integrative and Comparative Physiology*, 288(4), pp.R936–R946.

Grove, D., Genna, R., Paralika, V., Boraston, J., Hornyold, M.G. and Siemens, R., 2001. Effects of dietary water content on meal size, daily food intake, digestion and growth in turbot, Scophthalmus maximus (L.). *Aquaculture Research*, 32(6), pp.433–442.

Guillaume, J., 2001. *Nutrition and feeding of fish and crustaceans*. Springer Science & Business Media.

Guillaume, J., and Choubert, G., 2001. Digestive physiology and nutrient digestibility in fishes. In Guillaume, J., Kaushik, S., Bergot, P., & Metailler, R. (Eds.), *Nutrition and Feeding of Fish and Crustaceans*, UK, pp.27–58.

Guillaume, J., Kaushik, S., Bergot, P., Metailler, R., 2001. *Nutrition and Feeding of Fish and Crustaceans*. Springer London. 1: XXIV, p.408.

Harder W., (1975) *Anatomy of Fishes*. Schweizerbart'sche Verlagsbuchhandlung, Stuttgart, 132, p.612.

Harris, J. M., 1993. Widespread occurrence of extensive epimural rod bacteria in the hindguts of marine Thalassinidae and Brachyura (Crustacea, Decapoda). *Marine Biology*, 116(4), pp.615–629

Hassan, A.A., 2013. Anatomy and histology of the digestive system of the carnivorous fish, the brown-spotted grouper, *Epinephelus chlorostigma* (Pisces; Serranidae) from the Red Sea. *Life Science Journal*, 10(2), pp.2149–2164.

Hernandez-Blazquez, F. J., Guerra, R. R., Kfoury, J. R., Bombonato, P. P., Cogliati, B., and Silva, J. R. M. C. da, 2006. Fat absorptive processes in the intestine of the Antarctic fish Notothenia coriiceps (Richardson, 1844). *Polar Biology*, 29(10), pp.831–836.

Hidalgo, M.C., Urea, E. and Sanz, A., 1999. Comparative study of digestive enzymes in fish with different nutritional habits. Proteolytic and amylase activities. *Aquaculture*, 170(3–4), pp.267–283.

Hlophe, S.N., Moyo, N.A.G. and Ncube, I., 2014. Postprandial changes in p H and enzyme activity from the stomach and intestines of Tilapia rendalli (B oulenger, 1897), O reochromis mossambicus (P eters, 1852) and C larias gariepinus (B urchell, 1822). *Journal of Applied Ichthyology*, 30(1), pp.35–41.

Holst, J.J., Fahrenkrug, J., Stadil, F. and Rehfeld, J.F. (1996) Gastrointestinal endocrinology. *Scandinavian Journal of Gastroenterology*, 31(s216), pp.27–38.

Horn, M.H. (1997) Feeding and digestion. In Evans, D.H. (Ed.), *The Physiology of Fishes*, CRC Press, Boca Raton, FL, pp.43–63.

Horn, M.H., Gawlicka, A.K., German, D.P., Logothetis, E.A., Cavanagh, J.W. and Boyle, K.S. (2006) Structure and function of the stomachless digestive system in three related species of New World silverside fishes (Atherinopsidae) representing herbivory, omnivory and carnivory. *Marine Biology*, 149, pp.1237–1245.

Hossain A.M. and Dutta H.M. (1996) Phylogeny, ontogeny, structure and function of digestive tract appendages (Caeca) in teleost fish. In: *Fish Morphology*. Balkema Pub, Brookfield, VT, pp.59–76.

Izvekova, G.I., Solovyev, M.M., Kashinskaya, E.N. and Izvekov, E.I., 2013. Variations in the activity of digestive enzymes along the intestine of the burbot Lota lota expressed by different methods. *Fish Physiology and Biochemistry*, 39, pp.1181–1193.

Jany, K.D., 1976. Studies on the digestive enzymes of the stomachless bonefish Carassius auratus gibelio (Bloch): endopeptidases. *Comparative Biochemistry and Physiology Part B: Comparative Biochemistry*, 53(1), pp.31–38.

Jobling, M. (Ed.), 1995. *Environmental Biology of Fishes*. Fish and Fisheries Series 16. Chapman & Hall, New York.

Kapoor, B.G., Smit, H. and Verighina, I.A., 1976. The alimentary canal and digestion in teleosts. In *Advances in marine biology*, Vol. 13, Academic Press, pp.109–239.

Karila, P., Shahbazi, F., Jensen, J. and Holmgren, S. (1998) Projections and actions of tachykininergic, cholinergic, and serotonergic neurons in the intestine of the Atlantic cod. *Cell and Tissue Research* 291, pp.403–413.

Khayyami, H., Zolgharnein, H., Salamat, N. and Movahedinia, A., 2015. Anatomy and histology of the stomach and pyloric caeca in Mugilidae, *Liza aurata* (Risso, 1810), *L. abu* (Heckel, 1843) and *Mugil cephalus* (Linnaeus, 1758). 6(19), *Scientific Publication on the Persian Gulf*, pp.59–66.

Khojasteh, S.B., Sheikhzadeh, F., Mohammadnejad, D. and Azami, A., 2009. Histological, histochemical and ultrastructural study of the intestine of rainbow trout (*Oncorhynchus mykiss*). *World Applied Sciences Journal*, 6(11), pp.1525–1531.

Lee, S.M., Hwang, U.G. and Cho, S.H., 2000. Effects of feeding frequency and dietary moisture content on growth, body composition and gastric evacuation of juvenile Korean rockfish (Sebastes schlegeli). *Aquaculture*, 187(3–4), pp.399–409.

Lied, E. and Solbakken, R., 1984. The course of protein digestion in Atlantic cod (Gadus morhua). *Comparative Biochemistry and Physiology Part A: Physiology*, 77(3), pp.503–506.

Machado, M.R.F., de Oliveira Souza, H., de Souza, V.L., de Azevedo, A., Goitein, R. and Nobre, A.D., 2013. Morphological and anatomical characterization of the digestive tract of Centropomus parallelus and C. undecimalis. *Acta Scientiarum. Biological Sciences*, 35(4), pp.467–474.

Manjakasy, J.M., Day, R.D., Kemp, A. and Tibbetts, I.R. (2009) Functional morphology of digestion in the stomachless, piscivorous needlefishes Tylosurus gavialoides and Strongylura leiura forex (Teleostei: Beloniformes). *Journal of Morphology*, 270, pp.1155–11.

Martin, G.G., Natha, Z., Henderson, N., Bang, S., Hendry, H., Loera, Y., 2020. Absence of a microbiome in the midgut trunk of six representative Crustacea. *Journal of Crustacean Biology*, 40, pp.122–130. https://doi.org/10.1093/jcbiol/ruz087

Maynard, D.M., Dando, M.R., 1974. The structure of the stomatogastric neuromuscular system in *Callinectes sapidus, Homarus americanus* and *Panulirus argus* (Decapoda Crustacea). *Philosophical Transactions of the Royal Society B*, 268, pp.161–220.

Meister, M. F., Humbert, W., Kirsch, R., and Vivien-Roels, B., 1983. Structure and ultrastructure of the oesophagus in sea-water and fresh-water teleosts (Pisces). *Zoomorph*, 102, pp.33–51.

Mitra, A., Mukhopadhyay, P.K. and Homechaudhuri, S., 2015. Histomorphological study of the gut developmental pattern in early life history stages of featherback, Chitala chitala (Hamilton). *Fisheries & Aquatic Life*, 23(1), pp.25–35.

Mokhtar, D.M., Abd-Elhafez, E.A. and Hassan, A.H., 2015. Light and Scanning Electron Microscopic Studies on the Intestine of Grass Carp (*Ctenopharyngodon idella*): II-Posterior Intestine. *Journal of Aquaculture Research & Development*, 6(12), p.1.

Moriarty, D.J.W., 1973. The physiology of digestion of blue-green algae in the cichlid fish, Tilapia nilotica. *Journal of Zoology*, 171(1), pp.25–39.

Nelson, J.A., Wubah, D.A., Whitmert, M.E., Johnson, E.A. and Stewart, D.J., (1999) Wood-eating catfishes of the genus Panaque: gut microflora and cellulolytic enzyme activities. *Journal of Fish Biology* 54, pp.1069–1082.

Nelson, N. J., 1944. A photometric adaptation of the somogyi method for the determination of glucose. *Journal of Biological Chemistry*, 153, pp.375–380.

Nikolopoulou, D., Moutou, K.A., Fountoulaki, E., Venou, B., Adamidou, S. and Alexis, M.N., 2011. Patterns of gastric evacuation, digesta characteristics and pH changes along the gastrointestinal tract of gilthead sea bream (Sparus aurata L.) and European sea bass (Dicentrarchus labrax L.). *Comparative Biochemistry and Physiology Part A: Molecular & Integrative Physiology*, 158(4), pp.406–414.

Nordrum, S., Bakke-McKellep, A. M., Krogdahl, Å. and Buddington, R. K., 2000. Effects of soybean meal and salinity on intestinal transport of nutrients in Atlantic salmon (Salmo salar L.) and rainbow trout (Oncorhynchus mykiss). *Comparative Biochemistry and Physiology Part B: Biochemistry and Molecular Biology*, 125(3), pp.317–335.

Olsen, R.E. and Ringø, E., 1997. Lipid digestibility in fish: a review. *Recent Research Developments in Lipids*, 1, pp.199–264.

Pedersen, T. and Falk-Petersen, I.B., 1992. Morphological changes during metamorphosis in cod (Gadus morhua L.), with particular reference to the development of the stomach and pyloric caeca. *Journal of Fish Biology*, 41(3), pp.449–461.

Pegel', V. A., 1950. In *Physiology of Digestion in Fish*. Biological Series. Proc. Tomsk State University.

Preiser, H., Schmitz, J., Maestracci, D. and Crane, R.K., 1975. Modification of an assay for trypsin and its application for the estimation of enteropeptidase. *Clinica Chimica Acta*, 59(2), pp.169–175.

Purushothaman, K., Lau, D., Saju, J.M., Sk, S.M., Lunny, D.P., Vij, S. and Orbán, L., 2016. Morpho-histological characterisation of the alimentary canal of an important food fish, Asian seabass (*Lates calcarifer*). *PeerJ*, 4, p.e2377.

Ray, A. K., & Ringø, E. (2014). The gastrointestinal tract of fish. In Merrifield, D. & Ringø, E. (Eds.), *Aquaculture nutrition: Gut health, probiotics and prebiotics*, John Wiley & Sons ltd, West Sussex, UK, pp.1–13.

Reifel, C. W., and Travill, A. A. (1977). Structure and carbohydrate histochemistry of the esophagus in ten teleostean species. *Journal of Morph*, 152, pp.303–314.

Rieper, M. 1978. Bacteria as food for marine harpacticoid copepods. *Marine Biology*, 45, pp.337–345.

Roncari, G. and Zuber, H., 1969. Thermophilic aminopeptidases from Bacillus stearothermophilus. I. Isolation, specificity, and general properties of the thermostable aminopeptidase I. *International Journal of Protein Research*, 1(1–4), pp.45–61.

Rønnestad, I., Yufera, M., Ueberschär, B., Ribeiro, L., Sæle, Ø. and Boglione, C., 2013. Feeding behaviour and digestive physiology in larval fish: current knowledge, and gaps and bottlenecks in research. *Reviews in Aquaculture*, 5, pp.S59–S98.

Rust, M.B., 2002. Nutritional physiology. In Halver, J.E. and Hardy, R.W. (Eds.), *Fish Nutrition*, 3rd edn., Elsevier Science, New York, pp.367–452.

Saborowski, R., 2015. Nutrition and digestion. *The Natural History of the Crustacea*, 4, pp.285–319.

Sakata, T. and Taruno, N., 1987. Ecological studies on microflora of digestive tract of prawn Penaeus japonicus. *Suisan Yoshoku* (Aquaculture), 35, pp.147–151.

Schwenk, K. and Rubega, M., 2005. Diversity of vertebrate feeding systems. In Matthias Starck, J. and Wang, T. (Eds.), *Physiological and Ecological Adaptations to Feeding in Vertebrates*, Science Publishers. Enfield, USA.

Senarat, S., Kettratad, J., Jiraungoorskul, W. and Kangwanrangsan, N.. 2015. Structural classifications in the digestive tract of short mackerel, Rastrelliger brachysoma (Bleeker, 1851) from Upper Gulf of Thailand. *Songklanakarin Journal of Science & Technology*. 37(5), pp.561–567.

Shephard, K.L., 1994. Functions for fish mucus. *Reviews in Fish Biology and Fisheries*, 4(4), pp.401–429.

Sinha, G.M., 1983. Scanning electron microscopic study of the intestinal mucosa of an Indian freshwater adult major carp, Labeo rohita (Hamilton). *Zeitschrift für mikroskopisch-anatomische Forschung*, 97, pp.979–992.

Sklan, D., Prag, T., and Lupatsch, I., 2004. Structure and function of the small intestine of the tilapia Oreochromis niloticus×Oreochromis aureus (Teleostei, Cichlidae). *Aquaculture Research*, 35(4), pp.350–357.

Solovyev, M.M., Izvekova, G.I., Kashinskaya, E.N. and Gisbert, E., 2018. Dependence of pH values in the digestive tract of freshwater fishes on some abiotic and biotic factors. *Hydrobiologia*, 807, pp.67–85.

Solovyev, M.M., Kashinskaya, E.N., Izvekova, G.I. and Glupov, V.V., 2015. pH values and activity of digestive enzymes in the gastrointestinal tract of fish in Lake Chany (West Siberia). *Journal of Ichthyology*, 55, pp.251–258.

Somogyi, M., 1952. Notes on sugar determination. *Journal of Biological Chemistry*, 195, pp.19–23.

Sugiura, S.H. and Ferraris, R.P., 2004. Dietary phosphorus-responsive genes in the intestine, pyloric ceca, and kidney of rainbow trout. *American Journal of Physiology-Regulatory, Integrative and Comparative Physiology*, 287(3), pp.R541–R550.

Sullivan, M.X., (1907). The physiology of the digestive tract of elasmobranchs. *Bulletin of the Bureau of Fisheries*, 27, pp.1–27.

Suyehiro, Y., (1942) A study on the digestive system and feeding habits of fish. *Japanese Journal of Zoology*, 10, pp.1–303.

Tanaka, M., 1973. Studies on the structure and function of the digestive system of teleost larvae. Dr. thesis. Dep. Fish. Fac. Agricult., Kyoto University. pp.120.

Toledo-Solís, F.J., Uscanga-Martínez, A., Guerrero-Zárate, R., Márquez-Couturier, G., Martínez-García, R., Camarillo-Coop, S., Perales-García, N., Rodríguez-Valencia, W., Gómez-Gómez, M.A. and Álvarez-González, C.A., 2015. Changes on digestive enzymes during initial ontogeny in the three-spot cichlid Cichlasoma trimaculatum. *Fish Physiology and Biochemistry*, 41, pp.267–279.

van Nieuwenhoven, M.A., Kovacs, E.M., Brummer, R.J.M., Westerterp-Plantenga, M.S. and Brouns, F., 2001. The effect of different dosages of guar gum on gastric emptying and small intestinal transit of a consumed semisolid meal. *Journal of the American College of Nutrition*, 20(1), pp.87–91.

Vogt, G., 2021. Synthesis of digestive enzymes, food processing, and nutrient absorption in decapod crustaceans: A comparison to the mammalian model of digestion. *Zoology*, 147, p.125945.

Walford, J. and Lam, T.J., 1993. Development of digestive tract and proteolytic enzyme activity in seabass (Lates calcarifer) larvae and juveniles. *Aquaculture*, 109(2), pp.187–205.

Wilson, J. and Castro, L. In The multifunctional gut of fish Vol. 30 (eds Grosell, M., Farrell, A. & Brauner, C.) Ch. 1, 1–55 (Academic Press, 2011).

Wilson, R.W. and Grosell, M., 2003. Intestinal bicarbonate secretion in marine teleost fish—source of bicarbonate, pH sensitivity, and consequences for whole animal acid–base and calcium homeostasis. *Biochimica et Biophysica Acta (BBA)-Biomembranes*, 1618(2), pp.163–174.

Winkler, U.K. and Stuckmann, M.A.R.T.I.N.A., 1979. Glycogen, hyaluronate, and some other polysaccharides greatly enhance the formation of exolipase by Serratia marcescens. *Journal of Bacteriology*, 138(3), pp.663–670.

Yamamoto, M., and Hirano, T. (1978). Morphological changes in the esophageal epithelium of the eel, Anguilla japonica, during adaptation to seawater. *Cell and Tissue Research*, 192, pp.25–38.

Yoshikoshi, K. 1975. On the structure and function of the alimentary canal of Tigriopus japonicus (Copepoda; Harpacticoïda). I. Histological structure. *Bulletin of the Japanese Society of Scientific Fisheries*, 41, pp.929–935.

2 Fish Immune Response

An Overview

Ritam Guha, Sajna Beegum and Nandha Kumar
Cochin University of Science and Technology, Kochi, India

Abdullah Nazar
Kerala University of Fisheries and Ocean Studies, Kochi, India

Nayomi John
Mar Athanasius College (Autonomous), Kothamangalam, India

Sreeja Lakshmi
King Nandhivarman College of Arts and Science, Thellar, Vandavasi, India

Preetham Elumalai
Cochin University of Science and Technology, Kochi, India

Mehdi Soltani
Murdoch University, Perth, Australia
University of Tehran, Tehran, Iran

2.1 INTRODUCTION

Fish exhibit diverse shapes and sizes, sharing a common body structure characterized by scale, swim bladder, and fins for swimming. However, exceptions exist, as seen in the scale-less clingfish and fin-lacking eel. Notably, lungfish can survive for hours outside water (Firdaus-Nawi and Zamri-Saad 2016). The earliest vertebrate fossils, Ostracoderms, lacked jaws and paired fins, featuring a single dorsal nostril. Modern Cyclostomes like myxinoids and lampreys exhibit similarities with these primitive forms, especially in anatomical and cellular components of the fish immune system.

Approximately 400 million years ago during the Silurian period, jawed fishes emerged. Their descendants, the bony fish (Osteichthyes) and cartilaginous fish (Chondrichthyes) have become the predominant forms in present-day seas and freshwaters. These two fish groups diverged from distinct Placoderm lines, likely during the late Silurian or early Devonian period. Cartilaginous fishes, closely resembling their Devonian ancestors, symbolize the culmination of an evolutionary lineage. In contrast, bony fish constitute a progressive group encompassing two primary subdivisions: ray-finned fishes (Actinopterygii) and fleshy-finned fishes (Sarcopterygii). A subset of Sarcopterygii, the Crossopterygians, directly preceded amphibians in the evolutionary lineage. Another subset, the Dipnoi or lungfish, shares various morphological and embryological similarities with amphibians, yet they do not directly lead to the emergence of higher vertebrates (George 1997).

The immune system in fish serves to protect them from bacterial, viral, and parasitic threats. Current knowledge about the immune system, specifically the division into innate and adaptive immunity, is primarily derived from studies on humans and mice. However, recent research has shown a growing interest in understanding fish immunology. The exploration of immunology in

DOI: 10.1201/9781003503811-2

aquatic organisms began with Metchnikoff's observation of phagocytosis in wounded starfish larvae, revealing an early interest in defense mechanisms against environmental challenges (Graham et al. 2008). Fish immune system can be categorized into innate and adaptive components. The innate system includes physical barriers like scale, mucus, skin, gill, gut, and nostril epithelium, along with effector molecules such as antimicrobial peptides, interferons (IFNs), and cells like monocytes, macrophages, nonspecific cytotoxic cells (NCCs), and neutrophils. The adaptive system comprises lymphoid organs such as the thymus, spleen, head, kidney, and mucosa-associated lymphoid tissue (MALT). Effector molecules in the adaptive system include IgT, IgM, and IgD, with B and T lymphocytes playing crucial roles in adaptive immunity (Makesh et al. 2022).

2.2 OVERVIEW OF FISH IMMUNE SYSTEM

Fish depend on their immune system to protect against infectious agents. Although our comprehension of the immune system, particularly the distinction between innate and adaptive immunity, has predominantly originated from studies on humans and mice, recent investigations have increasingly focused on fish immunology. The examination of immunology in aquatic organisms traces back to Metchnikoff's early fascination, exemplified by his identification of phagocytosis in injured starfish larvae (Graham et al. 2008). The innate immune system encompasses physical barriers like mucus, skin, gill, gut, and nostril epithelium. Additionally, it includes effector molecules such as antimicrobial peptides, IFNs, and cells like monocytes, macrophages, NCCs, and neutrophils. Adaptive immunity, a sophisticated defense mechanism, delivers a specific and targeted response to pathogens. It entails a complex interaction of specialized cells, antibodies, and memory mechanisms to identify and eliminate specific threats. Crucial components of adaptive immunity involve lymphoid organs like the thymus, spleen, head kidney (HK), and MALTs, serving as centers for immune response development and coordination. Effector molecules, including immunoglobulins (Igs; IgT/Z, IgM, and IgD/W), are generated by B lymphocytes and play pivotal roles in neutralizing and marking pathogens for destruction (Makesh et al. 2022). B lymphocytes, or B cells, are responsible for producing antibodies that specifically target and bind to pathogens. T lymphocytes, or T cells, are another vital element of adaptive immunity, with diverse roles such as coordinating immune responses through cytokines or directly attacking infected cells, and promoting the development of memory cells for long-term immunity. The hallmark feature of adaptive immunity lies in its ability to "remember" prior encounters with pathogens. This immunological memory facilitates a quicker and more potent response upon re-exposure to the same pathogen, enhancing the level of protection.

The immune mechanism of fish and their ability to resist infections are significantly impacted by environmental temperature. Fish infections are influenced by microbial contamination, environmental stresses, and temperature, with specific immunity playing a crucial role in combating these challenges. Notably, specific immunity can be established even at lower temperatures (Ramy and William 2009).

2.2.1 Innate Immunity

In response to an infection, the innate immune system is the first and initial responder when foreign particles enter the cells. Parameters of the innate immune system can be broadly classified into physical, cellular, and humoral components.

2.2.1.1 Physical Parameters

The physical parameters crucial for fish defense encompass the protective fish scale and mucus layer, with epithelial cells covering the skin, scale, gills, and gut playing a vital role in combating infections (Adef et al. 2018). Goblet cells, responsible for secreting fish epidermal mucus, serve as the primary immune response against diverse pathogens, acting as a bridge between the fish and its immediate environment. The primary function of mucus involves the entrapment and elimination of microorganisms. Within the mucus, innate immunity is conferred by the presence of several antimicrobial peptides, lysozymes, lectins, and proteases (Dash et al. 2018). Maintaining osmotic balance and the removal of microbes are undertaken by the non-keratinized living cells in the fish epidermis. The gills, being a primary entry point for pathogens, are particularly susceptible to infections, and to safeguard the gills from infection, a protective layer of mucus is deployed (Sujata et al. 2021).

2.2.1.2 Cellular Parameters

The cellular parameters of the immune system include macrophages, granulocytes, dendritic cells, B cells, natural killer (NK) cells, and cytotoxic cells.

Macrophages – The important function of macrophage is the activation of lymphocytes and phagocytes/monocytes in specific immune responses. In fish, macrophages play a vital role in preventing the spread of infectious agents and the removal of phagocytosed foreign particles. In the distal intestine, macrophages act as antigen-presenting cells (APCs) which allow for the identification of antigens. Macrophages consist of different receptors on the surface of their cells such as toll-like receptors (TLRs), pattern recognition receptors (PRRs), and C-type lectin receptors (CLRs). They are also an important source of cytokines and chemokines connecting innate and adaptive immunities. Antigen presentation to T cells is mainly carried out by macrophages. Chemokine receptors, TLRs, and adhesion molecules are expressed on circulating monocytes consisting of CD14+, and at the site of inflammation or infection, it is recognized through a pathogen-associated molecular pattern (Doaa et al. 2023).

Granulocytes – Peripheral blood and certain organs of fish consist of neutrophils, eosinophils, and basophils. Based on the species and specific locations, granulopoiesis occurs in fish tissues such as spleen, kidney, epigonal organ, organ of Leydig, and other specialized tissues. The antiparasitic activity in certain species of fish is mainly carried out by eosinophils. The function of basophil is not discovered yet. Due to environmental pollutants, diseases, and other reasons, response toward pathogens by neutrophils and eosinophils can be varied (Mokhtar et al. 2023).

Dendritic cells – Dendritic cells are the major cells involved in APCs on the surface of mucosal and lymphoid tissues. The function is to present antigens to T cells to activate acquired immune response and regulate tolerance mechanism toward self-antigens. The dendritic cells present antigens through major histocompatibility complex (MHC) I and II (Fucikova et al. 2019).

B cells – Chondrichthyes fish is the first species to identify bonafide Ig-producing B cells. These B cells produce three types of antibodies including IgM, IgW, and IgNAR. Teleost B cells secrete three Ig isotopes including IgM, IgD, and IgT/Z. Teleost IgM and IgT/Z are involved in systemic and mucosal immune responses, respectively (Oriol Sunyer 2019).

NK cells – There are two types of NK cells: NCCs and NK-like cells. Adaptive cell-mediated cytotoxicity (CMC) consists of molecules indicated on cytotoxic T lymphocytes (CTLs) and target cells. CTLs kill host cells containing foreign bodies by binding T cell receptors (TCRs) and co-receptor CD8+ to MHC class I. Alternatively, APCs such as macrophages, dendritic cells, and B cells take up extracellular antigens to process them and present the resulting peptide to CD4+ T helper cells in class II MHC molecule (Uwe et al. 2013).

2.2.1.3 Humoral Parameters

The humoral parameters include many components such as transferrin, lysozymes, lectins, cytokines, complements, and natural antibodies. Transferrin is a non-cytokine serum protein molecule with activated macrophages as a primary antimicrobial activity. The substance released by necrotic cells contains transferring cleaving activity (James and Miodrag 2003).

Lysozyme is an essential element in the defense mechanism of fish. It is noted that fish lysozyme contains lytic activity against Gram-negative and Gram-positive bacteria. In marine and freshwater fish, it is contained in the mucus, lymphoid tissue, plasma, and other fluids. The level of lysozyme activity varies depending on the sex, age, and size of fish, water temperature, pH, toxicants, infections, nutrition quality, and degree of stress (Shailesh and Sahoo 2008).

Lectins in fish have many biological roles including phagocytosis, cell adhesion, complement activation, and innate immunity. Lectin is also present in skin mucus, stomach, intestine, liver, gills, eggs, skin, serum, and plasma. Lectins are involved in the identification of pathogen-associated molecular patterns (P-AMPs) acting in agglutination and neutralization of microbial infections (John et al. 2022). Cytokines are secreted in response to stimulation of an immune response. They can bind to receptors in an autocrine or paracrine manner. Cytokines are synthesized by macrophages, lymphocytes, granulocytes, dendritic cells, mast cells, and epithelial cells. They can also be divided into IFNs, interleukins, tumor necrosis factors, colony-stimulating factors, and chemokines. These are secreted in response to various pathogens, such as parasitic, bacterial, or viral agents. Acute-phase proteins such as mannose-binding lectin and C-reactive protein can regulate phagocytosis due to cytokine release in tissues (Sebastian Reyes-Cerpa et al. 2012).

Complement: There are more than 35 circulating proteins in the formation of the complement component in teleost fish. It is activated through alternative, classical, and lectin pathways; and its major functions are opsonization, lysis of pathogen, and regulation of innate and adaptive immune response (Lorena et al. 2022).

2.2.2 Adaptive Immunity

Adaptive immunity in fish is divided into cell-mediated and humoral responses. T cells are the main component of cell-mediated immune response. After activation, T cells differentiate into effector helper T cells and effector cytotoxic T cells (CTCs). Cytokines and master transcription factors are central elements involved in the differentiation of helper T cells. In teleost fish, CTCs contribute to cytotoxicity. B cells play a major part in the humoral-mediated immune response. B cells take part in opsonization, neutralization, and complement binding antibodies and promote antibody-dependent cellular cytotoxicity. So far, three major classes of antibodies IgM, IgD/W, and IgT/Z have been identified in teleost fish (Adef et al. 2021).

IgM was discovered to be the ancient antibody that has a similar function to that of gnathostomes. During the ontological development and adaptive humoral immunity, IgM is the one that is first expressed. Conventional B2 and "innate" B1 and marginal zone B cells have IgM as the surface receptors. In cartilaginous fish, 50% of the serum protein constitutes IgM. Chondrichthyan IgM loci acquire a multiple mini-cluster organization instead of a bony vertebrate class that contains a single translocon locus. There are five IgM subclasses. Only one C gene and a few gene segments are present in each cluster (Sara and Michael 2016).

IgD/W transcript is made by alternative splicing of Ig rearrangement indicated in the μ chain. Hence every B cell can respond to IgD or IgM. The anterior and posterior kidney, spleen, and gills contain IgD secreting cells. Serum contains 2–8 μg/mL of IgD that is a hybrid of the Cμ1 domain followed by a different number of Cδ domains, which is a different feature of teleost fish. Fish without Cμ1 have no IgD heavy chain (Sara and Michael 2016).

IgT/Z first identified in rainbow trout (IgT) and zebrafish (IgZ) is a chimeric protein consisting of Cμ1 domain and Cτ/ζ domain. It is important for the security of mucosal boundaries in fish. The ratio of IgT to IgM is greater than 60 in the gut mucus than in serum. Fish having infection of gut parasite *Ceratomyxa shasta* showed only parasitic-specific IgT in the gut but not in the serum. Secreted IgT in trout mucus is a tetramer, and in the serum, it is a monomer (Simon et al. 2013).

2.3 AIM AND SCOPE OF CHAPTER

This chapter provides a comprehensive understanding of the immune system in fish, highlighting its crucial role in defending against bacterial, viral, and parasitic threats. While our understanding of the immune system has been primarily derived from studies on mammals, this chapter focuses on recent advancements in fish immunology. It helps to elucidate the distinctions between innate and adaptive immunity in fish, with a particular focus on recent advancements in fish immunology research. Furthermore, it aims to explore the impact of environmental factors on the immune mechanisms of fish.

It provides an in-depth overview of the fish immune system, delineating the innate and adaptive components. This includes discussions on physical barriers, effector molecules, and immune organs crucial for the immune response in fish. The chapter delves into the molecular interactions between hosts and pathogens, examining the activation of receptors, production of immune signals, and the role of various elements such as antimicrobial peptides, complement proteins, and cytotoxic cells in the immune response. It explores the evolving trends in fish immunogenomics, emphasizing the significance of understanding the genetic aspects of the immune system. It discusses the generation of immunomodulating molecules and the role of Igs, B and T lymphocytes, and other genetic components in fish immunity. It also provides a detailed overview of innate immunity in fish, encompassing physical, cellular, and humoral components. It delves into the physical parameters, such as protective scales and mucus layers, emphasizing their role in combating infections. The chapter explores the cellular parameters of the fish immune system, introducing macrophages, granulocytes, dendritic cells, B cells, and NK cells. Each cell type functions, receptors, and contributions to innate immunity are discussed, offering a comprehensive understanding of the cellular defense mechanisms in fish. It also provides insights into the humoral aspects of adaptive immunity in fish, emphasizing the significant role of B cells in producing antibodies, i.e., IgM, IgD, and IgT/Z. The functions of these antibodies in opsonization, neutralization, and complement binding are elucidated. It also highlights how environmental conditions influence fish infections, microbial contamination, and the role of specific immunity in overcoming these challenges.

The chapter discloses the challenges faced by the aquaculture industry, with infectious diseases being a major contributor to economic losses, and also emphasizes scientifically proven disease management strategies, including vaccination, antibiotic usage, biological and chemical disease control methods, and biosecurity measures. It explores the use of probiotics and investigates immunostimulants, both natural and synthetic, and their role in boosting nonspecific cellular and humoral immune activity. It also highlights the complexity of disease control in aquaculture, especially due to the challenges posed by the aquatic environment and the difficulty in diagnosing fish diseases and the combination of immunology and genetics in understanding fish immunity. The chapter also explores the symbiotic relationship between fish hosts and gut microbiota, emphasizing its crucial role in immune system development and nutrient absorption.

2.4 ANATOMICAL COMPONENTS OF FISH IMMUNE SYSTEM

The immune system of fish is a complex network of interrelated components, each with a distinct role in defense. This intricate system is influenced by the immunological history and evolutionary development of an organism. Comprising both mucosal and systemic compartments, the vertebrate immune system's initial encounter and primary defense against pathogens occur through the mucosal immune system. In fishes, MALT is distributed in various anatomical locations, including gills, skin, and gut. Among these, the lymphoid system associated with the gut has been extensively studied. The immunological responses within the skin mucosa serve as a crucial anatomical and physiological barrier against external infections. MALT functions as a secondary immune system, and within MALTs, the substantial role of gut-associated lymphoid tissue (GALT) is particularly significant in immune actions (Hatten et al. 2001). The mucosal immune response plays a major role in both innate and adaptive immunity and they serve as the front line of defense against foreign particles. Since most pathogens attack or initiate infection in the mucous surface, the mucosal immune response plays an important role in the duration of infection. Recent studies concentrate on their cellular and molecular composition in different species. Skin, gut, and gills contain large areas for possible attack by pathogens and they are in direct contact with the surrounding aquatic environment

(Mogensen 2009). Understanding the defense mechanisms in fish is imperative for devising effective health management tools to sustain the growing global aquaculture industry. Research on the immune system and mucosal immunity is pivotal for the development of innovative vaccination strategies in fish.

2.4.1 Skin and Mucous Barrier

Skin acts as a physical barrier between the body and the external environment and the body surfaces are protected by epithelial tissue. Skin is the physical protective membrane that protects the body from pathogenic agents as well as prevents the leakage of water or nutrients. The physical barrier is achieved by a stratified epidermis layer. The external epidermal layer in fish is covered by mucus. Teleost skin is very unique and histologically diverse. Fish skin is different from mammals as they continuously secret mucus which has defensive activity against pathogens (Jenkins et al. 1994). Mucus provides structural and functional strength to adapt to the physical, chemical, and biological properties of the aquatic environment. The aquatic environment harbors a large number of pathogenic organisms; hence, the skin of aquatic organisms is very crucial as the first line of protection against infectious agents.

As fishes are in intimate contact with water environments, the occurrence of cutaneous diseases is more common in them compared to terrestrial vertebrates. The mucosal integument or skin serves as the protective barrier from the external environment. The skin barrier is present throughout the lining of the body openings as well as covers the fins. Apart from the mechanical barrier, skin contributes as a metabolically active tissue. The mucosal skin of fish integument serves as a multifunctional organ and plays a role in protection, sensory perception, communication, respiration, locomotion, ion regulation, thermal regulation, and excretion (Paludan et al. 2021). The complex structure and cell composition of the skin are responsible for various functions and protection. All these functions including immunity, osmoregulation, and excretion are very significant especially in the larval stages of fish because during early development stages, the surface-to-volume ratio is high, and cutaneous epithelial layers play a crucial role in protection. The structure of the larval skin of teleost fish consists of cells arranged in a bi-layer (including mucus cells and chloride cells or mitochondria-rich cells or monocytes) which rest on a basal membrane that is surrounded by an extensive network of the hemocoel. Mucus cell content in yolk sac larvae consists of exclusively neutral intracellular glycoproteins. Goblet cells at the larval development stages are known to contain *N*-acetyl glucosamine and sialic acid. Fish epidermis comprises numerous viable cell types among which the motile keratocyte is structurally important. Keratocytes have the potential to cover fish wounds with another new layer of cells due to the rapid migration of cells from the adjacent wound margins. Apart from the migratory activity, different cells on the fish epidermis are capable of neutralizing foreign particles and pathogens. This combined effect of the mucosal layer contributes to the innate immune response and protects them from microorganisms and other potentially toxic substances in the surrounding environment. The wound healing is a collective function of the keratocyte and the ability of the epithelial layer to repair wound lesions. After injury of the epidermis, immediate response is initiated which causes mucosal secretions and accumulation over the surface of adjacent cells to protect them from the entering pathogenic agents.

Fish mucosal surfaces, including their gills, skin, and gastrointestinal tracts, serve as significant points of microbial exposure and as a thin physical barrier separating the internal milieu from the exterior environment. A protective mucus layer covers the host defense mechanisms and associated epithelia (which contain live cells). The initial line of defense against infection through the skin epidermis is thought to be cutaneous mucus (Kunisawa 2005). The fish skin mucus functions as a natural, semipermeable barrier that is physical, biochemical, and dynamic and allows the flow of gametes, hormones, gases, odorants, and nutrients. Additionally, mucus serves as a biological barrier, which is an important part of how fishes defend themselves. Skin mucus has developed strong defenses that can be captured because it is impairment to the majority of bacteria and many pathogens; it immobilizes pathogens before they come into touch with epithelial surfaces. This happens as a result of the water circulation entrapping and expelling particles, bacteria, or viruses in this mucus layer. Additionally, the mucus in most fish species is constantly released and replaced, which hinders the invasion of pathogenic agents. Pathogens must migrate "upstream" since the mucus layer might occasionally be digested or removed. The ability of mucus to retain an undisturbed layer of mucus near epithelial surfaces despite severe shearing events (such as eating, coughing, intestinal peristalsis, and copulation) is a dynamic characteristic that is sometimes overlooked (Toda et al. 2011). The composition of mucus, a complex fluid, varies throughout the epithelial surface. The proteins in skin mucus must continue to function in extreme conditions like higher temperatures and hydraulic pressure because the skin mucus is exposed to the outside environment. Few surfaces are immune to the adhesion of mucus, which is a viscid (sticky) gel. Numerous creatures, including bacteria, barnacles, and snails, employ the adhesive properties of mucus to stick to the surfaces on which they dwell.

Small fishes also employ mucus to gather nutrients suspended in water. The gastrointestinal mucus of developed vertebrates has been examined to show that lipids in mucus secretions, particularly covalently bound fatty acids, contribute to fiber–fiber interactions that significantly increase the viscoelasticity of the gel. The balance between the rates of secretion, breakdown, and shedding determines how thick the mucus layer is. The thickness of the mucus blanket can be significantly increased by toxic and irritating agents that dramatically induce mucus secretion. Some fish, especially sharks, have small quantities of mucus on their skin naturally. The upkeep of the body's immune system depends greatly on the make-up and properties of skin mucus. Saliva and tear fluid have significantly reduced viscoelasticity and are easily permeable by motile bacteria simply by being somewhat more hydrated. Transporting mucus requires viscoelasticity to be well controlled, which is regulated by hydration. To control the viscoelasticity of secreted mucus gels, mucosal epithelia must somehow control the ionic environment (Hawkes 1974). Most mucosal epithelia probably do this in part by controlling the ionic environment to control mucus hydration and subsequently viscoelasticity. The airway mucus has been studied most thoroughly in this regard. The pH, calcium, non-mucin glycoproteins, and secreted lipids are just a few more elements that regulate mucus viscoelasticity. It has been thought about how fish mucus may serve a variety of purposes and how it may play a crucial part in fish immunity. Its first-line defensive and border roles in disease resistance have been investigated. Skin mucus also offers a medium in which an antibacterial mechanism may operate. Thus, fish skin mucus acts as a storage place for a wide range of physiologically active compounds as well as a large number of immune system defense components, both innate and acquired. In addition to preventing the invasion and growth of dangerous microorganisms, mucus also serves several other purposes, such as lubrication, ion regulation, and parental care behavior. Different fish species have shown epidermal mucus to have antimicrobial properties against infectious pathogens (bacteria and viruses), and increased expression of one or more of the aforementioned antimicrobial components in fish epidermal mucus has been seen following microbial stress. These findings support the idea that epidermal mucus plays a protective role in fish against infectious pathogens. High molecular weight, filamentous, highly glycosylated glycoproteins known as mucins are the most prevalent elements of the mucus layer and can contain up to 50% of their dry weight in carbohydrate chains. Mucins also contribute significantly to mucosae defense as a matrix for a variety of antibacterial compounds, and give mucosal layers viscoelastic and rheological qualities (Austin 2006).

2.4.2 Gills and Immune Interface

Gill as the respiratory organ consists of gill epithelium, glycocalyx layer, and mucous layer. Immune cells include macrophages, neutrophils, pillar cells, and eosinophilic granulocytes associated with lymphoid tissue which all prevent the entry of pathogens. Gills are regarded as mucosal surfaces, and they have developed components and structures that resemble those of the respiratory mucosa of mammals through convergent evolution. The upper respiratory tract of mammals has similar structures with the cuboidal or squamous cell epithelia of the gill in which primary and secondary lamella are covered by one to four layers, just like the local mucus-secreting cells, also known as goblet cells, which are typical of type I mucosal epithelia and serve as the first line of defense and cell protection. On pathogenic challenge or vaccination, several investigations have described the production of gill pathogen-specific IgM titers but antigen-specific responses mediated by other teleost Igs are as of yet unknown. IgT might be essential for gill immunity in fish since it is essential for mucosal immunity in the gut and skin. However, new evidence suggests that trout IgD may be crucial for gill immune reactions. Therefore, it is still unclear how the various teleost Igs function in the gill. Fish gills, which are frequently exposed to the outside environment, are well known to play a crucial role in fish immunology. These structures develop during ontogenesis from pharyngeal arches, and their medial epithelium comes from endoderm while their lateral epithelium comes from ectoderm. A variety of lymphoid structures, such as the tonsils, thymus, and other secondary lymphoid structures collectively known as Waldeyer's lymphatic ring, are produced in mammals as a result of the combination of pharyngeal endoderm and ectoderm. Tonsils in fish have not been described, although inter-branchial lymphoid tissue (ILT) in the salmonid gills was first found and described about ten years ago. Later, it was shown that these lymphoid aggregates might be seen in several additional teleost species in addition to salmonid species. The ILT was termed by Koppang et al. (2010) because it is located between the major lamellae at the distal stretch of the interlamellar septum. In the ILT, lymphoid cells, mostly T cells, are embedded in a network of epithelial cells, although there are also a few sporadic B cells and some highly MHC class cells. The thymic outer zone and its architecture appear to be related. Following its initial identification, the ILT has subsequently split into two parts: a proximal part (pILT), which corresponds to the lymphoid tissue that was first characterized, and a distal part (dILT), which extends along the major lamellae's trailing edges. Salmonid ILT CCL19 expression was discovered in 2020, supporting the idea that the ILT should be treated as a separate lymphoid organ. Salmonid ILT is entirely intraepithelial (IEL) and is separatable from the underlying tissue by a substantial basal membrane. Therefore, this structure is not vascularized, and it most likely is not innervated (Akira et al. 2006).

2.4.3 Gut-Associated Lymphoid Tissue (GALT)

The gastrointestinal tract of fish absorbs the nutrients while preventing pathogen invasion through its epithelium. If a pathogen enters, it is trapped and killed by GALT that is found in both bony and cartilaginous fish. The detailed mechanism of the GALT against infection in fish is unknown but it produces immune responses in both bony and cartilaginous fish (Akira et al. 2006).

2.4.4 Head Kidney and Spleen as Primary Immune Organs

In fish there are two stages of kidney development: pronephros and mesonephros; after the organogenesis, they persist together as cephalic/HK and endocrine/trunk kidney. The kidney has both exocrine functions as it produces urine and endocrine function to produce hormones. HK is the hematopoietic tissue which is equivalent to the bone marrow due to its structural similarity with higher vertebrates. Functional and transcriptional studies along with B cell development are reported in many studies. The HK also has antigen-sampling abilities and antigen-retention abilities. The sinusoidal endothelial cells, macrophages, and fibroblastic reticular cells form active phagocytic elements and it is common place in piscine pathological procedures to isolate the pathogens causing systemic infections from the HK (Zapata and Amemiya 2000). For many reasons, the HK is regarded as both a primary and secondary lymphoid organ. The HK contains proliferating B cell precursors and plasma cells; the trunk kidney contains plasma blasts and abundant B cells, some of which are activated. Thyroid cells arranged in follicles, the teleost adrenal counterpart, and the corpuscles of Stannius are also present in the kidney. There are three stages or generations in the development of the kidney in both birds and mammals: the pronephros, the mesonephros, and the metanephros. Here, the mesonephros and pronephros atrophy, leaving the metanephros as the kidney for life. The kidney performs endocrine tasks by producing hormones in addition to performing exocrine functions by producing urine. However, during organogenesis, the pronephros and mesonephros continue to exist as the cephalic or HK and the exocrine or trunk kidney, respectively, and in fish the metanephros do not form. Blood moves from the glomeruli to the sinusoids via efferent arterioles. There are several sinusoids in the kidney interstitium, and they form the renal portal system along with receiving blood from afferent arterioles and segmental veins from the lateral and dorsal musculature and the epidermis. Due to its anatomical resemblance to that of higher vertebrate bone marrow, the hematopoietic tissue found in the HK is referred to as the teleost counterpart of bone marrow. This idea is further supported by functional and transcriptional research, and B cell development has been the subject of numerous investigations. Additionally, various studies have shown that the HK is capable of retaining antigens. The active phagocytic components are composed of sinusoidal endothelial cells, macrophages, and fibroblastic reticular cells (Press and Evensen 1999). The HK is a common location for microorganisms producing systemic infections in piscine pathological treatments. Additionally, immunological genes in the HK are upregulated in response to a range of antigen challenges, which is a characteristic of secondary lymphoid organs. For instance, Koppang et al. (1998) showed that immunization causes MHC upregulation. For this reason, the HK is both a primary and secondary lymphoid organ as a result of these characteristics. MHC class II cells and T cells can be found scattered throughout the kidney, according to morphological investigations. B cell maturation occurs in an interesting gradient. The majority of the B cell precursors and plasma cells in the HK are proliferating, in contrast to the plentiful B cells, some of which are activated, and plasmablasts found in the trunk kidney (Zwollo et al. 2005). Most MHC class II cells in the renal tissue appear to be macrophages histologically. Melano-macrophages, however, are a significant cell population in the salmonid kidney in addition to B and T cells and macrophages. They were formerly believed to only store melanin. Salmonid head kidney (SHK-1) cell line, which had been categorized as a cell line made up of macrophage-like cells derived from the salmon HK, is able to synthesize melanin (Haugarvoll et al. 2006). In the kidney tissue, there exist many endocrine cells. They contain thyroid cells grouped in follicles in the HK, which, depending on the species, may be outside or inside the tissue (Geven and Klaren 2017). This region is also home to the teleost adrenal analog, and such tissue can be seen as islands between hematopoietic tissues.

The spleen is the primordial secondary lymphoid organ and most fish possess the spleen in which adaptive immune responses are generated. The primordium of the spleen arises from the mesenchyme derivative of mesoderm. Spleen filtrate peripheral blood, and visceral arteries supply blood to the spleen and the draining splenic veins join the hepatic portal system. Filtration occurs through leukocyte sheets of the white pulp, which will be released in red pulp consisting of peripheral blood. In the spleen melano-macrophages are found either in clusters or loosely dispersed in the white pulp. Enlargement of the spleen can occur due to different infections which clarifies its role as a secondary lymphoid organ. Retention of peripheral blood in the red pulp or proliferation of cells in the white pulp or both may result in splenomegaly. Almost all gnathostomes have a spleen which is thought to be the first secondary lymphoid organ and where adaptive immune responses are produced (Flajnik 2018). Local mesenchyme produced from mesoderm gives rise to the spleen's primordium, and a sophisticated vasculature develops inside this anlage. Leukocyte infiltration is followed by the production of the white pulp. The purpose of the spleen is to filter out peripheral blood. The organ is supplied by visceral arteries, and the splenic veins that drain it connect to the hepatic portal system. There are no signs of vagal innervation in the teleost spleen, which is innervated by sympathetic fibers. The arteries split off in the splenic tissue and come to an end in structures known as ellipsoids. Although their construction differs amongst different fish species, trout construction appears to be typical of the majority of species under study. In contrast to the more flattened ellipsoidal endothelial cells found in humans, the ellipsoids in this case are made up of cubical endothelial cells. Endothelial cells are spaced apart, and the basal lamina beneath them is shattered. This permits a thorough filtering of the structures. Macrophages and reticular cells surround the basal lamina. The white pulp is made up of these cells and structures put together. The white pulp is replete with MHC class II+ cells, which are almost certainly macrophages, and T cells. When compared to the scenario in mammals, the construction is essentially the same. The white pulp's leukocyte sheets serve as filters, and the final point of release is the red pulp, which is essentially just peripheral blood surrounding the white pulp. Fish have melanoma macrophages in their spleens, which, depending on the species, can be either grouped or more randomly distributed inside the white pulp. According to multiple studies on antigen retention, such locations are hypothesized to act as "dumping grounds" for various types of waste material. Thorsen et al. (2006) demonstrated the expression of genes related to the melanogenesis pathway in the spleen, indicating the possibility of de novo melanin synthesis in the organ. Spleen enlargement, or splenomegaly, can result from a variety of illnesses, which is in line with the organ's role as a secondary lymphoid organ. Splenomegaly can, in theory, result from either the retention of peripheral blood in the red pulp or the proliferation of cells in the white pulp or both.

2.4.5 Immune Cell Population (Immunocompetent Cells)

The immune cell populations that are found in fish are similar to those present in mammals. Immune cells present in fish consist of NK cells, granular leukocytes, dendritic cells, thrombocytes, cytotoxic cells, mast cells, eosinophilic cells, and melano-macrophage centers. The cell-mediated immune system includes leukocytes and NKs (type of lymphocytes); cytotoxicity and phagocytosis are the main processes seen in cellular immunity. The main cellular immune systems of fish are macrophages, granulocytes, dendritic cells, NK cells, and CTCs.

2.4.5.1 Natural Killer Cells

Fish processes two kinds of NK cells which include NCCs and NK cells. The NK cells have large and granular morphology, while NCCs are small agranular cells. The NK-like cells are also involved in destroying allogenic and virus-infected cells. The NK cell enhancement factors have been discovered which act as a cell marker. During a pathogenic infection, NK enhancement factor gene expression is activated in the tissue of fish which improves the cytotoxic effect of NCCs (Huang et al. 2021). NCCs were recognized first and represent the most investigated population of killer cells in teleosts. The functions of NCCs are similar to those of higher vertebrates, acting on target cells like virus-infected cells and tumor cells. The NCC also plays a role in antibacterial immunity by initiating the synthesis and release of cytokines. The NCCs found in tilapia and catfish can express components involved in the granule exocytosis pathway of CMC. In mammals, the nonspecific CMC reactions are carried out by NK cells. The NCC and NK-like cells are linked with the nonspecific CMC function in fish. Fish NK-like cells have been separated from blood leukocytes and are involved in functions like the destruction of virus-infected, allogeneic, as well as xenogenic target cells. In contrast, NCCs are found to target a variety of cells like some protozoans and tumor cells, and are also very active in the spleen and HK. The NCCs kill the affected cells spontaneously via necrotic and apoptotic mechanisms. Further studies show that granulysin, perforin, and serglycin gene encoding molecules which show lytic properties are expressed by NCCs. NCCs are found to express the NCC receptor protein 1 on their cell surface. Also, several NCC subsets are found in various immune compartments of fish (Jaso-Friedmann et al. 2000).

2.4.5.2 Macrophages

Macrophages have a wide-ranging role in specific immunity including lymphocyte activation and phagocytosis. Specific receptors are present on macrophage surfaces which can recognize β-glucan that initiates the bactericidal activity of leukocytes due to the production of reactive oxygen species (ROS). Another bactericidal activity is carried out by the production of nitric oxide catalyzed by nitric oxide synthase as demonstrated in the HK (Schoor & Plumb 1994). Observations on separating the molecules in crude leukocytes involved in nitric oxide production in macrophages of goldfish indicate the presence of transferring which acts as the mediator to activate fish granulocytes and macrophages. The major role in preventing the spread of pathogens and the destruction of phagocytosed pathogens is linked with neutrophils and macrophages. Macrophages act as an open environment in the form of antigen-representing cells where the interaction of antigen with the adaptive immune response is carried out. Several other granulocytes are also involved in innate clearance in the intestinal segments along with the resident macrophages. Macrophages are an important source of chemokines and cytokines which regulate immune action effectively. Moreover, macrophages act as an interacting site of antigens with T cells (Arango Duque and Descoteaux, 2014).

2.4.5.3 Neutrophils

Neutrophils are accommodated in blood, peritoneal cavity, and lymphoid organs. They are polymorphonuclear cells with the ability to phagocytose cells and foreign particles. They synthesize superoxide anions to act as a bactericidal component. Neutrophils exhibit inflammatory immune reactions against different pathogenic agents during infection. They are the primary granulocytes to be observed first at the injury site followed by macrophages. At the injury site, neutrophils regulate the phagocytic action by using antimicrobial peptides, proteolytic enzymes, and ROS. Neutrophils have the potential to release extracellular fibers referred to as neutrophil extracellular traps that are composed of DNA, histones, and proteins that can bind to and destroy pathogens. Moreover, the interaction of granular leukocyte-derived myeloperoxidase enzyme along with hydrogen peroxide produces hypochlorite which in turn is used to produce chloramines, oxidative substances for phagocytic activity (Biller-Takahashi and Urbinati 2014).

2.5 INNATE IMMUNE RESPONSE IN FISH

2.5.1 Toll-Like Receptors (TLRs) and Pattern Recognition Receptors (PRRs)

A class of receptors known as PRRs can identify chemical structures on the surface of pathogens, apoptotic host cells, and damage aged cells directly. PRRs serve as a link between specific and nonspecific immunity (Thaiss et al. 2016). PRRs can produce nonspecific anti-infection, anticancer, and other immune-protected actions by recognizing and attaching to ligands. Based on protein domain homology, the majority of PRRs in vertebrates' innate immune system can be divided into the following five categories. These include missing in melanoma-2 (AIM2)-like receptors (ALRs), TLRs, nucleotide oligomerization domain (NOD)-like receptors (NLRs), retinoic acid-inducible gene-I (RIG-I)-like receptors (RLRs), and CLRs (Gajewski et al. 2013). In principle, effector domains, intermediate domains, and ligand recognition domains make up PRRs. Through their effector domains, PRRs identify and bind their specific ligands and enlist adaptor molecules that share the same structure, thereby starting downstream signaling pathways that have an effect. The understanding of various PRR signaling pathways has been substantially advanced in recent years by the increased study on the identification and binding of PRRs and their ligands. This has also led to the development of new therapy concepts for immune-related disorders and even tumors. Innate immunity, which is split into two levels, is the first line of defense against pathogens that have progressively evolved in organisms. In the first level, the host's skin, mucosal tissue, blood–brain barrier, and chemical barrier (such as fatty acid, pH, enzyme, and complement system) can effectively keep away the invasion of general pathogenic microorganisms; in the second level, the innate immune system of vertebrates defends the organism through nonspecific immune defense supplemented by innate immune cells (Vidya et al. 2018).

Monocytes, neutrophils, macrophages, dendritic cells, NK cells, mast cells, eosinophils, and basophils are among the major types of innate immune cells. Innate immune cells lack specific antigen recognition receptors, in contrast to highly specialized T and B cells. PRRs induce immune-protective effects, such as anti-infection and antitumor effects and participate in the initiation and process of specific immune responses by recognizing and binding certain common molecules on the surface of pathogens, apoptotic host cells, and damaged cells (Fitzgerald and Kagan 2020).

One of the first PRRs identified in the innate immune system, which is crucial for inflammatory responses, is TLR. Membrane-bound signal receptors, or TLRs, are crucial PRRs in vertebrates' innate immune systems. These receptor molecules often have two roles: one is to convey signals and the other is to bind selectively to the ligand (Gajewski et al. 2013). To activate immune cells involved in the inflammatory response through gene transcription and to create and secrete a range of pro-inflammatory and antiviral substances, the relevant signal transduction will enhance the impact of anti-pathogen infection. TLRs are classified as type I transmembrane glycoproteins and consist of three distinct regions: the extracellular, transmembrane, and intracellular domains. Leucine-rich repeats (LRRs), which are found in the extracellular region, are in charge of extracellular pattern recognition and the recognition of particular ligands (Iwasaki and Medzhitov 2010).

2.5.2 Complement System and Its Role in Fish Immunity

Both the detection and removal of possible infections from the host are crucially dependent on the complement system. Furthermore, the development and coordination of an acquired immune response are greatly aided by the activation of the complement system. An essential component of the natural defense against common infections is the complement system. When complement is activated, strong and effective proteolytic cascades are created, which contribute to the pathogen's opsonization and lysis as well as the development of the classical inflammatory response by producing strong pro-inflammatory chemicals (Smith et al. 2019).

From its rudimentary form, the complement system has served as a crucial link between innate and adaptive immunity. Numerous complement proteins found in fish have been discovered and shown to share similarities with their mammalian counterparts. More than thirty proteins from plasma and cellular constituents make up the complement system.

Numerous elements of these activation pathways, such as factor Bf/C2, factor D (Df), and C3 (alternative pathway) and C1r/C1s (Fitzgerald and Kagan 2020) and C4 (traditional pathway), have been cloned in teleost fish. The presence of both classical and alternative activation pathways that are functionally well defined may be implied by the finding of these fish complement components. Nonetheless, there is evidence to support the theory that some of these elements might be functionally uninhibited in the sense that they might activate several pathways. However, it has been amply shown that fish complement is capable of opsonizing alien organisms for phagocyte destruction and lysing foreign cells. Fish have different isoforms of various complement proteins, including factor B and C3. It has been postulated that complement protein variety functions to increase the innate immune system's capacity for recognition and response. Complement cell involvement in host defense and the roles each particular protein, including its different isoforms, plays are crucial to comprehend both understanding how this system evolved and developing novel approaches for managing fish health (Walport 2001).

2.5.3 Interferons and Antiviral Response in Fish

Fish IFNs serve as a part of primary line of defense against infections. In 2003, the IFN genes for Atlantic salmon were discovered. According to Lutfalla et al. (2003), the IFNs were also recovered from pufferfish (*Tetraodon nigroviridis*) Figure 2.1. The primary cause of the increase of IFN gene expression in the body is virus infection. The primary cause of fish's antiviral status against pathogens is the upregulation of these genes during contact with the pathogens. Viperin, fish-specific protein kinase, and original antigenic immorality are the major genes involved in production. The viperin gene expression results in the innate immune system and antiviral defenses. As a result, they are crucial to the management of the immune system. Several viruses are affected by the antiviral activity of fish viperin. Primordial antigenic immortality is sometimes referred to as immunological imprinting, the Hoskins effect, or Fish-Specific Protein Kinase (PKZ) which initiates innate immune responses through the IRF3- and ISGF3-like mediated pathways (see Figure 2.2).

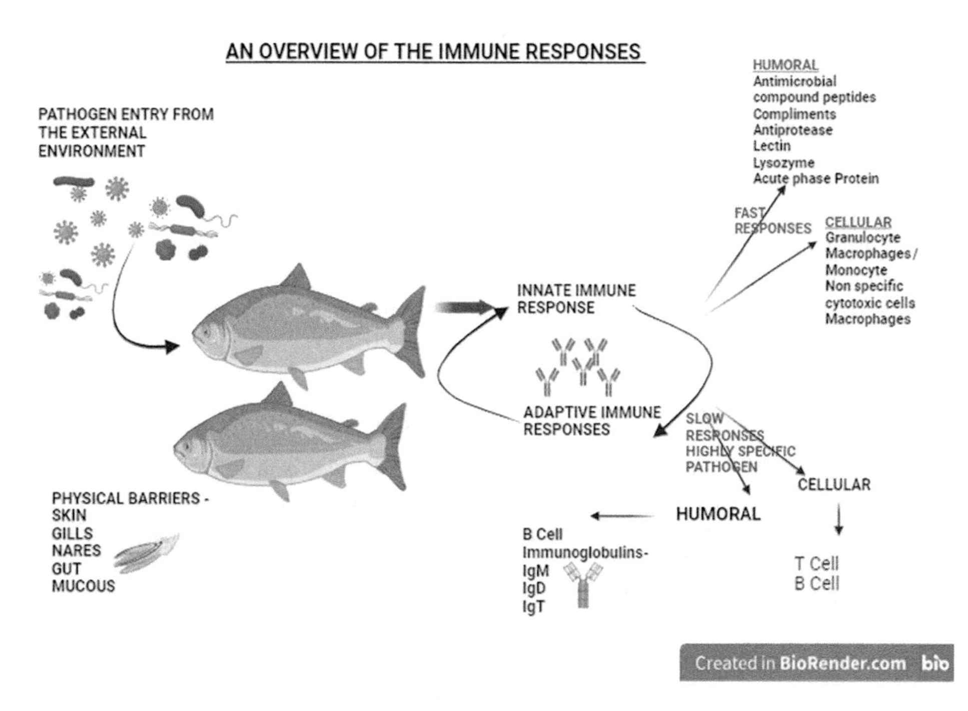

FIGURE 2.1 An overview of the immune response. This figure demonstrates elements of physical, cellular and humoral immunities of fish.

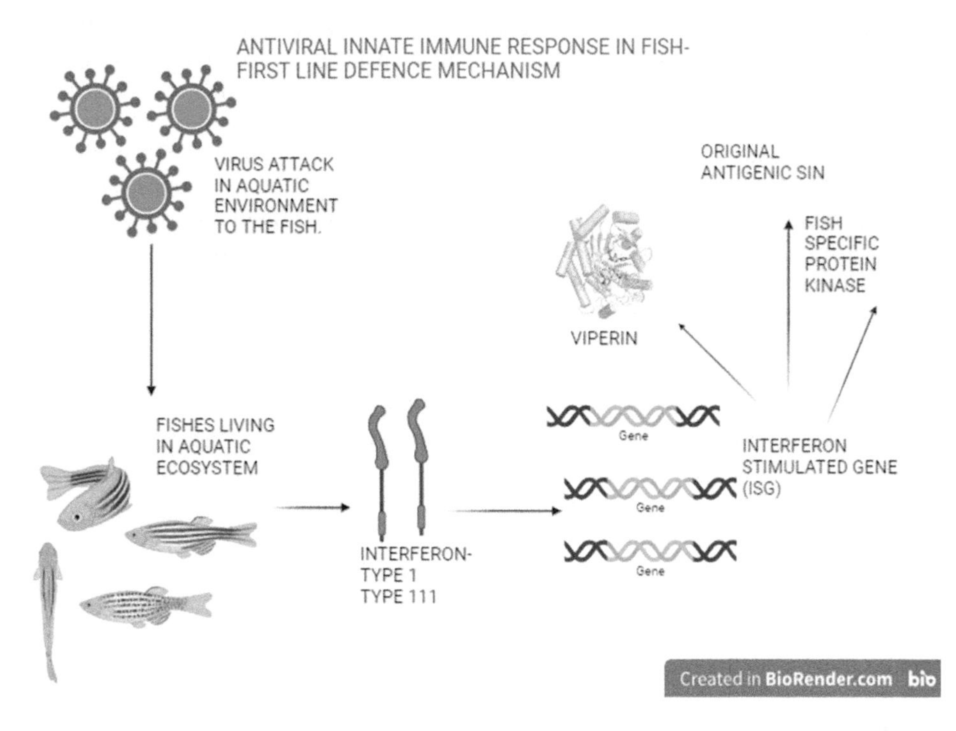

FIGURE 2.2 Fish can be vulnerable to a virus attack since they live in a more pathogenic environment. The virus infection on the fish body in an aquatic environment can lead to the development of IFNs type I and type II. Here, various gene developments can occur. The main genes created as a result of the release of IFNs into aquatic habitats include viperin, original antigenic sin, and IFN-stimulated genes. This system functions as a first-line defense mechanism against any viral disease that fish might come into contact with in an aquatic environment. This figure demonstrates the innate immunity in fish activated by IFNs against viral agents.

Due to the initial system of infection or difficulty meeting with the harmful organisms, the key immune system processes are linked to the immunological memory-based system. The pathogens are brought in by bacteria or viruses from other countries. Antiviral glycol, a proteinaceous component of IFNs that plays a significant role in protecting the body against sickness, makes up the majority of IFNs. It can make the body immune to all the illnesses. IFNs are proteins that are released and act as cytokines to make cells antiviral. Gene sequences, protein structures, and functional traits can be used to distinguish between the two IFN families (type I and type II). IFN genes have recently been cloned from various fish species including zebrafish, Atlantic salmon, spotted green pufferfish (*Tetraodon nigroviridis*), channel catfish, and Japanese pufferfish (*Takifugu rubripes*). Each fish's IFN gene had five exons and four introns (Zhang and Gui 2012).

2.5.4 Role of Antimicrobial Peptides (AMPs) in Defense

Fish inhabit an area where the pathogenicity of the aquatic habitats is more prevalent. When fish are on the defensive against any diseases, their bodies secrete mucus. Any marine organism that lives here creates antimicrobial peptides as its main product. When facing any of the pathogenic organisms present, AMPs are the first line of defense. Only after the infections are low metabolic costs necessary for the generation of AMPs. In terms of both function and specificity, AMPs are highly effective. They have an infectious memory-developing property. At a low metabolic cost, the AMPs are available after infection. From the fish, many of the AMPs have been separated. This primarily consists of *Pleurocidins* from winter flounder, American plaice, *Hippoglossoides platessoides*, and Atlantic halibut as well as Misgurin from loach and hagfish intestinal antimicrobial peptide (HFIAPs) from hagfish. Fish immune systems make it simple for infections to adapt and increase their chances of spreading in an aquatic habitat (Ravichandran et al. 2010).

Fish are the origin of disease evolution and have even played a significant role in the evolution of both major and minor food sources. Microbes that are pathogenic and saprophytic can more easily consume fish tissue when the immune system is inactive. The initial line of defense against any pathogens is the innate immune system. The N-terminal signal sequence, pro-region, and C-terminal cationic peptide make up synthetic AMPs. Nuclear magnetic resonance (NMR)methods are the main ones used to examine AMPs. The majority of the AMPs are from 5 groups. The majority of defense is cysteine-rich AMPs involved in harboring a consensus motif consisting of six cysteine residues that form three tri-molecular disulfide bridges. The β-sheet AMPs have a single hairpin structure, a length of about 20 residues, and a half-disulfide bond. Regular amino acid-rich AMPs like histidine, which is found in an AMP made from human saliva, are effective against *C. albicans*. Rare amino acids make up the antibiotic AMPs from modified amino acids – Nisin generated by *Lactococcus lactis*. Antimicrobial peptides play a vital role in a species-inherent host defense system. Antibacterial peptides showed a wide range of antibacterial activity in both *in vitro* and *in vivo* settings. The AMPs in fish species were not known in the early years. A dangerous peptide named pardaxin was taken from the *Pardachirus marmoratus* fish. The defensin, cathelicidin, and hepcidin families of peptides are discovered in various species. AMPs play a significant role in pharmaceuticals, antimicrobial peptides, and immunomodulatory effects.

The expression of AMPs known as piscidins, discovered in Atlantic cod and winter flounder, has been seen in the spleen, kidney, gill, skin, and even the gut of fish. Their chondrocytes, exocrine and endocrine glands, swim bladder, heart, oocytes, and other tissues all exhibit piscidins with an excellent antibacterial potential toward both Gram-negative and Gram-positive bacteria such as species of *Streptococcus, Pseudomonas, Bacillus*, and *Vibrio* genera. Piscidins also possess anti-viral, antifungal, and anti-parasitic activity. The winter flounder, *Pleuronectes americanus*, is home to the first piscidin discovered and isolated from its skin mucous secretions, known as pleurocidin. The mast cells of hybrid striped bass had their AMPs piscidin removed as well. Moronecidin, epinecidin, and dicentracin, as well as other related peptides, were also found in fish species.

Piscidin is particularly interesting because of its capacity for thermostability. Piscidin from the sea horse nesting pouch maintained complete activity after exposure from 20 to 80°C for 30 min, and only 20% activity loss during boiling at 100°C for 30 min to sustain antibacterial action at high salt concentrations. Anti-cancer action has also been discovered to be primarily important. The antimicrobial action of proteins generated from histones has recently gained significant attention. A new peptide was discovered in the Asian toad *Bufo Bufo gargarizans*, known as buforin I. It was discovered that this matched the N-terminal region of histone 2A exactly. Due to chrysophsin-1's powerful ability to eradicate the cariogenic bacterium *Streptococcus mutans*, piscidins are also promising for the treatment of oral diseases. Pleurocidin was also able to kill *S. mutans and S. sobrinus*, and the killing of biofilms was dose-dependent. This further proved pleurocidin's anti-cariogenic ability. A parasitic dinoflagellate was discovered as fragments of both termini, from histones H1, H2A, H2B, and H6. It also demonstrated that it retained its activity in physiological or higher salt concentration and was relatively stable in the presence of human saliva and no hemolysis was found. There are traits shared by all AMPs that encourage their development as therapeutic antimicrobials. These have broad-spectrum activity against a wide range of diseases and are particularly effective against pathogens that affect fish since they most likely co-evolved with those pathogens.

2.6 ADAPTIVE IMMUNE RESPONSES IN FISH

Innate and adaptive immunity form the two most important immune responses in jawed vertebrates. The building of memory cells is the major advantage of adaptive immunity. A quick and efficient interaction with the pathogens is possible through the adaptive immune system.

2.6.1 Overview of Fish Adaptive Immune System

The cartilaginous and teleost fish have a more advanced immune system like that of the mammals. Immunological memory cell development is the major event in the adaptive immune system. Memory T and B cells are also formed cells involved in the regulation of leukocytes or by evading the host-attacking cells. T cells form from the cell-mediated immune system. In the teleost, first involved in the development of the transcription factors' major components of the adaptive immune system are TCRs,Igs, MHC, and recombination activating gene (RAG). Cell-mediated immune responses and humoral responses are two categories of adaptive immunity and B cells are involved in cell-mediated immune responses. IgM, IgD, and IgT/Z are the groups under the humoral responses. MHCs I and II are mainly expressed in multiple tissues and MHC II is expressed in all professional APCs. Specific and nonspecific immunities are carried out separately. Macrophages are included in the specific immunity by capturing antigens inside the cell and killing them. Humoral-mediated immune systems mainly include the production of antibodies. They include mainly B cell receptors and the Igs. IgM is included and considered as one of the ancient classes of antibodies. IgD level is higher in the gills of rainbow trout than the level of IgM. IgT/Z is observed in the rainbow trout (IgT) and IgZ in zebrafish.

2.6.1.1 B Cells in Fish: Production of Antibodies

B cells immunity – The mucosal surface of many teleost fish enables protection from pathogens in the environment. Cellular and humoral defenses are equipped at the mucosal surface of the organisms. MALT includes Igs and B cells. The three main mucosal immune compartments found in bony fish are (Kordon et al., 2018a, 2018b) the GALT with the lamina propria (LP) and IEL compartments; the gill-associated lymphoid tissue; and the skin-associated lymphoid tissue (SALT). In teleost fish, the key hematopoietic organ, the HK forms mainly the primary lymphoid organ included in the production of B cells. Plasma blast cells form the spleen region and undergo the production of plasma cells. They further migrate to HK. The spleen has an effective function in trapping the antigens from the bloodstream. Determination for an effective vaccination and immuno competence is related to the ontogeny study of teleost fish. IgM-bearing and IgM-secreting cells exist. In most of the teleost fish, the LP and the intestinal epithelium are abundantly involved in the immune responses. The adaptive immune system presents for both the cartilaginous and teleost fish. Numerous lymphocytes are present in the skin epidermal layer of catfish and can secrete the IgM cells. Pathogen entry is highly possible through the mucosal surface of the gill region.

2.6.1.2 T Cells in Fish: Role in Cell-Mediated Immunity

Thymus possesses T cells. In the bony and cartilaginous fishes, lymphocyte development requires an important organ called thymus. T cells have been reported in the intestine gills. Cell-mediated immunity (TCR) is run by the T cells and regions such as MHC and CD3. CTCs and helper T cells form protection against pathogens. CTCs and helper T cells (Th cells) are the two categories among the T cells mainly. CTCs and helper T cells have been identified as CD8+ and CD4+, respectively. Fish T cells have been identified through the finding of antibodies against T cell-specific surface antigens. The primary lymphoid organ is the thymus which has two lobes, mainly cells that develop and mature from thymus. The thymus contains numerous lobules in each lobe. Peripheral cortex and central medulla are the two major regions in the lobes. The thymus is in the gill cavity. The located region can vary in their volume according to age, maturity, and sex. Some fish have a good differentiation between the central medulla and peripheral cortex of the lobes in the thymus. The rudiment of the thymus in *Sebastiscus marmoratus* (a teleost fish) was first visible 10–12 days post-hatch (seven days post-birth) at 20°C. Kidney and spleen consist of hematopoietic stem cells and the differentiated region is observed as well. The thymus is the essential organ for the development of T lymphocytes from early thymocyte progenitors to functionally competent T cells. Its high concentration was found in the anal region namely mAb DLT15+. It is also important in the antigen uptake at the posterior gut region and even for transportation.

2.6.1.3 Major Histocompatibility Complex (MHC) in Fish

Classical MHC class II functions in fish include expression of classical MHC class II, peptide presentation by classical MHC class II, and allograft rejection. Functions of classical MHC class II include expression of classical MHC class II, peptide presentation by classical MHC class II, and allograft rejection. Functions of classical MHC class I include expression of peptide presentation and allograft rejection. MHC gene studies are important in looking for resistance of the fish against disease pathogens. Investigation of the impact of linkage groups harboring MHC genes can be studied through the genome-wide quantitative trait loci (QTL). Extensive allelic polymorphism is shown in mammals by the MHC class II molecule and is expressed at the surface of professional APCs. Zebrafish and nurse sharks have shown the classical MHC class II polymorphism. Teleost fish have the old non-classical MHC class II and were named the "DB" and "DE." Atlantic salmon have shown the classical MHC class I molecule, identified by the peptide termini binding residues.

2.7 ENVIRONMENTAL AND STRESS-INDUCED IMMUNOMODULATION

2.7.1 Effect of Environmental Factors on Fish Immune Response

Aquatic environments are intricate ecosystems where various biotic and abiotic factors interact, influencing the health and well-being of aquatic organisms, including fish. Fish being ectotherms rely heavily on their surrounding environment to regulate their physiological processes, including immune responses. The immune system of fish plays a crucial role in protecting them from infections and diseases caused by pathogens. However, the effectiveness of fish immune responses is significantly influenced by environmental factors, which can either enhance or suppress their immune functions. This section explores the intricate relationship between environmental factors and fish immune responses, highlighting the key mechanisms involved. Water temperature is a critical environmental factor that profoundly affects fish immune responses. Temperature influences metabolic rates, enzymatic activities, and the speed of immune reactions. Studies have shown that higher temperatures often lead to increased metabolic demands, which may lead to greater production of ROS as by-products of metabolism. While ROS are essential for immune function, excessive production can lead to oxidative stress and immunosuppression, making fish more susceptible to infections (Reid et al. 2019). Zhao et al. (2023) on rainbow trout demonstrated that elevated water temperatures led to a decrease in lymphocyte proliferation and antibody production. Additionally, the expression of heat shock proteins (HSPs), known to be found in stress responses, elevates at higher temperatures. This suggests that fish activate stress-related pathways under thermal stress, which can influence immune functions.

Water quality parameters, such as dissolved oxygen levels, pH, and pollutants, can have profound effects on fish immunity. Poor water quality can stress fish, leading to immunosuppression and increased vulnerability to diseases (Portz et al. 2006). Pollutants like heavy metals and organic chemicals can directly damage immune cells and disrupt immune pathways (Boyd 2010). Furthermore, alterations in pH levels can affect the solubility and toxicity of various pollutants, ultimately impacting fish immune responses. Additionally, fluctuations in pH levels have been linked to changes in the immune gene expression of zebrafish.

Salinity fluctuations in aquatic environments, particularly in estuarine and coastal areas, can affect fish immune responses. Fish adapted to different salinity ranges exhibit varying immune reactions to sudden salinity changes. For example, euryhaline species are more resilient to salinity variations than stenohaline species, as they possess mechanisms to osmoregulate and maintain immune function under changing conditions (Whitney et al. 2016). The presence of pathogens and co-infections in aquatic ecosystems can significantly influence fish immune responses. Chronic exposure to low levels of pathogens may lead to immune suppression, making fish more susceptible to opportunistic infections (Derome et al. 2016). Furthermore, co-infections with multiple pathogens can lead to immune modulation, altering the outcome of infections and potentially enhancing disease severity.

Environmental pollutants, including heavy metals, pesticides, and industrial chemicals, pose significant threats to fish immune responses. These contaminants can accumulate in fish tissues, causing immunosuppression and making fish more susceptible to diseases. Exposure to lead resulted in decreased immune cell activity in grass carp. Similarly, exposure to polychlorinated biphenyls (PCBs) has been associated with altered immune gene expression in rainbow trout (Safe 1994). Human-induced habitat alterations, such as deforestation, urbanization, and dam construction, can disrupt fish habitats and subsequently impact their immune responses. Reduced habitat quality can lead to chronic stress, which suppresses immune function. The research by Maceda-Veiga (2013) demonstrated that habitat fragmentation negatively affected the immune system of the Iberian toothcarp, a freshwater fish species endemic to the Mediterranean region. Seasonal variations in environmental conditions can significantly affect fish immune responses. In colder months, fish metabolism slows down, affecting immune cell functions and antibody production. Research on common carp (*Cyprinus carpio*) (Tanck et al. 2000) showed that immune-related gene expression was downregulated during the winter months, potentially increasing the susceptibility of fish to pathogenic agents.

2.7.2 Impact of Water Quality and Pollutants on Fish Immunity

Water quality plays a critical role in the health and survival of aquatic organisms, including fish. Maintaining suitable water quality is essential for supporting various physiological processes, including immune system function. The immune system of fish, like other vertebrates, defends against pathogens and maintains overall health. However, water quality degradation due to pollutants can impair fish immune responses, leading to increased susceptibility to diseases. Water quality encompasses various physical, chemical, and biological parameters that collectively determine the suitability of aquatic habitats for fish and other aquatic organisms. Temperature, pH, dissolved oxygen levels, and nutrient concentrations are among the key factors affecting water quality. Changes in any of these parameters can have direct and indirect effects on fish immunity. For instance, water temperature fluctuations can influence the metabolic rate of fish and can alter the kinetics of immune cells and antibody production impacting their immune responses. This is because fish immune systems, particularly the humoral immunity, are temperature-dependent. Additionally, alterations in pH levels can affect the solubility of minerals and ions in water essential for immune system function, potentially leading to increased exposure of fish to toxic substances. Reduced dissolved oxygen levels due to pollution or eutrophication can stress fish and impair their immune systems. Changes in salinity can disturb osmoregulation, indirectly affecting immune responses.

2.7.2.1 Pollutants and Fish Immunity

A myriad of pollutants find their way into aquatic environments, including heavy metals, pesticides, industrial chemicals, and organic contaminants. They are released into aquatic environments, posing significant threats to fish immunity. These pollutants can directly suppress immune cell function, disrupt cytokine signaling pathways, and induce oxidative stress, weakening the immune system's ability to combat infections. For instance, heavy metals like mercury and lead are known to inhibit phagocytosis and reduce antibody production in fish. These substances can disrupt the delicate balance of fish immune responses. Heavy metals such as mercury, lead, and cadmium are known immune-toxicants that can suppress immune functions in fish. These metals interfere with immune cell development, cytokine production, and antibody responses, leaving fish vulnerable to infections.

Organic pollutants like polycyclic aromatic hydrocarbons (PAHs) and PCBs are also of concern. These compounds can accumulate in fish tissues, compromising immune cells' ability to recognize and eliminate pathogens. Additionally, pesticides that make their way into water bodies can disrupt fish endocrine systems, leading to immunosuppression and increased fish susceptibility to diseases.

Impact on Innate Immunity: Innate immunity as the first line of defense against pathogens is particularly vulnerable to waterborne pollutants. Pollutants can interfere with the proper functioning of innate immune cells, such as macrophages and neutrophils, impairing their ability to recognize and engulf pathogens. Altered innate immune responses weaken the overall immune defense, making fish more susceptible to infections.

Effects on Adaptive Immunity: Adaptive immunity, characterized by the production of specific antibodies and memory cells, is also influenced by water quality and pollutants. Pollutants can disrupt B cell and T cell proliferation, leading to decreased antibody production and compromised memory responses. This impairment reduces the fish's ability to mount effective immune responses upon subsequent exposure to pathogens.

Immunosuppression and Disease Susceptibility: The cumulative effects of water quality degradation and pollutant exposure can lead to immunosuppression in fish. Immunosuppressed fish are more susceptible to various diseases, including bacterial, viral, and parasitic infections. Reduced immune responses result in prolonged infection durations, increased disease severity, and higher mortality rates within fish populations.

Immunomodulation and Inflammation: Exposure to pollutants can trigger immunomodulatory responses in fish. This often involves a delicate balance between immune suppression and hyperactivation, leading to chronic inflammation. Prolonged inflammation can weaken the immune system over time, making fish more prone to infections. Furthermore, the stress response induced by exposure to pollutants can exacerbate immune dysfunction, as stress hormones can directly interfere with immune cell function.

Heavy Metal Exposure and Fish Immunity: A study by Dethloff and Bailey (1998) on effects of chronic cadmium exposure on the immune responses of rainbow trout showed that cadmium exposure led to a significant reduction in phagocytic activity of macrophages and decreased expression of immune-related genes. Additionally, antibody production was compromised, highlighting the adverse impact of heavy metal pollutants on both innate and adaptive immune functions in fish.

Impact of Microplastics on Fish Immunity: Microplastics, small plastic particles less than 5 mm in size, have emerged as a significant pollutant in aquatic ecosystems. These particles can adsorb and transport various contaminants, impacting both water quality and fish health. Recent research (Bhuyan 2022) has shown that microplastics can trigger an immune response in fish, leading to chronic inflammation. This immune activation, while initially protective, can lead to immune exhaustion and increased susceptibility to diseases over time.

Mechanisms of Impact: The impact of pollutants on fish immunity is often mediated through pathways involving the stress response and the hypothalamus–pituitary–interrenal (HPI) axis. Exposure to pollutants activates the stress response, leading to the release of stress hormones like cortisol. Chronic elevation of cortisol levels suppresses immune function by inhibiting immune cell production and altering immune cell activity. Furthermore, pollutants can disrupt the HPI axis, leading to dysregulation of immune responses. For instance, exposure to endocrine-disrupting chemicals can alter the production and release of hormones that modulate immune activity, leading to compromised immune function. Future research should focus on unraveling the complex interactions between different pollutants and their cumulative effects on fish immunity. Understanding the molecular mechanisms underlying immune responses in varying water quality conditions can pave the way for predictive models that estimate the impacts of environmental changes on fish health. These efforts will be crucial for the sustainable management of aquatic ecosystems and the preservation of fish populations.

2.7.3 Stress-Induced Immunosuppression and Fish Health

Stress is an inevitable part of life, affecting organisms across the animal kingdom, including fish. In the intricate environment of aquatic organisms, fish are particularly vulnerable to various stressors due to their sensitivity to environmental changes. Stress, in the context of fish health, can arise from a multitude of sources, such as changes in water temperature, poor water quality, mishandling, mis-transportation, over-crowding, over-feeding, and social interactions. Stress is known to have a profound impact on fish health, and one of the most significant consequences is the suppression of the immune system.

2.7.4 Stress and the Immune System

When a fish encounters a stressful situation, whether it's a sudden change in environmental conditions or the presence of a predator, its body responds by activating the stress response. Stress triggers a cascade of physiological responses collectively known as the stress response. This response involves the activation of the hypothalamic–pituitary–adrenal (HPA) axis and the release of stress hormones, such as cortisol, cortisone, and epinephrine, which prepare the organism for a "fight or flight" response. While this response is essential for survival in acute situations, chronic or repeated stressors can lead to negative health outcomes, including immunosuppression.

The immune system of fish, as in other vertebrates, plays a critical role in defending the organism against pathogens. It consists of innate and adaptive components that work in harmony to recognize and neutralize harmful invaders. However, chronic stress disrupts this delicate balance. Stress hormones like cortisol and cortisone can inhibit the immune response by suppressing the production and activity of immune cells, cytokines, and antibodies. This phenomenon, known as stress-induced immunosuppression, renders fish more susceptible to infectious agents.

2.7.5 Immunomodulation under Stress

One of the most noteworthy effects of chronic stress on fish health is the suppression of the immune system, known as stress-induced immunosuppression. Cortisol, the key stress hormone, plays a pivotal role in this process. While acute cortisol elevation can enhance immune responses, chronic elevation due to prolonged stress leads to a downregulation of immune-related functions. This includes reduced production of immune cells, impaired phagocytosis, and diminished cytokine production. Research by Campbell et al. (2021) provides compelling evidence for stress-induced immunosuppression in fish. The study exposed rainbow trout to various stressors and found that chronic stress significantly decreased the expression of immune-related genes, rendering the fish more susceptible to infections. This research underscores the complex interplay between stress and immunity in fish health.

2.7.6 Mechanisms of Stress-Induced Immunosuppression

The mechanisms behind stress-induced immunosuppression in fish are multifaceted and interconnected. Cortisol, the primary stress hormone, directly impacts immune cells by inhibiting their proliferation and function. Additionally, cortisol can disrupt the balance between pro-inflammatory and anti-inflammatory cytokines, leading to impaired immune responses. Glucocorticoid receptors, which mediate the effects of cortisol, are found on immune cells, enabling cortisol to directly modulate immune activities.

The interaction between the stress response and the immune system is complex and bidirectional. Stress hormones have been shown to impact various components of the immune system, such as reducing the production of pro-inflammatory cytokines, impairing the activity of immune cells, and suppressing the ability of fish to mount effective immune responses against pathogens. This interplay can be particularly concerning in aquaculture settings, where stressed fish may be more susceptible to infectious diseases. Furthermore, stress-induced changes in fish behavior can indirectly affect immune function. For instance, stressed fish might reduce their feeding, leading to malnutrition and further compromising their immune response. Aggressive behaviors triggered by stress can result in physical injuries, providing entry points for pathogens.

Implications for Fish Health: In aquaculture settings, stressed fish are more susceptible to infections, which can lead to disease outbreaks and economic losses. The consequences of stress-induced immunosuppression are far-reaching and have significant implications for both aquaculture and wild fish populations. Stress-induced immunosuppression has dire consequences for fish populations in terms of disease susceptibility. Fish compromised by chronic stress are more vulnerable to a range of pathogens, including bacteria, viruses, and parasites. Stress weakens the first line of animal defense, making fish less capable of responding to infections. Furthermore, the reduced effectiveness of immune responses prolongs the recovery time from infections.

A study by Waagbø (1994) focused on Atlantic salmon (*Salmo salar*) demonstrated a clear link between stress-induced immunosuppression and disease outbreaks. When subjected to chronic stress, the salmon exhibited higher mortality rates when exposed to a common fish pathogen, demonstrating compromised immune function under stress conditions. In wild fish populations, stress-induced immunosuppression can contribute to the spread of diseases and the decline of vulnerable species. Environmental stressors, such as pollution and habitat destruction, can exacerbate the effects of physiological stress, further compromising immune function. Additionally, the use of stress-reducing strategies, such as improved husbandry practices and the provision of enriched environments, has become imperative to ensure optimal growth and health in aquaculture operations. Understanding the connection between stress, immunosuppression, and fish health is crucial for the management of aquaculture systems and wild fish populations. Efforts to mitigate stress-induced immunosuppression in aquaculture settings are imperative to maintaining fish health and welfare. Several strategies have been proposed to minimize the negative impacts of stress on the immune system. Optimizing husbandry practices to reduce stressors, implementing gradual changes in environmental conditions, providing adequate space and proper handling techniques, and optimizing nutrition to support immune function can alleviate the impact of stress-induced immunosuppression.

2.8 HOST–PATHOGEN INTERACTIONS IN FISH

Economically important fish species are greatly affected by various pathogenic agents including Gram-negative and Gram-positive bacterial pathogens. Thus, efficient protective and preventive measures are required to reduce the economic losses. *Renibacterium salmoninarum*, *Nocardia seriolae*, *Mycobacterium marinum*, *Streptococcus* spp., *Lactococcus garvieae*, and *Vibrio* spp. are some of the bacterial pathogens infecting most of the economically important fish species. Knowledge of host–pathogen interactions and their interactions is related to bacterial virulence and fish immune response against the bacterial attack (Figure 2.3). Bacterial pathogens can invade fish tissues, proliferate the host system, and evade the fish defenses. Adherence and invasions of the bacteria to the fish host organism take place for a better survival and nutrient uptake. Fish as hosts are affected by the virulent and simultaneous attack by the bacterial population. This may lead to the killing of the fish species. The adaptive and innate immune system could be developed afterward by the host fish species.

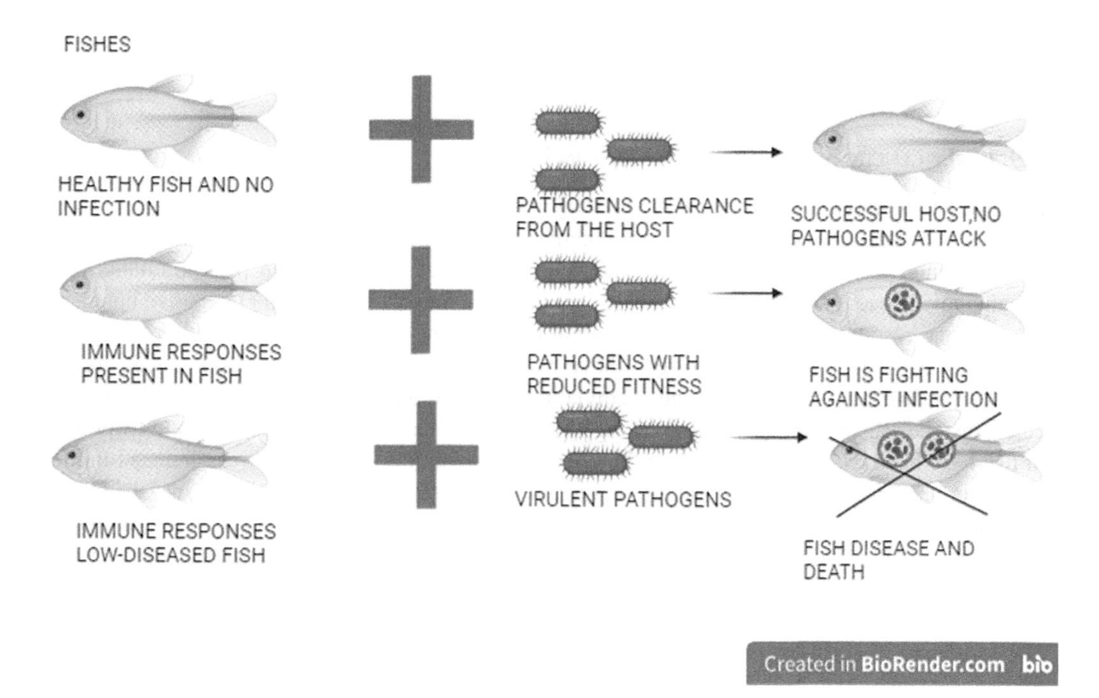

FIGURE 2.3 This figure shows differences in level of immune responses in fish with different levels of health status to the pathogens. The figure demonstrates the host fish and parasite interactions. The fish with different levels of immune responses are taken. In the first case, healthy and non-infected fish were taken. When the healthy fish were affected, the bacteria showed successful fish hosts. In the second case, sufficient immune response against the pathogens presents but not to a very high degree. Due to the immune responses against pathogens, the fish is fighting and surviving against infection. In the last case, fish hosts have a low immune response to the diseased fish. When the pathogens now infect the diseased fish, the result is the death of the fish. (Created with Bio render (https://biorender.com).

2.9 EMERGING TRENDS IN FISH IMMUNOLOGY

Novel findings in recent decades in fish immunology demonstrated that cells and molecules reflect on distinctive to either adaptive (or) innate system plays a specific role which is more complex than previouslyexplanations. Fish have more vulnerability against pathogens compared to higher vertebrates (Fuhrman 1999). The susceptibility and permissiveness of *Danio rerio* to cyprinid viruses (Streiff et al. 2023) reveal that they are less susceptible to the viruses due to upregulation of the immune genes. The understanding of fish immunology has more importance which serves as an essential link for understanding of higher vertebrates and is considered as a model for mammalians. *Danio rerio* resembles of many mammalian pathogens recognizing receptor, the cell type counterparts, and immune-competence which always attracts researchers. The teleost fish has been serving as the in vitro and in vivo model for evolutionary and immune-related studies in modern research. Bio-prospecting of immune genes (exploration of genetic information that could be developed into commercially valuable products and production of hybrid fish stocks with high immune response) laid a path for the future of fish immunology. Production of a hybrid of evolutionary divergent cyprinoid species with a diversity of MHC II B gene shows a hybrid advantage via potential heterosis. Fish immunology is a mimic of higher vertebrae. The specificity of species Igs has a higher level of biological and medical application of the fish immune system, exploring human infectious viruses in fish to evaluate the immune gene expression (Laghi et al. 2022). In biomedical applications, antimicrobial peptides and anti-viral genes have been in research levels for human application. Antimicrobial peptides from the teleost fish have shown potential against AMR (antimicrobial resistance) pathogenic microbes.

2.9.1 Epigenetic and Immune Response in Fish

Epigenetics modifiers have immunomodulatory properties, and the effect on immune response in fish has evolved in many cases; this epigenetics occurrence in fish immunity depends on many environmental factors, such as temperature, climate changes, alteration in pH, pollutants in the aquatic environment and disease outbreaks, creating the unusual positive (or) negative changes in the immune response. Epigenetic manipulation helps improve fish health and welfare. The environmental conditions and some triggers have been used in the regulation of epigenetic immune response in fish by β-glucan, histone methylation patterns, temperature-dependent cellular molecules, histone acetylation, etc., mainly using physical and chemical triggers. Behavioral fever drives in fish have been recently evaluated on the effect of epigenetic modulation, stress-induced immune response, and temperature dependent and vary in alteration of immune response in fish. In some cases of reinfection innate immunity mimics the adaptive-like memory response in fish immunity. The response relies on the temperature, environmental stress, and microbial contamination. Induced and non-induced epigenetic immune response in fish helps in better response in the disease outbreak condition. The adaptive immune system includes lymphocytes, Igs, (TCRs, and products of MHC to allow the clonal selection of B and T cells. Innate immunity is the first line of defense that reacts to pathogens in a very short time and does not provide long-lasting protection. These are the simple forms of immune response in the fish (Turvey and Broide 2010).

2.9.2 Immunogenomics and Advances in Fish Vaccine Development

Immunogenomics is the combination of immunology and genetics, in terms of study of genes that are involved in the immune system of fish. Immunogenomics studies in fish laid a path for the understanding of immunity and related regulation and the mechanism behind innate and adaptive immune responses in fish. This study reveals the genetic factors involved in the disease-resistance nature of the fish (Matsuo and Nonaka 2004). Researchers are focused on various kinds of vaccine development for aquaculture diseases caused by bacteria, parasites, and viruses. Different types of fish vaccines were developed such as live attenuated (whole-cell vaccine), inactivated vaccine, recombinant vaccine, subunit vaccine, DNA vaccine, and biofilm vaccine. Advances have been made with the evaluation of the immune gene expression in the vaccinated fish. Early studies on vaccine development were focused mainly on the whole vaccine, a whole-cell vaccine produced from the disease-causing virulent microorganisms that can cause infection. Based on these, researchers are interested in the production of recombinant vaccines in the field of aquaculture. The delivery method and target, using adjuvant and nanoparticles for efficacy of vaccines are evaluated by the immune gene up and downregulation compared with the control group or unvaccinated group. The HK, spleen, and liver are the common target organs for the expression of the immune gene which has been considered the primary lymphoid organs of fish. Interesting immune genes are MHC I, MHC II, IL-1 to IL-8, CD4, CD8, IgT, IgM, IgD, etc., which have specific roles in the immune response of the fish. The vaccinated fish are evaluated to see if it is an innate or adaptive immune response against the produced vaccine. Most likely the immunogenomics studies are carried out using the novel research model *Danio rerio* for toxicological, and environmental factors, drug discovery, and development of vaccines. Advancements in vaccine development also focus on the delivery of the vaccine using immunostimulant, adjuvant, and nanoparticles; they improve the immune response and the stability of the vaccine delivered into the host. Some vaccines have been commercialized and some are in field-level trials. The effectiveness of vaccines is also based on the route of delivery, which includes oral, intra-peritoneal, intra-muscular injection, and immersion routes (Jiema et al. 2019). Fish immunization was carried out for as long as 50 years and is generally accepted as an effective method for preventing a wide range of bacteria, parasites, and virus infections. Vaccines also help fish build up an adaptive immunity to accelerate this response in later infections by targeting disease-causing organisms.

2.9.3 Role of Gut Microbiota in Fish Immune Regulation

Microbiome and interaction between the gut microbiota with the body function are one of the leading research focuses on gut microbiota-producing compounds that influence the behavior of the host immune cells. The gut microbiota plays a crucial role in the development of fish immune systems by helping them with optimal nutrient absorbance (Talwar et al. 2018). They also influence the digestion of feed in the fish gut. The colonization of fish gut starts early in the larval stage and it continues driven toward the achievement of a complex assemblage of gut-associated microbes (Nayak 2010). The host gut exerts selective pressure on the microbiota which in turn influences the host immune response. The major concept of gut microbiota is a symbiosis of the relationship between both animal and gastrointestinal microbiota (CGMI) that is a key factor for development, immunity, and nutrient conversion, rendering them for bio-availability for various uses. Germ-free has provided opportunities to increase our knowledge of the nature of host–microbe interaction interplay in health and disease. The ability to form biofilm helps microbiota to survive in the environment offered by the GI tract of the animal. They play an important role in epithelial renewal and maturation, which in turn regulate immune response. Understanding of the intestinal microbiome of fish will guide the development of probiotics, and prebiotics, and hope full effectors as novel findings the microbiome presents in the fish produced a particular antibody in their gill in response to pathogenic microbes. They also include physiological functions such as growth, development, and most importantly the digestion and immune response in fish.

2.10 CONCLUSION

Fish immunology has gained interest in research and science for decades. Immune responses are evoked from the body when the surrounding circumstances are unsuitable for the organism's health. Unlike other vertebrates, fish immunity comprises innate and adaptive immunity with humoral and cellular responses. Host–pathogen–environment interactions play a crucial role in the immune response pattern, where the system cells recognize the antigens with different receptors such as TLR and PRR, and direct the cellular or humoral immunity to nullify the pathogenic effect. Primarily MALT, GALT, SALT, mucus, and skin act as physical barriers followed by B cell- or T cell-mediated response. There are cells like macrophages, NK cells, neutrophils, and monocytes that help to phagocyte or opsonize the antigens. Infections are often natural in aquaculture practices, where a cure is much more needed. Thus, herd immunity is important in sustainable aquaculture practices. Antibody-mediated or humoral immune responses play a major role in herd immunity where memory cells of the fish will recognize the antigen and will produce specific antibodies to neutralize them. In recent trends, passive immunization has been one of the great successes in science where fish vaccines are developed to combat several pathogens. Therapeutic treatment vaccines are highly effective when they trigger the fish's immune system to produce immunity against the specific antigen. Immunogenomics and epigenetics are thrusting areas for the achievement in vaccinology and to understand the immune response pattern of fish. It is also important to understand the physical as well as molecular aspects of immune responses against different pathogens, which will enhance the scope to prevent disease outbreaks in aquaculture. Through only understanding the signaling pathways and response mechanisms, future prospects will open the development of different vaccines and other treatment measures for a sustainable aquaculture.

BIBLIOGRAPHY

Akira, S., Uematsu, S., & Takeuchi, O. (2006). Pathogen recognition and innate immunity. *Cell*, 124(4), pp.783–801.

Ainsworth, J. (2003). *Fish Granulocytes: Morphology, Distribution and Function* (pp.123–148). Volume 2,.

Alvarez-Pellitero, P. (2008). Fish immunity and parasite infections: From innate immunity to immunoprophylactic prospects. *Veterinary Immunology and Immunopathology*, 126(3–4), pp.171–98.

Arango Duque, G., & Descoteaux, A. (2014) Macrophage cytokines: Involvement in immunity and infectious diseases. *Frontiers in Immunology*, 5, p.491.

Austin, B. (2006) The bacterial microflora of fish, revised. *The Scientific World Journal*, 6, pp.931–945.

Avtalion, R. R., & William Clem, L. (2009). Environmental control of the immune response in fish. *C R C Critical Reviews in Environmental Control*, 11(2), pp.105–162.

Bailey, C., Strepparava, N., Wahli, T., & Segner, H. (2019). Exploring the immune response, tolerance and resistance in proliferative kidney disease of salmonids, *Developmental & Comparative Immunology*, 90, pp.165–175.

Bavia, L., Santiesteban-Lores, L. E., Carneiro, M. C., & Prodocimo, M. M.. (2022). Advances in the complement system of a teleost fish, *Oreochromis niloticus. Fish and Shell Immunology*, 123, pp.61–74.

Bhuyan, M. S. (2022). Effects of microplastics on fish and in human health. *Frontiers in Environmental Science*, 10, p.250.

Biller-Takahashi, J.D., & Urbinati, E.C. (2014) Fish Immunology. The modification and manipulation of the innate immune system: Brazilian studies. *Anais da Academia Brasileira de Ciências*, 86, pp.1484–1506.

Bjørgen, H., & Koppang, E. O. (2022). Anatomy of teleost fish immune structures and organs. *Immunogenetics*, 73(1), pp.53–63.

Boyd, R. S. (2010). Heavy metal pollutants and chemical ecology: Exploring new frontiers. *Journal of Chemical Ecology*, 36, pp.46–58.

Bruce, T. J., & Brown, M. L. (2017). A review of immune system components, cytokines, and immunostimulants in cultured finfish species. *Open Journal of Animal Sciences*, 7(3), pp.267–288.

Campbell, J. H., Dixon, B., & Whitehouse, L. M. (2021). The intersection of stress, sex and immunity in fishes. *Immunogenetics*, 73(1), pp.111–129.

Consuegra, S., Webster, T.U., & Anka, I., (2023). Microbiome, epigenetics and fish health interactions in aquaculture. *Epigenetics in Aquaculture*, pp.245–262.

Cuesta, A., Esteban, M. A., & Meseguer, J. (2003). In vitro effect of chitin particles as the innate cellular immune system of gilthead seabream (*Sparusaurata* L.). *Fish & Shellfish Immunology*, 15, pp.1–11.

Dash, S., Das, S. K., Samal, J., & Thatoi, H. N. (2018). *Epidermal mucus, a major determinant in fish health: A review*. 19(2): pp.72–81.

Derome, N., Gauthier, J., Boutin, S., & Llewellyn, M. (2016). Bacterial opportunistic pathogens of fish. *The Rasputin effect: When commensals and symbionts become parasitic*, pp.81–108.

Dethloff, G. M., & Bailey, H. C. (1998). Effects of copper on immune system parameters of rainbow trout (Oncorhynchus mykiss). *Environmental Toxicology and Chemistry: An International Journal*, 17(9), pp.1807–1814.

Fillatreau, S., Six, A., Magadan, S., Rosario, C., Sunyer, J. O., & Boudinot, P. (2013a). The astonishing diversity of Ig classes and B cell repertoires in teleost fish: A Review. *Frontiers in Immunology*, 4, p.28, http://doi.org/10.3389/fimmu.2013.00028

Fillatreau, S., Six, A., Magadan, S., Castro, R., Sunyer, J. O., & Boudinot, P. (2013b). The astonishing diversity of Ig classes and B cell repertoires in teleost fish. *Frontiers in immunology*, 4, p.28.

Firdaus-Nawi, M., & Zamri-Saad, M. (2016). Tropical agricultural science. Major components of fish immunity: A Review.

Fischer, U., Koppang, E. O., & Nakanishi, T. (2013). Teleost T and NK cell immunity. *Fish Shell Immunology*, 35(2), pp.197–206.

Fitzgerald, K. A. & Kagan, J. C. (2020) Toll-like receptors and the control of immunity. *Cell* 180, pp.1044–1066.

Flajnik, M. F. (2018) A cold-blooded view of adaptive immunity. *Nature Reviews Immunology*, 18(7), pp.438–453.

Frías-Lasserre, D., & Villagra, C.A., (2017). The importance of ncRNAs as epigenetic mechanisms in phenotypic variation and organic evolution. *Frontiers in Microbiology*, 8, p.2483.

Fucikova, J., Palova-Jelinkova, L., Bartunkova, J., & Spisek, R. (2019). Induction of Tolerance and Immunity by Dendritic Cells: Mechanisms and Clinical Applications. *Frontiers in Immunology*, 10, 2393. doi: http://doi.org/10.3389/fimmu.2019.02393

Fuhrman, J.A., (1999). Marine viruses and their biogeochemical and ecological effects. *Nature*, 399(6736), pp.541–548.

Gajewski, T. F., Schreiber, H., & Fu, Y.-X. (2013) Innate and adaptive immune cells in the tumor microenvironment. *Nature Immunology*, 14, pp.1014–1022.

Geven, E. J., & Klaren P. H. (2017) The teleost head kidney: Integrating thyroid and immune signalling. *Developmental & Comparative Immunology*, 66, pp.73–83.

Gnanagobal, H., & Santander, J. (2022). Host–pathogen interactions of marine gram-positive bacteria. *Biology*, 11(9), p.1316.

Gong, H., Wang, Q., Lai, Y., Zhao, C., Sun, C., Chen, Z., ... Huang, Z. (2021). Study on immune response of organs of Epinephelus coioides and Carassius auratus after immersion vaccination with inactivated Vibrio harveyi vaccine. *Frontiers in Immunology*, 11, p.622387.

Graham, N. A. J., McClanahan, T. R., MacNeil, M. A., Wilson, S. K., Polunin, N. V. C., Jennings, S., Chabanet, P., Clark, Susan, Spalding, M. D., Letourneur, Y., Bigot, Lionel, Galzin, R., Öhman, M. C., Garpe, K. C., Edwards, A. J., and Sheppard, C. R. C. (2008) Climate warming, marine protected areas and the ocean-scale integrity of coral reef ecosystems. *PLoS ONE*, 3(8), pp.30–39.

Guo, C. J., He, J., & He, J. G. (2019). The immune evasion strategies of fish viruses. *Fish & Shellfish Immunology*, 86, pp.772–784.

Hatten, F., Fredriksen, Å., Hordvik, I., & Endresen, C. (2001) Presence of IgM in cutaneous mucus, but not in gut mucus of Atlantic salmon, Salmo salar. Serum IgM is rapidly degraded when added to gut mucus. *Fish & Shellfish Immunology*, 11, pp.257–268.

Haugarvoll, E., et al. (2006) Melanogenesis and evidence for melanosome transport to the plasma membrane in a CD83+ teleost leukocyte cell line. *Pigment Cell & Melanoma Research*, 19(3), pp.214–225.

Hawkes, J.W. (1974) The structure of fish skin—I. General organization. *Cell and Tissue Research*, 149, pp.147–158.

Hitzfeld, B. (2005). Encyclopedic Reference of Immunotoxicology. Chapter: Fish immune system.

Hoar, W. S., Randall, D. J., Iwama, G., & Nakanishi, T. (1997). *The fish immune system: Organism, pathogen, and environment*. Academic Press.

Huang, Y., Zheng, Q., Wang, Z., Tang, J., Lu, Y., Qin, Q., Cai, J., & Jian, J. (2021) Fish natural killer enhancing factor-A (NKEF-A) enhance cytotoxicity of nonspecific cytotoxic cells against bacterial infection. *Molecular Immunology*, 133, pp.184–193.

Iwama, G. (1997). *The fish immune system: Organism, Pathogen and Environment*. Academic Press.

Iwasaki, A., & Medzhitov, R. (2010) Regulation of adaptive immunity by the innate immune system. *Science*, 327, pp.291–295.

Jaso-Friedmann, L., Ruiz, J., Bishop, G., & Evans, D. (2000) Regulation of innate immunity in tilapia: Activation of nonspecific cytotoxic cells by cytokine-like factors. *Developmental & Comparative Immunology*, 24, pp.25–39.

Jenkins, P., Wrathmell, A., Harris, J., & Pulsford, A. (1994) Systemic and mucosal immune responses to enterically delivered antigen in Oreochromis mossambicus. *Fish & Shellfish Immunology*, 4, pp.255–271.

John, N., Salim, R., Ramesh, S., & Paul, N. M. (2022). Fish Lectins: History, Types, and Structural Classification. In *Aquatic Lectins: Immune Defense, Biological Recognition and Molecular Advancements* (pp.45–61). Singapore: Springer Nature Singapore.

Kawakami, H., Shinohara, N., & Sakai, M. (1998) The Non-specific Immunostimulation and Adjuvant Effects of *Vibrio anguillarum* Bacterin, M-glucan, Chitin and Freund's Complete Adjuvant against *Pasteurellapiscicida* Infection in Yellowtail. *Fish Pathology*, 33, pp.287–292.

Kiron, V. (2012). Fish immune system and its nutritional modulation for preventive health care: *Animal Feed Science and Technology*, 173(1–2), pp.111–133.

Kodama, H., Hirota, Y., Mukamoto, N., Baba, T., & Azuma, I. (1993) Activation of rainbow trout (*Oncorhynchus mykiss*) phagocytes by muramyl dipeptide. *Developmental & Comparative Immunology*, 17, pp.129–140.

Koppang, E. et al. (1998) Differing levels of Mhc class II β chain expression in a range of tissues from vaccinated and non-vaccinated Atlantic salmon (*Salmosalar* L.). *Fish & Shellfish Immunology*, 8(3), pp.183–196.

Koppang, E. O., Fischer, U., Moore, L., Tranulis, M. A., Dijkstra, J. M., Köllner, B., ... Hordvik, I. (2010). Salmonid T cells assemble in the thymus, spleen and in novel interbranchial lymphoid tissue. *Journal of anatomy*, 217(6), pp.728–739.

Kordon, A. O., Karsi, A., & Pinchuk, L. (2018a). Innate Immune Responses in Fish. *Antigen Presenting Cells and Professional Phagocytes*, 18(9), pp.1123–1139.

Kordon, A. O., Karsi, A., & Pinchuk, L. (2018b). Innate immune responses in fish: antigen presenting cells and professional phagocytes. *Turkish Journal of Fisheries and Aquatic Sciences*, 18(9), pp.1123–1139.

Kordon, A. O., Pinchuk, L., & Karsi, A.. (2021a). Turkish journal of fisheries and science: Adaptive immune system in fish.

Kordon, A. O., Pinchuk, L., & Karsi, A. (2021b). Adaptive Immune System in Fish. *Turkish Journal of Fisheries and Aquatic Sciences*, 22(4), http://doi.org/10.4194/TRJFAS20235

Kunisawa, J., & Kiyono, H. (2005) A marvel of mucosal T cells and secretory antibodies for the creation of first lines of defence. *Cellular and Molecular Life Sciences*, 62, pp.1308–1321. http://doi.org/10.1007/s00018-005-5035-1

Laghi, V., Rezelj, V., Boucontet, L., Frétaud, M., Da Costa, B., Boudinot, P., Salinas, I., Lutfalla, G., Vignuzzi, M., & Levraud, J.P. (2022). Exploring zebrafish larvae as a COVID-19 model: Probable abortive SARS-CoV-2 replication in the swim bladder. *Frontiers in Cellular and Infection Microbiology*, 12, pp.790–851.

Lieschke, G. J., & Trede, N. S. (2008). Fish immunology. *Current Biology*, 19(16). http://doi.org/10.1016/j.cub.2009.06.068

Lorena, B., Santiesteban-Lores, L. E., Carneiro, M. C., & Prodocimo, M. M. (2022), Advances in the complement system of a teleost fish, Oreochromis niloticus, *Fish & Shellfish Immunology*, 123, pp.61–74, https://doi.org/10.1016/j.fsi.2022.02.013

Maceda-Veiga, A. (2013). Towards the conservation of freshwater fish: Iberian Rivers as an example of threats and management practices. *Reviews in Fish Biology and Fisheries*, 23, pp.1–22.

Makesh, M., Megha, K. B., & Rajendran, K. V. (2022). *Overview of Fish immune system*. Springer, https://doi.org/10.1007/978-981-19-1268-9_1

Mashroof, S., & Criscitiello, M. F. (2016). Fish immunoglobulins: A review. Biology (Basel), 5(4): p.45. https://doi.org/10.3390/biology5040045

Masso-Silva, J. A., & Diamond, G. (2014). Antimicrobial peptides from fish. *Pharmaceuticals*, 7(3), pp.265–310.

Matsuo, M.Y., & Nonaka, M. (2004). Repetitive elements in the major histocompatibility complex (MHC) class I region of a teleost, medaka: Identification of novel transposable elements. *Mechanisms of Development*, 121(7–8), pp.771–777.

Ma, J., Bruce, T., Sudheesh, P., Knupp, C., Loch, T., Faisal, M., & Cain, K.. (2018). *Journal of Fish Diseases-Front Cover*. https://doi.org/10.1111/jfd.12947

Mogensen, T. H. (2009) Pathogen recognition and inflammatory signaling in innate immune defenses. *Clinical Microbiology Reviews*, 22, pp.240–273.

Mokhtar, D. M., Zaccone, G., Alesci, A., Kuciel, M., Hussein, M. T., & Sayed, R. K. (2023). Main components of fish immunity: An overview of the fish immune system. *Fishes*, 8(2), p.93.

Nakanishi, T., Shibasaki, Y., & Matsuura, Y. (2015). T cells in fish. *Biology*, 4(4), pp.640–663.

Nayak, S.K., (2010). Probiotics and immunity: A fish perspective. *Fish & Shellfish Immunology*, 29(1), pp.2–14.

Ninawe, A. S., Hameed, A. S. S., Selvin, J. (2017). Advancements in diagnosis and control measures of viral pathogens in aquaculture: An Indian perspective. *Aquaculture International*, 25(1), pp.251–264.

Oriol Sunyer, J. (2019). Evolutionary and Functional Relationships of B Cells from Fish and Mammals: Insights into their Novel Roles in Phagocytosis and Presentation of Particulate Antigen. 12(3), 2012.

Pais, R., Khushiramani, R., Karunasagar, I., & Karunasagar, I. (2008) Effect of immunostimulants on the haemolymph hemagglutinins of tiger shrimp *Penaeus monodon*. *Aquaculture Research*, 39, pp.1339–1345.

Paludan, S.R., Pradeu, T., Masters, S.L., & Mogensen, T.H. (2021) Constitutive immune mechanisms: Mediators of host defence and immune regulation. *Nature Reviews Immunology*, 21, pp.137–150.

Peeler, E. J., & Taylor, N. G. (2011) The application of epidemiology in aquatic animal health -opportunities and challenges. *Veterinary Research*, 42(1), p.94.

Portz, D. E., Woodley, C. M., & Cech, J. J. (2006). Stress-associated impacts of short-term holding on fishes. *Reviews in Fish Biology and Fisheries*, 16, pp.125–170.

Powell, D. (2000) Microscopic functional anatomy: Immune system. In: Ostrander, G. (ed) *The Laboratory Fish*. Academic Press, San Diego, pp.441–449.

Press, C.M., & Evensen, Ø. (1999) The morphology of the immune system in teleost fishes. *Fish & Shellfish Immunology*, 9(4), pp.309–318.

Rauta, P. R., Nayak, B., & Das, S. (2012). Immune system and immune responses in fish and their role in comparative immunity study: A model for higher organisms. *Immunology Letters*, 148(1), pp.23.

Ravichandran, S., Kumaravel, K., Rameshkumar, G., & Ajithkumar, T. T. (2010). Antimicrobial peptides from the marine fishes. *Research Journal of Immunology*, 3(2), pp.146–156.

Reid, G. K., Gurney-Smith, H. J., Flaherty, M., Garber, A. F., Forster, I., Brewer-Dalton, K., & De Silva, S. (2019). Climate change and aquaculture: Considering adaptation potential. *Aquaculture Environment Interactions*, 11, pp.603–624.

Reyes-Cerpa, S., Maisey, K., Reyes-López, F., & Toro-Ascuy, D., Sandino, A. M., & Imarai, M. (2012). Fish cytokines and immune response: A review. Chapter 1 Fish Cytokines and Immune Response.

Robertsen, B. (2006). The interferon system of teleost fish. *Fish & Shellfish Immunology*, 20(2), pp.172–191.

Rodger, H. D., Henry, L., & Mitchell, S. O. (2011). Non-infectious gill disorders of marine salmonid fish. *Reviews in Fish Biology and Fisheries*, 21, pp.423–440.

Safe, S. H. (1994). Polychlorinated biphenyls (PCBs): Environmental impact, biochemical and toxic responses, and implications for risk assessment. *Critical Reviews in Toxicology*, 24(2), pp.87–149.

Sahoo, S., & Banu, H., Prakash, A., & Tripathi, G.. (2021). Immune system of Fish: An Evolutionary Perspective. A Review.

Salinas, I., Zhang, Y. A., & Sunyer, J. O. (2011). Mucosal immunoglobulins and B cells of teleost fish. *Developmental & Comparative Immunology*, 35(12), pp.1346–1365.

Saurabh, S. & Sahoo, P. K. (2008). Lysozyme: An important defense molecule of fish innate immune system. *Aquaculture Research*, 39, pp.223–239. http://doi.org/10.1111/j.1365-2109.2007.01883.x

Schoor, W.P., Plumb, J.A. (1994) Induction of nitric oxide synthase in channel catfish Ictalurus punctatus by Edwardsiella ictaluri. *Diseases of Aquatic Organisms*, 19, pp.153–155.

Šimková, A., Řehulková, E., Rasoloariniaina, J.R., Jorissen, M.W., Scholz, T., Faltýnková, A., Mašová, Š. and Vanhove, M.P., (2019). Transmission of parasites from introduced tilapias: A new threat to endemic Malagasy ichthyofauna. *Biological Invasions*, 21, pp.803–819.

Siwicki, A.K. (1989). Immunostimulating influence of levamisole on non-specific immunity in carp (*Cyprinuscarpio*). *Developmental & Comparative Immunology*, 13, pp.87–91.

Smith, N.C., Rise, M.L., Christian, S.L. (2019) A comparison of the innate and adaptive immune systems in cartilaginous fish, ray-finned fish, and lobe-finned fish. *Frontiers in Immunology*, 10, p.2292.

Soleto, I., Granja, A. G., Simon, R., Morel, E., Diaz-Rosales, P., & Tafalla, C. (2019). Identification of CD8α+ dendritic cells in rainbow trout (Oncorhynchus mykiss) intestine. *Fish and Shell Immunology*, 89, pp.309–318.

Stafford, J. L., & Belosevic, M. (2003). Transferrin and the innate immune response of fish: Identification of a novel mechanism of macrophage activation. *Developmental & Comparative Immunology*, 27(6–7), pp.539–554. http://doi.org/10.1016/S0145-305X(02)00138-6

Streiff, C., He, B., Morvan, L., Zhang, H., Delrez, N., Fourrier, M., Manfroid, I., Suárez, N.M., Betoulle, S., Davison, A.J., & Donohoe, O., (2023). Susceptibility and permissivity of zebrafish (Danio rerio) larvae to cypriniviruses. *Viruses*, 15(3), p.768.

Talwar, C., Nagar, S., Lal, R., & Negi, R.K., (2018). Fish gut microbiome: Current approaches and future perspectives. *Indian Journal of Microbiology*, 58, pp.397–414.

Tanck, M. W. T., Booms, G. H. R., Eding, E. H., Bonga, S. W., & Komen, J. (2000). Cold shocks: A stressor for common carp. *Journal of Fish Biology*, 57(4), pp.881–894.

Thaiss, C. A., Zmora, N., Levy, M. & Elinav, E. (2016) The microbiome and innate immunity. *Nature*, 535, pp.5–74.

Thorsen, J., Høyheim, B., & Koppang, E.O. (2006) Isolation of the Atlantic salmon tyrosinase gene family reveals heterogeneous transcripts in a leukocyte cell line. *Pigment Cell & Melanoma Research*, 19(4), pp.327–336.

Toda H., Saito Y., Koike T., Takizawa F., Araki K., Yabu T., Somamoto T., Suetake H., Suzuki Y., Ototake M., et al. (2011) Conservation of characteristics and functions of CD4 positive lymphocytes in a teleost fish. *Developmental & Comparative Immunology*, 35, pp.650–660.

Tort, L., Balasch, J. C., & Mackenzie, S. (2003). Fish immune system. A crossroads between innate and adaptive responses. *Inmunología*, 22(3), pp.277–286.

Turvey, S.E., & Broide, D.H., (2010). Innate immunity. *Journal of Allergy and Clinical Immunology*, 125(2), pp.S24–S32.

Verma, M., Rogers, S., Divi, R.L., Schully, S.D., Nelson, S., Joseph Su, Ross, S.A., Pilch, S., Winn, D.M., & Khoury, M.J., 2014. Epigenetic research in cancer epidemiology: Trends, opportunities, and challenges. *Cancer Epidemiology, Biomarkers & Prevention*, 23(2), pp.223–233.

Vidya, M. K. et al. (2018) Toll-like receptors: Significance, ligands, signaling pathways, and functions in mammals. *International Reviews of Immunology*, 37, pp.20–36.

Waagbø, R. (1994). The impact of nutritional factors on the immune system in Atlantic salmon, Salmo salar L.: A review. *Aquaculture Research*, 25(2), pp.175–197.

Walport, M. J.. (2001) Complement. Second of two parts. *The New England Journal of Medicine*, 344, pp.1140–1144.

Wang, B., Thompson, K. D., Wangkahart, E., Yamkasem, J., Bondad-Reantaso, M. G., Tattiyapong, P., … Surachetpong, W. (2023). Strategies to enhance tilapia immunity to improve their health in aquaculture. *Reviews in Aquaculture*, 15, pp.41–56.

Whitney, J. E., Al-Chokhachy, R., Bunnell, D. B., Caldwell, C. A., Cooke, S. J., Eliason, E. J., … Paukert, C. P. (2016). Physiological basis of climate change impacts on North American inland fishes. *Fisheries*, 41(7), pp.332–345.

Yamaguchi, T., & Dijkstra, J. M. (2019). Major histocompatibility complex (MHC) genes and disease resistance in fish. *Cells*, *8*(4), pp.378.

Yu, F., Yang, C., Zhu, Z., Bai, X. & Ma, J. (2019). Adsorption behavior of organic pollutants and metals on micro/nanoplastics in the aquatic environment. *Science of the Total Environment*, 694, p.133643.

Zapata, A., Amemiya, C. (2000). Phylogeny of lower vertebrates and their immunological structures. In: *Origin and evolution of the vertebrate immune system*. Springer, pp.67–107. https://link.springer.com/chapter/10.1007/978-3-642-59674-2_5

Zhang, Y. B., & Gui, J. F. (2012). Molecular regulation of interferon antiviral response in fish. *Developmental & Comparative Immunology*, 38(2), pp.193–202.

Zhao, G., Liu, Z., Quan, J., Sun, J., Li, L., & Lu, J. (2023). Potential role of miR-8159-x in heat stress response in rainbow trout (Oncorhynchus mykiss). *Comparative Biochemistry and Physiology Part B: Biochemistry and Molecular Biology*, p.110877.

Zwollo, P., et al. (2005). B cell heterogeneity in the teleost kidney: Evidence for a maturation gradient from anterior to posterior kidney. *The Journal of Immunology*, 174(11), pp.6608–6616.

3 Biology and Life Cycle of *Bacillus* Probiotics

Mehdi Soltani
Murdoch University, Perth, Australia
University of Tehran, Tehran, Iran

Einar Ringø
UiT The Arctic University of Norway, Tromsø, Norway

3.1 INTRODUCTION

The physiological life habitat of *Bacillus* bacteria is remarkable. They are degraders of most substrates derived from animal and plant sources including cellulose, starch, proteins, agar, and hydrocarbons; antibiotic producers; heterotrophic nitrifiers; denitrifiers; nitrogen fixers; iron precipitators; selenium oxidizers; oxidizers and reducers of manganese; facultative chemolithotrophs; acidophiles; alkalophiles; psychrophiles, thermophiles; and others (Slepecky & Hemphill, 2006). *Bacillus* species such as *Bacillus licheniformis* and *B. subtilis* are widely distributed in the marine flora as a part of dominant primary inhabitants of the marine environments (Priest, 1993). *Bacillus* bacteria generally more frequently exist in sediments than in the water column, and their spores may account for up to 80% of the total heterotrophic flora mainly in the sediments (Siefert et al., 2000) and therefore, they can be naturally ingested by aquatic animals, e.g. shrimps and some demersal fish that feed in or on the sediments (Rengpipat et al., 1998). To understand some biological functions of *Bacillus* such as spore resistance to physical and chemical substances and identification of the spore DNA repair process during spore germination and outgrowth, it is fundamentally important to know about their cellular structure and life cycle.

3.2 CELLULAR SURFACE STRUCTURE

The vegetative cell wall of *Bacillus* species is made up of peptidoglycan (PG) containing meso-diaminopimelic acid or lysine (Bartlett & White, 1985). All *Bacillus* species contain large amounts of an anionic polymer, such as teichoic acid or teichuronic acid, that is bonded to muramic acid residues. The levels of phosphate and magnesium in the growth medium determine the type of anionic polymer. The lipoteichoic acids are associated with the cell membranes, and glycerol teichoic acids vary a great deal between *Bacillus* species and within species (Kaya et al., 1984). *Bacillus* bacteria have peritrichous flagella that in most species such as *B. subtilis* present chemotaxis effect (Slepecky & Hemphill, 2006). Capsule of *Bacillus* as an important factor involved in the bacterial virulence is made up of protein or glycoprotein, but some *Bacillus* species, e.g. *B. circulans* and *B. pumilus*, produce carbohydrate capsules. Members of genus *Bacillus* possess an S-layer (glycoprotein subunits or crystalline surface proteinous layers) with different molecular weight, 40–200 kDa. However, some strains of *Bacillus* species may lack S-layer. The actual function of S-layer is unknown, although it may play a function in bacteria–metal interactions (Slepecky & Hemphill, 2006). In *Bacillus* bacteria, macro-fibers with helical shapes are multicellular and multistranded structures (Mendelson, 1978) that may have some roles in cell-surface molecular organization and force interactions in the cell wall. The establishment and maintenance of these macro-fibers are dependent on

DOI: 10.1201/9781003503811-3

the physiological and genetic conditions (Surana & Mendelson, 1988). The membranes of *Bacillus* species have been studied extensively, especially as a model of membrane structure in Gram-positive bacteria, to assess their roles in sporulation, germination, and thermophilic status. For example, at high temperatures, the physical properties of the membrane of thermophilic species, e.g. *B. stearothermophilus*, can be changed due to changes in the composition of the membrane lipid (Slepecky & Hemphill, 2006). Lipids of *Bacillus* membranes are great in diversity with an identified wide variation in fatty acids (Minnikin & Goodfellow, 1981). The main phospholipids in the membrane are phosphatidylglycerol, di-phosphatidylglycerol, and phosphatidyl-ethanolamine. Menaquinone is the major isoprenoid quinones, and most species contain menaquinones with seven isoprenoid units (Slepecky & Hemphill, 2006). Some enzymes including malate, glycerate-3-phosphate, lactate, nicotinamide adenine dinucleotide hydrogen (NADH), and succinate dehydrogenases have been identified in bacill membranes. Buchanan (1987) reported six penicillin-binding proteins in *B. subtilis*, but one of these factors is absent in some strains of *B. subtilis*.

3.3 LIFE CYCLE

The processes of resting cell formation and the change back to the vegetative cell in *Bacillus* bacteria make them excellent bacteria for manipulation, fast growth, and mass production. *Bacillus* life cycle contains three different physiological phases of vegetative growth, sporulation, and germination phases, and transition from one phase to another is operated by the level of nutrient availability (Figure 3.1) (Moir, 2006; Rosenberg et al., 2012). Pathways with multiple signaling play a significant role in the transitional functions for nutritional and growth rate information directly to the life cycle of the cells and would permit the cells to constantly assess their environments and cell cycle process (Wang & Levin, 2009).

Vegetative growth: Vegetative growth phase is featured by a binary symmetric fission cell growth under the available required nutrients. Replication of bacterial genome is closely tied to the vegetative cell division cycle (Wang & Levin, 2009), and a total separation of sister cell by cleavage of the cell wall may not occur under some conditions forming a long chain (Chai et al. 2010). During vegetative stage, a transcriptional variability related to nutrient availability has been demonstrated among different bacterial cells (Rosenberg et al. 2012) that may represent an additional strategy to enhance population.

Germination: A detailed description of the germination process in *Bacillus* has been reported by various researchers (e.g. Moir, 2006; Plomp et al., 2007; Zhang et al., 2010). Spores of *Bacillus* can remain dormant for a long time period with a remarkable resistance to environmental damages including heat, radiation, toxic chemicals, and pH extremes. Under favorable environmental conditions, the spore breaks dormancy and restarts growth in a process called spore germination and outgrowth. Higgins and Dworkin (2012) reported that the morphological changes during the sporulation process are accompanied by the activation of different specific transcription factors in each spore compartment. The process of germination contains three stages:

Stage I: The initial process in response to nutritional replenishment, the so-called activation, occurs when some germinating molecules, e.g. amino acids, sugars, and purine nucleosides, are sensed by germination receptors located in the inner membrane of the spore. Releasing of hydrogen, potassium, sodium and calcium ions causes an increase in the spore core pH (from 6.5 to 7.7). Activation of enzymes of cortex lytic results in the degradation process of the protective spore PG cortex. It has been shown that the activation process is reversible, thus it is not necessarily that the spore switch toward the germination and outgrowth process (Setlow et al., 2003; Paredes-Sabja et al., 2011; Stewart & Setlow, 2013).

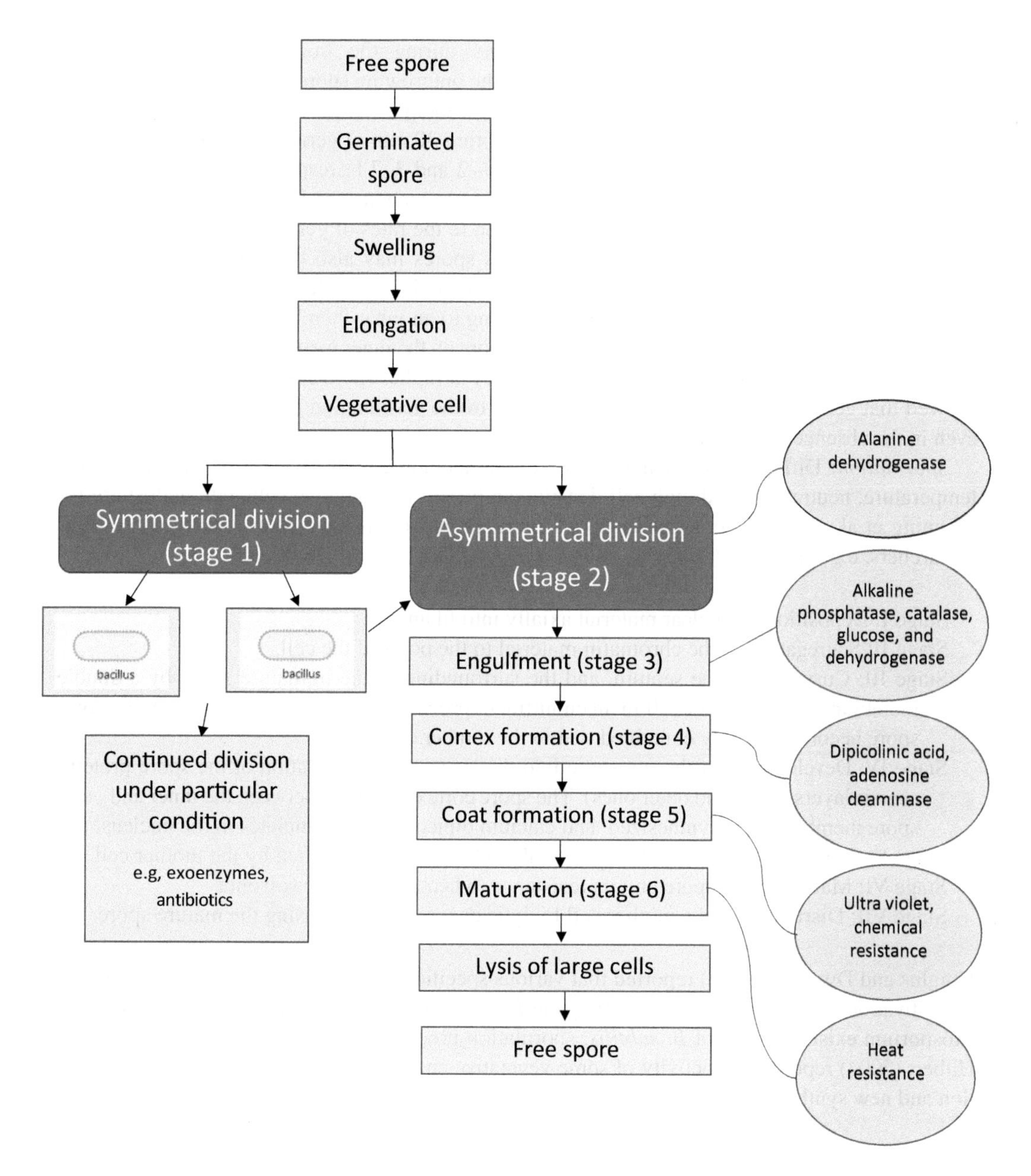

FIGURE 3.1 General life cycle of *Bacillus* bacteria (germination, outgrowth, and sporulation) and effects of some biochemical and physical incidents related to different stages. Designed by Prof. M. Soltani and Dr M. Bromand.

Stage II: Degradation and releasing of pyridine-2,6-dicarboxylic acid (PDA) occur before the spore core rehydration happens. Zhang et al. (2010) demonstrated that dehydration can cause the initiation of protein mobility and reactivation of biochemical processes during outgrowth.

Stage III: Spore coat hydrolysis provides a disclosure of the developing vegetative cell. The outgrowth is the transition of the germinated spore to a growing cell during the time that cell division occurs.

Adenosine triphosphate (ATP) generation occurs during the first stage via conversion of 3-phosphoglycerate (stored in the spore core) and the outgrowing spore changes to the use of extracellular nutrients at later stage (Zhang et al., 2010). Plomp et al. (2007) reported that swelling of spore in some species, i.e. *B. atrophaeus*, occurs within 30 min of germinant contact but the initiation of coat layer and outgrowth elongation took 1–2 and 3–7 h, respectively. Chen et al. (2006) demonstrated that germination process is highly heterogeneous for individual spores, and some factors such as the gene expression make a variation in the rates of germination of different spores (Zhang et al., 2013). The germination of *Bacillus* spores may also occur by very low levels of muropeptides produced via degradation of PGs. Shah et al. (2008) demonstrated that cell wall PG fragments can activate spore germination via binding to an inner membrane-bound eukaryotic-like protein kinase, whereas nutrients act through receptors on the inner membrane of the spore. This can cause an increase in the spore germination efficiency at higher spore concentrations. Setlow (2008) showed that germination in a few spores may lead to the germination of the total spore population even in the absence of nutrients.

Sporulation: Different environmental conditions including high mineral rate, nutrient distress, temperature, neutral pH, and high cell density can bring the vegetative cells into endospore forms (Veening et al., 2009). The seven stages of sporulation process have been described by different researchers, e.g. Errington (2003), Setlow (2007), and Higgins and Dworkin (2012).

Stage I: Deposition of nuclear material axially into filaments.
Stage II: Segregation of the chromatin material to the poles of the cell.
Stage III: Curvation of the septum, and the surrounding of the immature spore by a double membrane of the mother cell in an engulfment process like phagocytosis; the smaller forespore becomes entirely contained within the mother cell.
Stage IV: Development of the forespore into the spore, and assimilation of the spore proteinaceous layers (inner and outer ones). The spore cortex (PG layer between the inner and outer spore membranes) is synthesized, and calcium dipicolinate accumulates in the nucleus.
Stage V: Synthetization of spore coat with about 80 proteins deposited by the mother cell.
Stage VI: Maturation of spore and its resistance to heat and organic solvents.
Stage VII: Disruption of the mother cell by lytic enzymes and releasing the mature spore.

Higgins and Dworkin (2012) reported that various specific transcription factors are associated with the activation of sporulation process. In some *Bacillus* species, an additional protein layer named exosporium exists. In case of *B. subtilis*, sporulation process takes 6–8 h at 37°C, and Piggot and Hilbert (2004) reported the activity of some vegetative enzymes i.e. stopping, remaining, modification and new synthetization enzymes change.

3.4 SPORE RESISTANCE FACTORS

Several factors have been reported to be associated with *Bacillus* spore resistance, and some environmental/chemical factors able to affect these are given in Table 3.1.

Spore coat: The spore coat as the first barrier to penetration of large molecules (Nicholson et al., 2000) plays a role in spore resistance to attack by bacteriovorus (Laaberki & Dworkin, 2008) and to some chemical biocide actions such as oxidizing substances (Young & Setlow, 2004). Protection of resting spores from enzyme activities, e.g. lysozyme and mechanical disruption, is also a function of spore coat (Gould, 2003). Riesenman and Nicholson (2000) demonstrated that the coat, particularly the inner coat layer, influences spore resistance to relevant UV wavelengths. Additionally, Setlow (2006) reported that proteins of spore coat can react with and detoxify some chemical substances. However, spore coat has little or no role in the resistance of the spore to some other chemical agents, heat, and radiation.

TABLE 3.1
Some Environmental and Chemical Factors That Can Affect *Bacillus* Spore Resistance Factors

Environmental/ Chemical Factor	Spore Resistance Factor	Way of Affecting the Spore (Damage/Disrupt)
Dry heat	Alpha/beta-type small acid-soluble proteins (ABSASPs) and DNA repair	DNA
Wet heat	ABSASPs, core mineralization, core water content, and dipicolinic acid (DA)	Essential spore proteins
Ionizing radiation	ABSASPs, spore core components, core dehydration, and DA	DNA and/or RNA proteins
UV radiation	ABSASPs, DA, core water content, and DNA repair	DNA
Microwave radiation	Core water content	Essential spore proteins, inner membrane, DNA, and coat
Chemical substance	ABSASPs, spore coat, permeability of inner membrane, and DNA repair	Chemicals such as nitrous acid, formaldehyde, and alkylating agents, e.g. ethylene oxide, affect DNA Chemicals such as ozone, hypochlorite, chlorine, dioxide, ethanol, and strong acid affect inner membrane Lytic enzymes affect spore cortex

Source: (Modified from Sella et al., 2014).

Cortex PGs of spore: Cortex contains PG the thickness of which is an important factor in maintaining spore resistance to heat and core water content via reaching in protoplast dehydration (Atrih & Foster, 1999). However, Popham (2002) observed that the spore PGs could vary with no significant effect on spore core dehydration. Additionally, Zhang et al. (2012) demonstrated that the cortex of PG structure affects the level of spore germination enhancement leading to a more rapid completion of germination.

Spore maturation: It has been shown that those spores that released first from their mother cells during sporulation stage are less resistant than the total spores formed during the whole sporulation phase that may in part be due to the modifications that occurred during the spore maturation (Sanchez-Salas et al., 2011).

Core water content: Rupley and Careri (1991) reported that dehydration of spore core may increase spore resistance to wet heat via stabilizing proteins to thermal denaturation. Indirect core proteins immobilization by the dehydration of spore core may also protect irreversible protein aggregation (Sunde et al., 2009). Kaieda et al. (2013) suggested that the water of spore core is in a gelatinous form with high water mobility and plays a major factor for the spore resistance to wet heat and hydrogen peroxide heat. Setlow (2006) reported that core proteins in highly dehydrated spore core have higher resistance to wet heat due to a decrease in molecular motion.

Core mineralization: Mineralization of spore is associated with accumulation of high level of some cations (Ca^{2+}, Mg^{2+}, and Mn^{2+}) and DNA in the spore core can enhance the resistance of spores via dehydration of the protoplast and other mechanisms. Both mineralization and demineralization of spores affect the heat resistance in some *Bacillus* species, e.g. *B. atrophaeus*. Different mineral ions have different impacts on the wet heat protection, with a greater protection made by calcium, magnesium, manganese, potassium, and sodium cations (Bender & Marquis, 1985; Beaman & Gerhardt, 1986; Yamazaki et al., 1997). About 10% of the dry weight of *Bacillus* spores is dipicolinic acid or PDA that is chelated with divalent cations mainly by calcium ions. Setlow et al. (2006) reported that dipicolinic acid is highly important in spore resistance and stability, and also has some specific roles in protecting of spore DNA from damage. Desiccation, gamma irradiation, and ultraviolet radiation are some of the killing treatments affecting the generation of reactive oxygen species causing cell death via protein damage, e.g. enzymes repairing oxidative DNA damage. The protection of spore proteins to these killing treatments may be due to the high levels of dipicolinic acid conjugated to ions, especially the available calcium cations, in spore as reported by Granger et al. (2011).

Alpha/beta-type small acid-soluble spore proteins (ABSASPs): In the resting *Bacillus* spores, DNA is saturated with a group of nonspecific DNA-binding ABSASPs that are synthesized only during sporulation. It has been demonstrated that these proteins have a protective function that prevents DNA damage (Setlow & Setlow, 1996). The protection of DNA of spore from ultraviolet, free radical damage (Setlow, 1995; Moeller et al., 2013) as well as resistance to dry heat, formaldehyde, and hydrogen peroxide is associated with these proteins being inside the resting spore (Paidhungat & Setlow, 2001; Setlow, 2000).

DNA repair: Metabolically the resting spores are inactive; however, the repairing of damaged DNA proteins occurs during germination period when spores reactivate and return to vegetative form (Moeller et al., 2013). Repairing of DNA proteins is highly important in spore outgrowth that occurred during spore resting. (Setlow, 1995). Nicholson et al. (2000) demonstrated that both enzymes of spore can repair DNA damage and the damaged DNA accumulated in the resting spore can induce the synthesis of DNA repair proteins upon subsequent spore germination. Moeller et al. (2013) showed that DNA repair is the most important mechanism for spore resistance to ionizing radiation. Details of functions of DNA repair and mutagenesis are outlined in Lenhart et al. (2012).

Permeability of inner membrane: Spore resistance is also associated with the inner membrane permeability that is a significant barrier to chemical agents as demonstrated by Paidhungat and Setlow (2001). Rose et al. (2007) also reported that a higher permeability of this membrane is associated with spore resistance under a lower acid condition.

Chemical status: The spores of *Bacillus* bacteria possess a reversible cation exchange system loaded with calcium ions, providing both high wet heat and high dry heat resistances for the spore, but under acidified condition they become less resistant (Alderton & Snell, 1969). Exposing of spores to low pH conditions and some heat results in loss of minerals, i.e. calcium, manganese, magnesium, potassium, and sodium, as well as a reduction in heat resistance. Conversely, if the spores are subjected to heated condition together at alkaline condition, e.g. pH 8, they absorb mineral ions and become more resistant to the heat environment (Beaman & Gerhardt, 1986).

Sporulation conditions: Both nutritional and environmental factors are necessary when studying their impacts on spore resistance, and growth medium and temperature are the major factors affecting the spore resistance (Melly et al., 2002). For instance, Rose et al. (2007) reported that *B. subtilis* spores prepared on solid medium gave a higher wet heat and chemical resistance than spores prepared in liquid medium. Some divalent ions, particularly calcium and manganese, can induce sporulation and increase the heat resistance (Hornstra et al., 2009; Minh et al., 2011). Minh et al. (2008) demonstrated that *B. subtilis* spores produced at lower water activity were smaller and more sensitive to heat condition than water with high activity. Moeller et al. (2011) demonstrated that use of cysteine and thio-proline in the sporulation medium enhanced the spore resistance to ultraviolet radiation and H_2O_2. Lindsay et al. (1990) reported that heat resistance in *B. subtilis* spores was increased at 23–45°C, but the resistance declined above 45°C. Cortezzo and Setlow (2005) also demonstrated that the produced spores at lower temperatures were more sensitive to DNA damage by some chemical substances, e.g. formaldehyde and nitrous acid, than spores produced at higher temperatures. A higher sporulation temperature associated with higher level of spore mineralization has also been reported by Palop et al. (1999).

The prepared *B. subtilis* spores at 22–48°C provided similar values of dipicolinic acid and ABSASPs at lower and higher temperatures (Melly et al., 2002). These spores were more resistant to wet heat than the spores prepared at lower temperatures. However, Kaieda et al. (2013) were unable to show a decrease in core water content under increasing temperature for sporulation. At higher temperatures, a similar resistance to ultraviolet radiation, dry heat, and oxidizing agents, e.g. formaldehyde and glutaraldehyde, occurs (Melly et al., 2002).

Other factors influencing spore resistance: Spore resistance can also be affected by some other factors such as osmolarity, pH, and presence of germinating substances, e.g. alanine, Ca-dipicolinic acid, and dodecylamine (Setlow et al., 2003; Pinzón-Arango et al., 2009). Some organic materials including proteins and carbohydrates are able to increase spore resistance to heat, while glucose may reduce the spore resistance (Molin & Svensson, 1976). During spore resistance assessment, the interactions between the inherent parameters and environmental factors are necessary. For example, Moeller et al. (2012) reported that the survival and mutagenesis of spore under ultraviolet radiation were highly dependent on structural components of spore such as small acid-soluble spore proteins, coat layers, and dipicolinic acid, being the major protective components.

BIBLIOGRAPHY

Alderton G, Snell N. 1969. Chemical states of bacterial spores: dry-heat resistance. *Appl. Microbiol.*, 17(5): pp.745–9. doi: 10.1128/am.17.5.745-749.1969.

Atrih A, Foster SJ. (1999) The role of peptidoglycan structure and structural dynamics during endospore dormancy and germination. *Anton Van Leeuw*, 75, pp.299–307.

Bartlett AT, White MPJ (1985) Species of *Bacillus* that make a vegetative peptidoglycan containing lysine lack diaminopimelate epimerase but have diaminopimelate dehydrogenase. *J Gen Microbiol*, 131, pp.2145–2152.

Beaman TC, Gerhardt P (1986) Heat resistance of bacterial spores correlated with protoplast dehydration, mineralization and thermal adaptation. *Appl Environ Microbiol*, 52, pp.1242–1246.

Bender GR, Marquis RE (1985) Spore heat resistance and specific mineralization. *Appl Environ Microbiol*, 50, pp.1415–1421.

Buchanan, C. E. (1987) Absence of penicillin-binding protein 4 from an apparently normal strain of *Bacillus subtilis*. *J Bacteriol*, 169, pp.5301–5303.

Chai Y, Norman T, Kolter R, Losick R (2010) An epigenetic switch governing daughter cell separation in *Bacillus subtilis*. *Genes Dev*, 24, pp.754–765.

Chen D, Huang SS, Li YQ (2006) Real-time detection of kinetic germination and heterogeneity of single *Bacillus* spores by laser tweezers Raman spectroscopy. *Anal Chem*, 78, pp.6936–6941.

Cortezzo DE, Setlow P (2005) Analysis of factors that influence the sensitivity of spores of *Bacillus subtilis* to DNA damaging chemicals. *J Appl Microbiol*, 98, pp.606–617.

Errington J (2003) Regulation of endospore formation in *Bacillus subtilis*. *Nat Rev Microbiol*, 1, pp.118–126.

Gould GW (2003) Mechanisms of resistance and dormancy. In: Hurst A, Gould GW (editors) *The bacterial spore*. London: Academic Press, pp.173–209.

Granger AC, Gaidamakova EK, Matrosova VY, Daly MJ, Setlow P (2011) Effects of Mn and Fe levels on *Bacillus subtilis* spore resistance and effects of Mn2, other divalent cations, orthophosphate, and dipicolinic acid on protein resistance to ionizing radiation. *Appl Environ Microbiol*, 77, pp.32–40.

Higgins D, Dworkin J (2012) Recent progress in *Bacillus subtilis* sporulation. *FEMS Microbiol Rev*, 36, pp.131–148.

Hornstra LM, Beek AT, Smelt JP, Kallemeijn WW, Brul S (2009) On the origin of heterogeneity in (preservation) resistance of *Bacillus* spores: Input for a 'systems' analysis approach of bacterial spore outgrowth. *Int J Food Microbiol*, 134, pp.9–15.

Kaieda S, Setlow B, Setlow P, Halle B (2013) Mobility of core water in *Bacillus subtilis* spores by 2H NMR. *Biophys J*, 105(9): pp.2016–2023.

Kaya SK, Yokoyama Y Araki EI, 1984. N-*acetylmannos-aminyl* (14) acetylglucosamine, a linkage unit between glycerol teichoic acid and peptidoglycan in cell walls of several *Bacillus* strains. *J Bacteriol*, 158,990 p.996.

Laaberki MH, Dworkin J (2008) Role of spore coat proteins in the resistance of *Bacillus subtilis* spores to *Caenorhabditis elegans* predation. *J Bacteriol*, 190, pp.6197–6203.

Lenhart JS, Schroeder JW, Walsh BW, Simmons LA (2012) DNA repair and genome maintenance in *Bacillus subtilis*. *Microbiol Mol Biol Rev*, 76, pp.530–564.

Lindsay JA, Barton LE, Leinart AS, Pankratz HS (1990) The effect of sporulation temperature on sporal characteristics of *Bacillus subtilis*. *Curr Microbiol*, 21, pp.75–79.

Little SF, Ivins BE (1999) Molecular pathogenesis of *Bacillus anthracis* infection. *Microbial Infection*, 1(2): pp.131–139.

Melly E, Genest PC, Gilmore ME, Little S, Popham DL, Driks A, Setlow P (2002) Analysis of the properties of spores of *Bacillus subtilis* prepared at different temperatures. *J Appl Microbiol*, 92, pp.1105–1115.

Mendelson NH 1978. Helical *Bacillus subtilis* microfibers: morphogenesis of a bacterial multicellular microorganism. *Proc Natl Acad Sci USA*, 75, pp.2478–2482.

Minh NTH, Durand A, Loison P, Perrier-Cornet JM, Gervais P (2011) Effect of sporulation conditions on the resistance of *Bacillus subtilis* spores to heat and high pressure. *Appl Microbiol Biotechnol*, 90,pp.1409–17.

Minh NTH, Guyot S, Perrier-Cornet JM, Gervais P (2008) Effect of the osmotic conditions during sporulation on the subsequent resistance of bacterial spores. *Appl Microbiol Biotechnol*, 80, pp.107–14.

Minnikin, D. E. M. Goodfellow, (1981). Lipids in the classification of *Bacillus* and related taxa. In Berkeley, R. C. and M. Goodfellow (Eds.), *The aerobic endospore forming bacteria* (pp.59–103). London: Academic Press.

Moeller R, Raguse M, Reitz G, Okayasu R, Li Z, Klein S, Setlow P, Nicholson WL (2013) Resistance of *Bacillus subtilis* spore DNA to lethal ionizing radiation damage relies primarily on spore core components and DNA repair, with minor effects of oxygen radical detoxification. *Appl Environ Microbiol*. AEM-03136.

Moeller R, Vlašić I, Reitz G, Nicholson WL (2012) Role of altered rpoB alleles in *Bacillus subtilis* sporulation and spore resistance to heat, hydrogen peroxide, formaldehyde, and glutaraldehyde. *Arch Microbiol*, 194(9): pp.759–67.

Moeller R, Wassmann M, Reitz G, Setlow P (2011) Effect of radioprotective agents in sporulation medium on *Bacillus subtilis* spore resistance to hydrogen peroxide, wet heat and germicidal and environmentally relevant UV radiation. *J Appl Microbiol*, 110, pp.1485–1494.

Moir A (2006) How do spores germinate? *J Appl Microbiol*, 101,pp.1–5.

Molin G, Svensson M (1976) Formation of dry-heat resistant *Bacillus subtilis* var. *niger* spores as influenced by the composition of the sporulation medium. *Anton Van Lee J M S*, 42, pp.387–395.

Nicholson WL, Munakata N, Horneck G, Setlow P (2000) Resistance of *Bacillus* endospores to extreme terrestrial and extraterrestrial environments. *Microbiol Mol Biol R*, 64, pp.548–572.

Paidhungat M, Setlow P (2001) Spore germination and outgrowth. In: Hoch JA, Losick R, Sonenshein A, editors. Bacillus subtilis *and Its Relatives: From Genes to Cells*. Washington, DC: American Society for Microbiology, pp.537–548.

Palop A, Sala FJ, Condon S (1999) Heat resistance of native and demineralized spores of *Bacillus subtilis* sporulate at different temperatures. *Appl Environ Microbiol*, 65, pp.1316–1319.

Paredes-Sabja D, Setlow P, Sarker MR (2011) Germination of spores of *Bacillales* and *Clostridiales* species: mechanisms and proteins involved. *Trends Microbiol*, 19, pp.85–94.

Piggot PJ, Hilbert DW (2004) Sporulation of *Bacillus subtilis*. *Curr Opin Microbiol*, 7, pp.579–586.

Pinzón-Arango PA, Scholl G, Nagarajan R, Mello CM, Camesano TA (2009) Atomic force microscopy study of germination and killing of *Bacillus atrophaeus* spores. *J Mol Recogniti*, 22, pp.373–9.

Plomp M, Leighton TJ, Wheeler KE, Hill HD, Malkin AJ (2007) *In vitro* high-resolution structural dynamics of single germinating bacterial spores. *Proc Natl Acad Sci USA*, 104, pp.9644–9649.

Popham DL. (2002) Specialized peptidoglycan of the bacterial endospore: The inner wall of the lockbox. *Cell Mol Life Sci* 59, pp.426–433.

Priest F. G. (1993) Systematics and ecology of *Bacillus*. In: Sonenshein, A. L. editor. Bacillus subtilis *and Other Gram-Positive Bacteria, Biochemistry, Physiology, and Molecular Genetics*. Washington, DC: American Society for Microbiology Press, pp.3–16.

Rengpipat S., Phianphak W, Piyatiratitivorakul S, Menasveta P. (1998) Effcts of a probiotic bacterium on black tiger shrimp *Penaeus monodon* survival and growth. *Aquaculture* 167, pp.301–313.

Riesenman PJ, Nicholson WL (2000) Role of the spore coat layers in *Bacillus subtilis* spore resistance to hydrogen peroxide, artificial UV-C, UV-B, and solar UV radiation. *Appl Environ Microbiol*, 66, pp.620–626.

Rose R, Setlow B, Monroe A, Mallozzi M, Driks A, Setlow P (2007) Comparison of the properties of *Bacillus subtilis* spores made in liquid or on agar plates. *J Appl Microbiol*, 103, pp.691–699.

Rosenberg A, Sinai L, Smith Y, Ben-Yehuda S (2012) Dynamic expression of the translationalmachinery during *Bacillus subtilis* life cycle at a single cell level. *PLoS ONE*, 7, p.e41921.

Rupley JA, Careri G. (1991) Protein hydration and function. *Adv Protein Chem*, 41, pp.37–172.
Sahin O, Yong EH, Driks A, Mahadevan L. (2012) Physical basis for the adaptive flexibility of *Bacillus* spore coats. *J R Soc Interface*, 9(76): pp.3156–3160
Sanchez-Salas J-L, Setlow B, Zhang P, Yong-Qing Li Y, Setlow P (2011) Maturation of released spores is necessary for acquisition of full spore heat resistance during *Bacillus subtilis* sporulation. *Appl Environ Microbiol*, 77, pp.6746–6754.
Sella SRBR, Vandenberghe LPS, Soccol CR (2014) Life cycle and spore resistance of spore-forming *Bacillus atrophaeus*. *Microbiol Res*, 169, pp.931–939. http://doi.org/10.1016/j.micres.2014.05.001
Setlow B, Setlow, P (1996) Role of DNA repair in *Bacillus subtilis* spore resistance. *J Bacteriol*, 178, pp.3486–3495.
Setlow B, Cowan A E, Setlow P (2003) Germination of spores of *Bacillus subtilis* with dodecylamine. *J Appl Microbiol*, 95,pp.637–648.
Setlow B, Atluri S, Kitchel R, Koziol-Dube K, Setlow P (2006) Role of dipicolinic acid in resistance and stability of spores of *Bacillus subtilis* with or without DNA-protective alpha/beta-type small acid-soluble proteins. *J Bacteriol*, 188, pp.3740–7.
Setlow P (1995) Mechanisms for the prevention of damage to the DNA in spores of *Bacillus* species. *Annu Rev Microbiol*, 49, pp.29–54.
Setlow P (2000) Resistance of bacterial spores. In: Storz G, Hengge-Aronis R (Eds.), *Bacterial Stress Responses*. Washington, DC: *American Society for Microbiology*, pp.217–230.
Setlow P (2006) Spores of *Bacillus subtilis*: Their resistance to and killing by radiation, heat and chemicals. *J Appl Microbiol*, 101, pp.514–525.
Setlow P (2007) I will survive: DNA protection in bacterial spores. *Trends Microbiol*, 15, pp.172–180.
Setlow P (2008) Dormant spores receive an unexpected wake-up-call. *Cell*, 135, pp.410–412.
Shady TSM (1997) Studies on the application of *Bacillus subtilis* lipase in detergency. *Ann Agric Sci* (Cairo), 42(1): pp.73–80.
Shah IM, Laaberki M-H, Popham DL, Dworkin J (2008) A eukaryotic-like *ser/thr* kinase signals bacteria to exit dormancy in response to peptidoglycan fragments. *Cell*, 135,pp.486–496.
Siefert JL, Larios-Sanz M, Nakamura LK, Slepecky RA, Paul JH, Moore ERB, Fox GE & Jurtshuk P Jr (2000) Phylogeny of marine *Bacillus* isolates from the Gulf of Mexico. *Curr. Microbiol.* 41, pp.84–88.
Slepecky R, Hemphill E (2006) The genus *Bacillus*. Nonmedical. In: Dworkin M, Falkow S, Rosenberg E, Schleifer K-H & Stackebrandt E, editors. *The Prokaryotes*, Vol. 4, pp.530–562. New York: Springer.
Stewart KAV, Setlow P (2013) Numbers of individual nutrient germinant receptors and other germination proteins in spores of *Bacillus subtilis*. *J Bacteriol*, 195(7): pp.1484–1491.
Sunde EP, Setlow P, Hederstedt L, Halle B (2009) The physical state of water in bacterial spores. *Proc Natl Acad Sci*, 106, pp.19334–19339.
Surana UAJW, Mendelson NH 1988. Regulation of *Bacillus subtilis* macrofibe twist development by Dalanine. *J. Bacteriol.* 170, pp.2328–2335.
Veening JW, Murray H, Errington JA (2009) Mechanism for cell cycle regulation of sporulation initiation in *Bacillus subtilis*. *Genes Dev*, 23, pp.1959–1970.
Wang JD, Levin PA (2009) Metabolism, cell growth and the bacterial cell cycle. *Nat Rev Microbiol*, 7, pp.822–827.
Yamazaki K, Kawai Y, Inoue N, Shinano H (1997) Influence of sporulation medium and divalent ions on the heat resistance of *Alicyclobacillus acidoterrestris* spores. *Lett Appl Microbiol*, 25, pp.153–156.
Young SB, Setlow P (2004) Mechanisms of killing of *Bacillus subtilis* spores by hypochlorite and chlorine dioxide. *J Appl Microbiol*, 95, pp.54–67.
Zhang JQ, Griffiths KK, Cowan A, Setlow P, Yu J (2013) Expression level of *Bacillus subtilis* germinant receptor determines the average rate but not the heterogeneity of spore germination. *J Bacteriol*, 195 (8): pp.1735–40.
Zhang P, Garner W, Yi X, Yu J, Li Y-G, Setlow P (2010) Factors affecting variability in time between addition of nutrient germinant and rapid dipicolinic acid release during germination of spores of *Bacillus* species. *J Bacteriol*, 192, pp.3608–3619.
Zhang P, Thomas S W, Li Y, Setlow P (2012) Effects of cortex peptidoglycan structure and cortex hydrolysis on the kinetics of Ca2+ dipicolinic acid release during *Bacillus subtilis* spore germination. *J Bacteriol*, 194,pp.646–652.

4 Physiology of Spore Formation for *Bacillus* Probiotic Production

Lokesh Pawar
College of Fisheries (Central Agricultural University- Imphal), Agartala, India
University of Algarve, Faro, Portugal

Arya Singh
ICAR – Central Institute of Fisheries Education, Mumbai, India

Nayan Chouhan
College of Fisheries (Central Agricultural University- Imphal), Agartala, India

Hadeer A. Amer
Kafer El-Sheikh University, Kafer Elsheikh, Egypt

K. Likitha
Kerala University of Fisheries and Ocean Studies, Kochi, India

Naga Prasanthmadduluri
Andhra Pradesh Fisheries University, Muttukuru, India

Soibam Khogen Singh
College of Fisheries (Central Agricultural University- Imphal), Agartala, India
Krishi Vigyan Kendra, Ukhrul, ICAR -Research Complex for NEH Region, Hungpung, India

Mehdi Soltani
Murdoch University, Perth, Australia
University of Tehran, Iran

4.1 INTRODUCTION

The intricate process of spore formation in *Bacillus* species has long captivated the scientific community due to its remarkable physiological and industrial significance. Probiotics, known for their beneficial effects on human health, have gained immense popularity in recent years. Understanding the physiology behind spore formation is crucial for optimizing the production of these valuable microbial products, as the spore state enhances their stability and viability, allowing for easy storage and transportation. By unravelling the intricate interplay between spore formation and probiotic

 DOI: 10.1201/9781003503811-4

production, this chapter aims to provide valuable insights for researchers and industry professionals alike, ultimately contributing to the advancement of *Bacillus*-based probiotic technologies. In this chapter, we explore the underlying mechanisms and regulatory pathways that govern spore formation in *Bacillus* and shed light on the physiological factors influencing the production of probiotics and its intricate connection to the production of probiotics by *Bacillus* organisms with an overview of the physiology of *Bacillus* spp., exploring their morphology and cell structure, motility, as well as their metabolic and nutrient utilization capabilities.

4.2 PHYSIOLOGY OF *BACILLUS* SPP.

The phylum *Firmicutes* includes a varied collection of Gram-positive, rod-shaped bacteria known as the *Bacillus* genus. These microorganisms exhibit a remarkable range of physiological and metabolic capabilities, rendering them highly significant in various fields, including agriculture, medicine, and biotechnology. Their adaptability and utility have led to their exploitation in numerous industrial and medical applications.

4.2.1 Morphology and Cell Structure

Bacillus spp. typically present themselves as rod-shaped bacteria, existing either as individual cells with lengths ranging from 2 to 5 μm and widths of 0.5–1.0 μm or as chains resembling bamboo sticks, characterized by concave ends. These bacteria possess a cytoplasmic membrane, which acts as a protective barrier. Furthermore, they possess a thick peptidoglycan cell wall that retains the Gram-positive stain, resulting in their characteristic purple appearance under a microscope. Notably, *Bacillus* spp. are known to form endospores, which are resilient structures enabling survival under harsh conditions.

4.2.2 Motility

Bacillus spp. display motility through the presence of peritrichous flagella. These flagella enable the bacteria to move either towards favourable environmental conditions or away from unfavourable ones, such as light or chemical gradients. Flagellar movements allow *Bacillus* spp. to actively explore their surroundings and locate sources of nutrients. Additionally, some species within this genus exhibit gliding motility, a mechanism independent of flagella.

4.2.3 Metabolism and Nutrient Utilization

Bacillus spp. is metabolically versatile bacteria. Primarily aerobic organisms, they require oxygen for growth and metabolism. These bacteria possess the utilization capacity towards a wide array of carbon and energy sources, including sugars, amino acids, and organic acids. To achieve this, extracellular enzymes are produced that facilitate the breakdown and assimilation of these compounds. *Bacillus* spp. harbour various enzymes involved in aerobic respiration, such as cytochromes and oxidases. In the absence of oxygen, certain species can also ferment specific substrates. Their capability to produce diverse enzymes, such as lipases, proteases, and amylases, contributes to their efficient nutrient utilization capabilities, allowing them to thrive in diverse environments.

4.2.4 Metabolic Pathways

Bacillus spp. possesses an extensive repertoire of metabolic pathways. Key pathways include the glycolysis, pentose phosphate pathway, and tricarboxylic acid (TCA) cycle. These metabolic routes enable efficient processing and breakdown of various compounds, leading to the generation of energy. *Bacillus* bacteria's proficiency in metabolic processes makes them highly valuable in industrial applications, such as the production of biofuels and bioremediation.

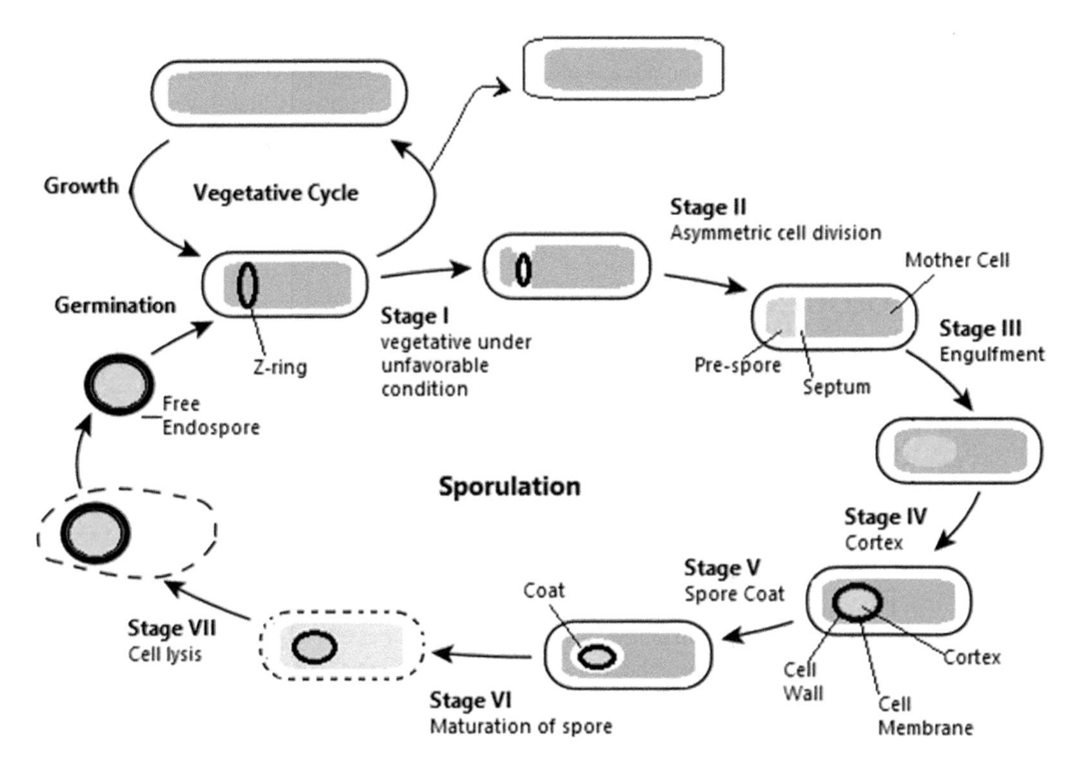

FIGURE 4.1 Life cycle of *Bacillus subtilis* (From Wikipedia, the free encyclopaedia (Adapted from https://commons.wikimedia.org/wiki/File:Sporulation_in_Bacillus_subtilis.jpg).)

4.2.5 Sporulation and Germination

Bacteria exhibit the ability to persist in two distinct forms: spores and vegetative forms (Figure 4.1). When the surrounding environment provides favourable conditions for bacterial metabolic activities, they predominantly exist in their vegetative forms. In this state, bacteria actively absorb nutrients, undergo growth, and reproduce. However, in harsh and unfavourable environments, bacteria face extinction as they become unable to withstand the conditions. Nonetheless, certain bacteria have evolved the capability to survive in such adverse conditions by transforming into highly resilient, metabolically dormant forms known as spores. These microorganisms are referred to as "spore-forming bacteria." It is worth noting that specialized methods, such as autoclaving, can effectively destroy these spores (Madigan et al., 2018). Spores are classified into two groups according to how they form in connection to the bacterial cell:

4.2.5.1 Exospore

Exospores are produced externally to the vegetative cell through budding at one end of the bacterial cell. They are subsequently released outside as spores, which enable them to survive adverse conditions. Unlike endospores, exospores lack dipicolinic acid (DPA), an important component found in endospores. Examples of exospore-forming bacteria include methane-oxidizing bacteria of the genus *Methylosinus* and *Methylocystis*.

4.2.5.2 Endospore

Endospores are formed within the bacterial cell and are subsequently released outside as spores to endure adverse conditions. Endospores are the most commonly observed spores and are often used synonymously with the term "spore." The structure of an endospore is considerably more complex than that of a vegetative cell, comprising multiple layers. These layers, arranged from the outermost to the innermost, include the exosporium (outermost layer), spore coat (middle layer), cortex (inner layer), and core (innermost layer). The core of an endospore contains a compacted, inactive chromosome. Additionally, more than one layer of proteinaceous material, known as the spore coat, surrounds the core, along with a cortex rich in peptidoglycan. Together, these layers provide protection against factors such as ultraviolet (UV) radiation, extreme temperatures (up to 80–85°C for most species), solvents, hydrogen peroxide, and lysozyme enzymes (Henriques & Moran, 2007; Nicholson et al., 2000; Cutting, 2011).

4.2.6 Pathogenicity

Although many *Bacillus* species are harmless, certain strains can cause infection in human and animal. One notable example is *B. anthracis*, causing anthrax, a severe and potentially fatal disease. Other species, such as *Bacillus cereus*, have been associated with food poisoning (Madigan et al., 2018).

4.2.7 Production of Enzymes and Secondary Metabolites

Bacillus species are renowned for their ability to produce a varied range of extracellular enzymes, with proteases, amylases, lipases, and cellulases. These enzymes play a crucial role in breaking down cell wall components of pathogenic bacteria (James et al., 2021). Moreover, they have the potential to neutralize "anti-nutritional" components in feed ingredients (Popov & Algburi, 2021). Furthermore, these enzymes aid in the digestion of indigestible starch polysaccharides, thereby enhancing nutrient absorption in the intestines. One notable example is *Bacillus subtilis*, which produces enzymes such as subtilisin, catalase, and lactic acid. The presence of such enzymes promotes the development of beneficial microbes like *Lactobacillus* species. It is essential to have stable probiotics capable of colonizing the gastrointestinal tract of fish to ensure optimal digestion (Senok et al., 2005) In addition to enzyme production, *Bacillus* species are also renowned for their capacity to generate a wide range of secondary metabolites. These metabolites contribute to their ecological and physiological characteristics and possess various properties, including antimicrobial, antifungal, antiviral, and anticancer activities. Additionally, these metabolites aid in competition with other microorganisms and adaptation to different environments. *Bacillus*-produced antimicrobials exhibit a broad-spectrum effect contrary to a wide range of pathogens, including yeasts, fungi, and both Gram-negative and Gram-positive bacteria. Secondary metabolite production typically occurs during the stationary growth phase and is influenced by nutrient limitations, such as nitrogen and phosphorus, as well as an abundant carbon and energy source. For instance, *Gibberella fujikuroi* or *Fusarium moniliforme* produces the secondary metabolite gibberellic acid (GA3), which is a plant growth hormone (Peng et al., 2020). Bacteriocins, which are non-toxic peptides, display a bactericidal mechanism by binding to receptors on bacterial surfaces. They are generally more susceptible to proteases related to antibiotics (Yang et al., 2014; Mokoena, 2017).

4.2.8 Biofilm Formation

Biofilm formation is a common characteristic among many *Bacillus* species. The process begins with individual *Bacillus* cells attaching to a solid substrate using structures such as pili, fimbriae, or other surface appendages. Over time, these individual cells multiply, forming microcolonies that are rooted within a matrix of extracellular polymeric substances (EPS) which was self-produced. This EPS matrix plays a vital role in providing structural stability to the biofilm, aiding in adhesion to the surface, and acting as a physical barrier that inhibits the penetration of the immune system and antibiotics. *Bacillus* biofilms are highly resilient due to the presence of persister cells, which are metabolically inactive and exhibit extreme resistance to antimicrobial agents. As the biofilm matures, the microcolonies create a three-dimensional structure, and channels and water-filled voids develop within the biofilm, facilitating nutrient and waste exchange.

Bacillus species use quorum sensing to control biofilm creation. This involves making and sensing signalling molecules (autoinducers) for bacterial communication within the biofilm group. This coordination manages activities like sporulation, biofilm formation, and virulence. Grasping biofilm creation's mechanisms is vital in medicine, industry, and environmental microbiology. Biofilms impact chronic infections, industrial system fouling, and the environment. These species are recognized for enzyme and secondary metabolite production. Their enzymes break cell walls, counter anti-nutrients, and enhance nutrient absorption in intestines. Moreover, they generate diverse secondary metabolites with antimicrobial and bioactive traits. *Bacillus* species make strong biofilms that affect biological and ecological processes. Comprehending enzyme, metabolite, and biofilm production in *Bacillus* is valuable, from refining feed quality to inventing new antimicrobials.

4.2.9 Market Potential for Spore-Forming Probiotics

According to Poshadri et al. (2022), the probiotics industry is expected to achieve a substantial milestone, reaching $73.9 billion by 2030, with a consistent annual growth rate of 8.6%. Probiotics, as defined by the OIE and WHO, are living microorganisms that offer a wide range of health advantages when consumed in suitable amounts. These benefits encompass detoxification, cholesterol reduction, improved digestion and mineral absorption, as well as safeguarding against gastroenteritis. For probiotics to be effective, it is crucial that they are viable and administered in sufficient quantities, with at least 10^7 CFU/g on a regular basis. This ensures a positive impact on the gastrointestinal microbiota and the body's defence system. Two of the most popular probiotic bacterial species are *Lactobacillus* and *Bifidobacterium*. However, these bacteria may lose their efficacy during production and storage due to their sensitivity to stomach acid and high temperatures. In contrast, *Bacillus* species demonstrate distinct advantages, since *Bacillus* species provide valuable antimicrobial benefits, fortify the immune system, promote growth, enhance nutritional value, and furnish protection against gastrointestinal diseases (Poshadri et al., 2022).

4.2.10 Optimization of Spore-Forming Probiotic Bacteria

The production of spores with high efficiency is a critical step in bio product manufacturing. *Bacillus* spp. starts forming heat-resistant spores when the exponential stationary growth phase ends and nutrient depletion occurs, typically taking about 8 hours. The cell density of *Bacilli* ranges from 1×10^8 to 1.5×10^{10} spores/mL, with sporulation efficiency varying between 30% and 80% (Rao et al., 2007). Various methods, such as mathematical and statistical approaches, artificial neural networks, genetic algorithms, and fuzzy logic, can be used to improve the cultivation environment and culture media. Experimental statistical techniques, including the Plackett–Burman design and the response surface methodology [RSM; also known as central composite design (CCD); Box and Wilson 1951], have demonstrated efficacy in improving the nutritional environment for submerged bacterial cultivation (Chen et al., 2009). Allowing sporulation to occur in conditions that support high cell density can significantly enhance the output of probiotics (Elisashvili et al., 2018).

4.3 GROWTH AND SPORE PRODUCTION

4.3.1 Growth of *Bacillus*

Bacillus species exhibit wide temperature and pH tolerance and thrive aerobically. They grow on nutrient-rich substrates like agar plates and broth. Most strains prefer temperatures of 30–40°C, but can survive from 4–15°C to 35–55°C (Andersson et al., 1995). The optimum pH range for growth is 4.9–9.3; however, pH inhibits less in foods, evident by limited meat growth at pH 4.35 (Ochoa & O'Ryan, 2017). *Bacillus cereus* group includes various Gram-positive species producing spores found in nature as cells and spores (Liu et al., 2017). Spores are vital for the wide presence of *Bacillus* due to resilience against temperature, pH, UV radiation, enzymes, and toxins. Favourable conditions trigger dormant spores to germinate into new cells, which multiply through sporulation and binary fission (McKenney et al., 2013).

4.3.2 Spore Production of *Bacillus*

Chapter 3 provides a detailed description of the sporulation, vegetative growth, and germination phases of the life cycle of spore-forming *Bacillus*. The microbe's sense of nutrient availability controls transitions between these phases. Through a variety of signalling pathways, the cell cycle machinery receives this information about growth rate and nutrition, allowing for constant environmental sensing and cell cycle process fine-tuning (Moir, 2006; Wang & Levin, 2009).

4.4 EFFECT OF CARBON SOURCE

The media containing a combination of casamino acids, yeast extract, peptones, and minerals are used in spore formation. The bacteria which form spore will strive until the nutrients in the medium are depleted and on depletion, the sporulation initiated. However, the resultant shows the heterogenous sporulation.

The carbon source has a very crucial role in the spore production of the *Bacillus* sp. and the most common source of carbon used is glucose. The different carbon source undergoes a different metabolic pathway. The studies conducted on *B. cereus* ATCC 14579 with lactate as one of the main carbon source have shown that the lactate neither caused catabolite repression nor repressed the TCA cycle which leads to the activation of Spo0A-P resulting in sporulation unlike glucose and also sporulation not necessarily being induced by nutrition depletion (De Vries et al., 2005).

4.4.1 Case Studies

Bacillus subtilis strain carries out carbon metabolism through TCA cycle which specifically identifies the reaction between the bottom part of glycolysis and TCA cycle as metabolic subsystem which is catalysed by the enzymes, PEP carboxy-kinase and malic enzyme consuming 23% of the metabolized glucose. The *B. subtilis* used for the production of riboflavin undergoes two major pathways for glucose catabolism which are the pyruvate shunt and the pentose-phosphate pathway (Papagianni, 2012).

For industrial synthesis of poly-γ-glutamic acid (γ-PGA), *B. licheniformis* can be harnessed by augmenting carbon utilization within the TCA cycle and boosting glutamic acid generation. This is achieved through increased expression of citrate synthase and pyruvate dehydrogenase, reconfiguring the glyoxylate shunt via isocitrate lyase upregulation and removing the pyruvate formate lyase gene. Employing these metabolic engineering strategies resulted in heightened γ-PGA production (Li et al., 2021).

Four different carbon sources such as sucrose, glucose, starch, and corn flour were used. Among the sources, glucose is found to be the least performing with a spore yield of 4.18×10^3 spores/mL and corn flour being the most productive one with a yield of 3.74×10^7 spores/mL of *Bacillus subtilis*. The low performance of glucose is assumed to be because of catabolite repression as glucose inhibits the induction enzymes partially involved in the sporulation process (Shi and Zhu, 2007a). In particular, at a concentration of 2 g/L, a spore yield of 2.3×10^9 spores/mL was found, and the spore yield gradually decreased as the glucose concentration rose. On the other hand, supplying the established carbon source at a rate of 0.5% boosted spore production, reaching a peak of 2.3×10^9 spores/mL in 72 hours. Moreover, it was discovered that sucrose and xylose were acceptable carbon sources for the spore formation. According to the studies' overall findings, the lowest concentration of 0.2% glucose produced the greatest number of spores, 6×10^9 spores/mL (Khardziani et al., 2017a).

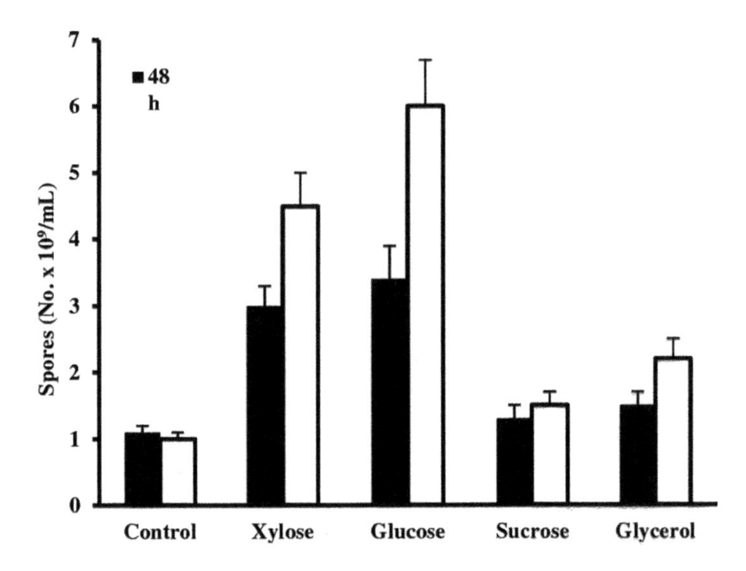

FIGURE 4.2 Spore production of *B. amyloliquefaciens* B-1895 with different carbon sources over 48 h and 72 hours (Khardziani et al., 2021).

Khardziani et al. (2021) observed that the medium without carbon source has produced the least number of spores, 11 × 10^8 spores/mL, of *B. amyloliquefaciens* B-1895 by utilizing the peptone and the yeast extract present in the media. Then enhancement of the medium with any source of carbon at the rate of 0.5% has shown raise in the spore to different extent based on the source. The glucose medium is observed with the highest number of spore count of about 60 × 10^8 spores/mL on submerged cultivation for 72h while sucrose is the one with poorest performance, this may be due to low sucrose activity of *B. amyloliquefaciens* B-1895 as shown in Figure 4.2.

In their study, Monteiro et al. (2014) employed *B. subtilis* to determine the optimal medium for spore development using several chemically defined media and compared the results with DSM (Difco Sporulation Media). The results concluded that the spore production in chemically defined media was lower than DSM. Among the chemically defined media, the medium containing glucose and citric acid, glycerol, or both as carbon source has shown the best results. Glucose was the main carbon source used and similar to the previous studies done by the author, glucose showed inhibitory effects on a concentration above 20 g/L.

From all the studies discussed above, the importance of carbon supply in this context is apparent for the homogenous sporulation of *Bacilli* probiotics. The reduction of carbon source acts as main stimulus to initiate the process of sporulation. Also, different carbon sources undergo different metabolic pathways, depending on which the sporulation period can be increased or decreased during practical application. The carbon source concentration also has a crucial role where above certain levels of concentration result in declining production.

4.5 EFFECT OF LIGNOCELLULOSIC SUBSTRATE

Lignocellulose is a rich and renewable source of carbon; it is the heterogenous mixture of hemicellulose (20%–35%), lignin (10%–25%), and cellulose (35%–50%) bounded in a matrix (Saritha, 2012; Wei, 2017). These polysaccharides on fermentation are converted to simple fermented sugars. Agricultural wastes such as corn stover, wheat straw, waste wood, and rice straw are the most commonly available lignocellulosic source (Balwinder Singh et al., 2023); other sources with their origin are shown in Table 4.1. The lignocellulosic substrate is mostly used as the replacement for carbon source in order to lessen the cost of production.

TABLE 4.1
Different Sources of Lignocellulose (Abo, Bodjui et al., 2019)

Agricultural leftovers	Foliage, crop residues, plant stems, and bovine excrement
Forest sources	Wood like cedar, spruce, and willow, along with sawdust and wooden fragments
Industrial source	Chemically derived pulps
Municipal waste sources	Leftover food, kraft paper

The study on agro-industrial lignocellulosic biomass with solid-state fermentation improved the performance of *Bacillus amyloliquefaciens* B-1895. The sunflower oil mill and mushroom spent substrate are observed with poor results whereas the corncobs gave the high number of spores (4.7×10^{11} spores g^{-1} biomass). The maximum spore formation (8.2×10^{11} spores g^{-1} biomass) was observed when the 0.06/L of nutrient enriched media of 2g KH_2PO_4, 1 g $MgSO_4.7H_2O$, 1 g NaCl per litre of distilled water and 10 g peptone was used to moisten the 0.015 kg of corncobs (Berikashvili, 2017). Likewise, on the other investigation on *B. amyloliquefaciens* B-128 by Koteswara Rao et al. (2007), the combination of tapioca with lactose at the concentration of 16.7 g/L and 12.7 g/L, respectively, gave the maximum yield of 5.92×10^8 spores/mL when submerged was observed. Some of the *Bacillus* species and their spore yield are listed in Table 4.2.

Khardziani et al. (2021) replaced the lignocellulosic materials in the place of glucose medium. The lignocellulosic substrate of milled corncobs was optimized with casein hydrolysate at a concentration of 40 mM N. The spore formation of 2.8×10^{10} spores/mL was observed as maximum in the shake flasks of *Bacillus amyloliquefaciens* B-1895. Also, the study effectively explains the importance of chemical composition of lignocellulosic source in order to be effectively used by the bacterial species for sporulation and to achieve high cell density for large-scale production.

The research by Wangka Orm et al. (2014) demonstrates that utilizing aqueous extracts of cassava root, sticky rice, sweet potato, and rice, along with the addition of 20 g/L dextrose, for submerged-type fermentation by *Bacillus* KKU03 and *Bacillus* KKU02 strains, was compared to a control group using nutritional broth as the carbon source. The batches containing cassava root and sweet potato exhibited superior spore formation. The highest spore concentration, reaching 1.3×10^8 and 8.32×10^8 spores/mL for *Bacillus* KKU03 and *Bacillus* KKU02, respectively, was achieved when cooked cassava root was employed as the carbon source. Additionally, supplementing cassava root with 2.0 g of $(NH_4)_2SO_4$ and 0.1 g of $MgSO_4$ resulted in a 1.75-fold increase in total spore count, with *Bacillus* KKU02 reaching 6.61×10^7 spores/mL and *Bacillus* KKU03 achieving 1.62×10^8 spores/mL.

Among the various sources used in the study of *Bacillus subtilis*, corn flour in combination of 4.54% $(NH_4)_2SO_4$ gave the most positive result (7×10^8 spores/mL) as the corn flour is hydrolysed to glucose at a slower rate to uptake glucose (Shi and Zhu, 2007a). The experiment was conducted

TABLE 4.2
Spore Yields of *Bacillus* spp. in Media with Different Lignocellulosic Substrates

Bacillus Species	Lignocellulose Substrate	Spores/mL	References
B. amyloliquefaciens B-1895	15 g corncob with nutrient enrichment	82×10^{10}	Berikashvili et al. (2017)
B. subtilis KATMIRA 1933	40 gL^{-1} milled mandarin peels	7×10^{10}	Khardziani et al. (2017b
B. amyloliquefaciens B-1895	Milled corncobs and casein hydrolysate	2.8×10^{10}	Khardziani et al. (2021)
B. amyloliquefaciens B-128	Tapioca with lactose	0.059×10^{10}	Koteswara et al. (2007)
Bacillus amyloliquefaciens BS-20	Corn meal, glucose, soyabean meal, beef extract, Mn^{2+}, Fe^{2+}, and Ca^{2+}	0.805×10^{10}	Ren et al. (2018)

TABLE 4.3
Influence of Diverse Lignocellulosic Growth Materials on Spore Production in *B. subtilis* KATMIRA, 1933

Lignocellulosic Substrate	Spores (No. × 10^{10} Spores/mL)
Control	0.08
Mandarin peels	5.7
Corn cobs	0.44
Soyabean	0.08
EPR from corn grains	2.92
Banana peels	0.63
EPR from wheat grains	0.72
Extraction cake of sunflower	0.13
Wheat bran	0.61
Sunflower oil cake	0.06
Wheat straw	1.62

with corn meal, soyabean meal, wheat bran, and molasses also with metal ion-optimized RSM. The metal ion-optimized RSM expressed $(3.10 \pm 1.41) \times 10^9$ (CFU/mL). The medium containing 9.0 g/L of corn meal, 8 g/L of glucose, 9.5 g/L of soyabean meal, and 7.2 g/L of beef extract combined with minerals like Mn^{2+}, Fe^{2+}, and Ca^{2+} at a concentration of 1.0 mM, 3.0 mM, and 2.1 mM, respectively, has shown the spore production of 8.8-folds of 8.05×10^9 spores/mL. The study indicated that the soyabean meal and corn meal have a positive effect on the spore production of *B. amyloliquefaciens* BS-20; however, the molasses and wheat bran show no remarkable effects (Ren et al., 2018).

An experiment with ten different lignocellulosic sources for the submerged fermentation experiment with *Bacillus subtilis* KATMIRA 1933 for 3 days was performed, and the results were evaluated by Khardziani et al. (2017). Amongst all the substrates, successive use of mandarin peels and EPR (ethanol production residue) derived from corn grains as shown in Table 4.3 has shown a better result. When 0.01 g/L of mandarin peels was introduced, only 4×10^9 spores/mL were produced; however, with increase in the substrate concentration up to 0.04 g/L result revealed 7- to 10-folds production of spores as enough levels of glycosyl hydrolases resulted in the relaxed hydrolysis of mandarin peel polysaccharide to simpler sugar acting as both carbon ad energy source. Even though the substrates like sunflower oilcake and soyabean meal have shown positive result for growth and accumulation of vegetative cells, there was no mass sporulation (Table 4.3).

Several researches demonstrated the efficacy of some defined culture medium producing high yields on the *Bacillus* sp. production. Nevertheless, these cannot be used in the large-scale production as they are very expensive. Hence, with adequate optimization, the cheaper lignocellulosic materials which are generally the by-products of agro-industry can be used to partially or totally replace the carbon source which is highly priced. The lignocellulosic materials have high potential to be utilized in production of *Bacilli* probiotic and the studies are relatively scarce with a huge scope for further studies and research.

4.6 EFFECT OF NITROGEN SOURCE

Bacillus species, renowned for their spore formation capabilities and stability, have gained wide recognition in the field of bioremediation for their ability to improve water quality and reduce nitrogen levels in aquaculture systems (Qiao et al., 2020 & Hong et al., 2005). Studies conducted by Hoffmann et al. (1998) and Nakano et al. (1998) have demonstrated that several *Bacillus* species can utilize nitrite and nitrate as alternative nitrogen and electron sources, effectively eliminating nitrite and enhancing water quality. Notably, *B. subtilis* (Chen and Hu, 2011), *B. lichenformis* (Meng et al., 2009), *B. amyloliquefaciens* (Xie et al., 2013), and *B. cereus* (Lalloo et al., 2007) have shown a strong ability to remove nitrite and improve water conditions. These bacteria also exhibit proficiency in absorbing dissolved organic matter (DOM) and total nitrogen ammonia (TAN), making them suitable for wastewater reuse in prawn farming. Additionally, they release compounds that stimulate microalgal growth (Watanabe et al., 2008). The use of *B. subtilis* in bio-floc of tilapia systems has been found to significantly reduce nitrogenous waste, promote development, enhance immunity, and protect against damage induced by lipopolysaccharides (James et al., 2021c).

Research endeavours carried out by various authors, including Chen et al. (2009), have aimed at optimizing the medium composition to enhance spore production in *Bacillus* species. To increase spore yields, numerous experiments using various carbon and nitrogen sources have been conducted. Based on the Plackett–Burman design, corn steep liquor, yeast extract, and soyabean flour were the best nitrogen sources for spore production, while starch (corn), wheat bran, and flour (corn) were found to be the most effective carbon sources. Along with examining the impact of different nitrogen-based nutrients on spore yields, the researchers maintained D-glucose monohydrate as the consistent carbon source. Soyabean flour and yeast extract were the next best sources of nitrogen for spore formation, after corn steep liquor. Conversely, when L-glutamic acid, beef peptone, or soy peptone served as the sole nitrogen source, low spore yields were observed. Furthermore, organic substances and inorganic salts were tested as nitrogen supplements to corn cobs. The researchers discovered that corn cobs alone yielded significant spore growth, yielding 7.2×10^9 spores/mL without any additional nitrogen sources. Moreover, the choice of nitrogen source significantly influenced spore production in *Bacillus* strains. The addition of organic nitrogen sources, such as casein hydrolysate, to the medium enhanced spore yields compared to control media. Among the inorganic compounds tested, KNO_3 was found to be the most beneficial nitrogen source, resulting in a more than 3-fold increase in spore output. Conversely, ammonium sulphate and ammonium nitrate tended to inhibit sporulation. Other studies by Shi and Zhu (2007a) found ammonium sulphate (4.54%) and corn meal (1.2%) to be effective sources of nitrogen for spore production in *B. subtilis*.

In summary, these studies highlight the significance of optimizing the medium composition for spore production in *Bacillus* species. Corn flour and starch, and wheat bran are favourable carbon sources, while soyabean flour, yeast extract, and corn steep liquor serve as optimal nitrogen sources. These findings contribute to the understanding of the factors influencing spore yields and can guide the development of efficient cultivation strategies for *Bacillus* species. Exactly, as discussed above, several factors influence spore yields and sporulation efficiency in *Bacillus* species, including nitrogen and carbon sources and their ratio (C:N ratio), as well as mineral supplementation.

In a study conducted by Qingshan and Zengguo et al. (2019), it was found that *Bacillus subtilis* M7-1 showed the greatest potential for denitrification in culture systems. The researchers screened isolates using chemically defined media (SCDDM) containing sodium succinate, $NaNO_2$, KH_2PO_4, $FeSO_4 \cdot 7H_2O$, $CaCl_2$, $MgSO_4 \cdot 7H_2O$, and distilled water. Within 24 hours, M7-1 effectively broke down nitrite-N without needing excessive dissolved oxygen when a medium containing shrimp meal as the primary carbon source and nitrite as the nitrogen source were employed. This medium mimics the nutritional conditions found in shrimp farming ponds.

A study by Yuniarti et al. (2019b) investigated the influence of the carbon-to-nitrogen (C:N) ratio on sporulation and spore formation in *Bacillus* spp. SB 4. The selection of nitrogen and carbon sources and their ratio significantly affected spore production and viability. The researchers calculated the sporulation efficiency (%) based on the ratio between spore production and the highest number of vegetative cells. It was observed that when the medium's C:N ratio is maintained at 11.9, the cessation of vegetative cell development and the initiation of spore production coincided. Maintaining balanced C:N ratio was crucial to prevent the pH from dropping below 5.6, as it could affect spore formation and cell proliferation. The study recorded a maximum spore production of 5.13×10^7 cells per mL, while other investigations reported higher values for *B. coagulans* (3.8×10^{11} cells mL^{-1}) (Monteiro et al., 2014) and *B. subtilis* (3.6×10^{10} cells mL^{-1}) (Vakil et al., 2016).

Spore formation may be enhanced by raising the carbon concentration and C:N ratio; an average increase of about 21.41% was seen for every three units increase in the C:N ratio.

Tian et al. (2022) conducted a study to explore the spore medium ingredients for *B. subtilis* BSNK-5. The effect of soyabean flour at different levels on spore production was investigated, and it was found that at a concentration of 12 g/L, it generated a total cell density of 1.2×10^9 CFU mL^{-1}, a spore cell density of 9.8×10^8 CFU mL^{-1}, and a sporulation proficiency of 82.6%.

In summary, these studies provide valuable insights into the species-specific requirements for nitrogen sources and their concentrations in the medium, considering the unique physiological characteristics of *Bacillus* species and the carbon source. The findings emphasize the importance of managing nitrogen content and promoting healthy culture systems in aquaculture.

4.7 EFFECT OF MEDIUM pH

The development and spore formation of *Bacillus* are impacted by environmental conditions, including pH levels. The highest spore yield occurs at the optimal pH. The pH, in conjunction with temperature, affects the specific growth rate and spore formation rate. The subsequent studies elaborate on the role of medium pH in the production of *Bacillus* spores. In uncontrolled conditions, a significant pH variation is noticed: pH drops to 6.5 from 6.7 at the start of exponential growth and then abruptly rises to 8.1 during the rest of this phase. As sporulation begins, the pH slowly increases to 9.0, resulting in a final production of 4.2×10^8 spores/mL of *B. subtilis*. Maintaining a consistent pH enhances sporulation efficiency. Lowering the pH to 5.0 reduces sporulation by 6%, while it is maintained at 50% efficiency within the pH range of 6.0–9.0. The peak spore output of 7.5×10^9 spores/mL is achieved at a pH of 7.5, with a 50% sporulation efficiency. The final cell concentration reached 3.6×10^9 spores/mL, nine times higher than batches without pH control when pH was maintained (Monteiro et al., 2005).

Tzeng et al. (2008) used shake flask experiments to investigate the pH's impact on *B. amyloliquefaciens* B-128. Comparing the initial alkaline pH of 8.0 to an uncontrolled pH, the spore production increased from 7.96×10^8 spores/mL to 10.30×10^8 spores/mL, whereas the acidic pH of 5.0 and 6.0 repressed the spore production. Moreover, the culture ended with medium pH of 7.34 and 7.08 when started at 9.0 and 8.0, respectively. On the other hand, no such decrease in pH was observed in cultures that started at pH 7.0 or lower. Two controlled batches of *B. amyloliquefaciens* B-128 in a 20 L airlift bioreactor were contrasted with an uncontrolled batch. The controlled batches had pH levels of 7.0 and 8.0, and the aeration rate was set at 1.5 vvm. Out of these three, the uncontrolled pH batch showed 6-fold increase up to 8.6×10^9 spores/mL of maximum spore concentration compared to the pH-controlled batches. The impact of pH on the production of *B. subtilis* EA-CB0575 spores was investigated in a 14 L bioreactor employing a single factorial design. No discernible impact on the spore cell density ($P = 0.285$) and sporulation efficiency ($P = 0.895$) 48 hours after fermentation was observed when the pH was continuously maintained between 6.5 and 7.0 and with uncontrolled pH with an initial pH of 5.5. Nonetheless, in uncontrolled conditions, 2.27×10^9 spores/mL with 92.9% sporulation efficiency were noted, and pH fluctuated between 5.5 and 7.0 throughout the fermentation period (Posada-Uribe et al., 2015).

Neutral pH was found to be paramount for the growth and spore production of *Bacillus* sp. Also, alkaline pH promotes sporulation. On the contrary, the acidic medium suppresses the growth and sporulation efficiency. Medium pH is found to vary through the process of sporulation.

4.8 EFFECT OF STIRRING AND AIR SUPPLY ON MICROELEMENT IMPACT

During the fermentation process, stirring primarily induces a mixture and cutting action that effectively blends oxygen, heat, and nutrients within the fermentation mixture. This action disperses air into small bubbles, thereby increasing the interface between gas and liquid and preventing the clustering of mycelia to enhance oxygen absorption (Mantzouridou et al., 2002). The effects of agitation include shear forces, which lead to various outcomes in microorganisms such as changes in physical structure, fluctuations in growth and production of metabolites, and potential damage to cellular structures (Mantzouridou et al., 2002). Excessive agitation speed can lead to shear forces and non-uniform mixing that may harm sensitive microorganisms, disrupt the final product's formation, and escalate energy consumption (Giavasis et al., 2006). Conversely, inadequate agitation speed can elevate the viscosity of the fermented liquid broth, impeding efficient mass transfer (Bandaiphet & Prasertsan, 2006).

Especially under conditions of gentle mechanical stirring, aeration plays a critical role in oxygenating the fermentation procedure and facilitating the blending of the fermentation mixture. Alongside supplying necessary oxygen for cellular proliferation, aeration also eliminates the waste gases generated during fermentation (Mantzouridou et al., 2002). Nevertheless, an elevated aeration rate leads to a decrease in the fermentation broth's volume.

Primary Use

- Aeration is to ensure that the submerged culture of microorganisms has enough oxygen for metabolic needs.
- Agitation guarantees that microbial cells are suspended uniformly in a homogeneous nutritional solution.
- Fermentation dictates the need for aeration and agitation.
- Aeration furnishes the culture with oxygen.
- Eliminates undesired volatile by-products of metabolism.
- Agitation is achieved through stirring or as a result of aeration.

4.9 EFFECT OF MICROELEMENTS

Bacillus probiotics are affected by nutrients, especially trace metals like Zn^{2+}, Cu^{2+}, I^{-}, and Fe^{2+}, in their growth and metabolism. Bacterial growth would be hampered by these trace element in a low or a high concentration (Finney & O'Halloran, 2003). While divalent cations, especially calcium, help the spore become dehydrated and mineralized, calcium carbonate, magnesium sulphate, and phosphate all encourage the sporulation process.

According to Monteiro et al. (2018), pH values between 6.9 and 9.0 showed no effect on B. *subtilis* sporulation efficiency.

4.10 EFFECT OF THE CULTIVATION METHOD

There are several types of cultivation methods used for spore-forming probiotics, each with its advantages and limitations. The choice of the best cultivation method depends on various factors, including the specific strain of spore-forming probiotics, which significantly impact the growth, sporulation, and overall production of *Bacillus* probiotics (Fedorova et al., 2019). While there isn't a single universally "best" method for maximum production of spore-forming probiotics, certain methods have been widely studied and proven effective. Here are some common cultivation methods and their potential benefits which are also shown in Figure 4.3.

Mechanical stirred tanks utilized for cellular cultivation come with three main operational modes: batch, fed-batch, and continuous. These reactors come with built-in sensors that oversee a range of cultural attributes, including pH, temperature, concentrations of gases like CO_2 and O_2, biomass levels, cell structure, nutrient components within the growth medium, protein levels, protein activity, and concentrations of metabolites (Chisti, 2001). To ensure precise management of these attributes, control systems are employed. The categorization of reactor operation depends on whether cells are suspended within a liquid medium (submerged cultivations – SF) or affixed to a solid substrate (solid-substrate fermentation – SSF). Additionally, the reactors contain a gas phase that serves to

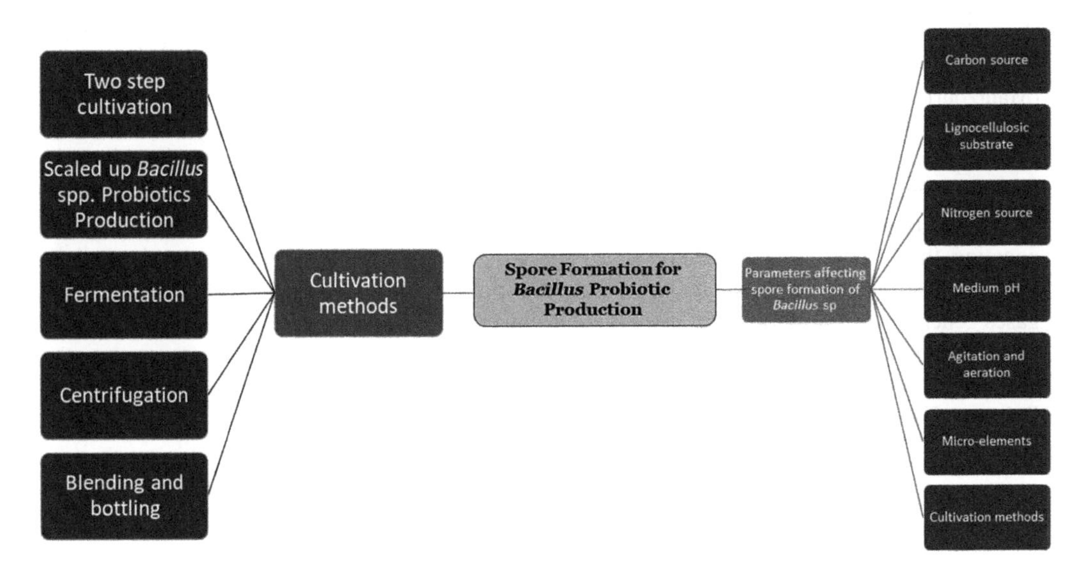

FIGURE 4.3 Different cultivation methods and sources for spore formation of *Bacillus* sp. probiotic production with the factors affecting the sporulation.

deliver oxygen during aerobic processes (for aerobic bacteria), an inert gas such as nitrogen during anaerobic processes (for anaerobic bacteria), and eliminates the carbon dioxide generated through cellular metabolism. Irrespective of the culture mode, the gas phase is consistently operated in a continuous manner, with gas consistently entering and exiting the reactor (Chisti, 2001).

For probiotics to be developed commercially, mass production is essential, and fermentation processes have a big impact on production costs. Studies have indicated that SSF can be a more economical and ecologically friendly method of producing probiotics than submerged cultivations and SSF is widely used in industry because of its automation benefits (Soccol et al., 2017).

In a batch culture operation, no biomass, fresh nutrient medium, or culture broth is added or withdrawn, except for the gas phase. The process begins by adding a small amount of cell culture, called the "inoculum," to a sterilized nutrient medium. This inoculum is derived from previous batch cultures. Some culture conditions like pH and dissolved oxygen levels are typically left uncontrolled during each subculture. Although a typical batch culture may involve some additions (such as acid/base for pH control and antifoam) and withdrawals (for culture assessment), efforts are made to minimize net volume changes (Soccol et al., 2010).

In terms of cellular proliferation, the batch cultivation method goes through distinct phases. Initially, there is a lag phase where cells acclimate to their surroundings and synthesize necessary enzymes. This is succeeded by an active growth phase marked by an exponential rise in both cell count and mass. Following this stage, cell growth decelerates due to dwindling nutrients, leading to a phase of non-exponential growth. In this period, specific secondary metabolites are either hastened or triggered. Subsequently, the growth phase transitions into a stationary phase, characterized by near depletion of vital nutrients, prompting the synthesis of secondary metabolites and the production of spores. As the final stage, the death phase sets in, resulting in a decline in cell numbers (Soccol et al., 2010). Within industrial bioprocesses, serial batch cultures are frequently employed, where a portion of the previous batch's culture is utilized as an inoculum for the ensuing batch. This practice is termed repeated batch operation with recycling. In cases where no transfer of culture happens between batches, it is referred to as repeated batch operation without recycling, or parallel batch cultures (Cinar, 2003a).

In conclusion, the cultivation method chosen for spore-forming probiotics has a significant impact on their growth, sporulation, and overall production. Mechanically agitated reactors offer various operational modes, such as batch, fed-batch, and continuous, allowing for precise control of various cultural characteristics. Both SF and SSF have their advantages, with the former being commonly used in industry for its automation advantages, while the latter proves to be cost-effective and environmentally friendly. In batch cultures, the growth phases, such as lag phase, exponential growth phase, non-exponential growth phase, stationary phase, and death phase, play essential roles in the production of secondary metabolites and spores. Understanding the cultivation methods and their effects on spore-forming probiotics is crucial for optimizing production processes in the probiotic industry.

Probiotic research typically emphasizes a single strain using submerged fermentation (SMF) with pure cultures, a method less common and more expensive in animal production (Zhao et al., 2008). An economical and simpler alternative is solid-state fermentation (SSFer) for producing probiotics (Ying et al., 2009). SSFer-prepared probiotics have demonstrated superior effects on animal growth and intestinal microbiota balance compared to SMF-prepared ones (Zhang, 2013). In SSFer, multiple factors influence probiotic production, such as moisture, initial pH, inoculation volume, flask dry content, sterilization, and fermentation time. Optimizing these parameters revealed 50% moisture as optimal for both *B. subtilis* MA139 and *L. reuteri* G8-5. While *B. subtilis* MA139 thrived at this level, *L. reuteri* G8-5 required slightly more moisture. Moisture below 50% hindered *L. reuteri* G8-5 growth, while above 50% hindered *B. subtilis* MA139. Consequently, a 50% moisture level was chosen for further optimization (Popov & Algburi, 2021).

Studies comparing *Bacillus* spp. probiotic production in SMF and SSFer conditions are limited. *B. licheniformis* B36 succeeded in SSFer using plant raw materials with abundant lignocellulosic substrates. The highest spore yield achieved was 1×10^{11} spores/g, by using 15 g of wheat bran and 5 g of rice straw powder. The presence of additional glucose or sucrose resulted in a 35% or 25% increase in spore output, respectively. Similarly, the addition of peptone and yeast extract led to a 16% and 24% increase in spore yield (Popov & Algburi, 2021). Interestingly, cultivation method affected bacterial substrate preferences. Wheat straw declined *B. amyloliquefaciens* B-1895 sporulation in SMF method but became a preferred nutrient source in SSFer. Utilizing SSFer proved advantageous for *Bacillus* culture across media, significantly boosting spore production compared to SMF in the same timeframe (Popov & Algburi, 2021).

In a fed-batch culture process, nutrients are carefully introduced to a setup similar to a batch system. This allows precise control over nutrient availability, cell growth, uptake, and targeted metabolite production. Secondary metabolites like antibiotics and enzymes are usually generated when cell growth is restricted. Introducing nutrients in a fed-batch setup can increase the overall culture mass and volume, unless highly concentrated nutrient forms are used. The feeding rate can be predetermined or adjusted through feedback control, and nutrient supply stops when the maximum allowed volume is attained. Following a fed-batch process, a terminal batch operation might be carried out to utilize any remaining nutrients. Typically, fed-batch operations are both preceded and followed by batch processes. This sequence (batch–fed-batch–batch) can be repeated several times, resulting in consecutive fed-batch operations. It is common to transfer culture from one sequence to the next for inoculation (Cinar, 2003b).

In order to improve the probiotic *B. coagulans* and its spores' cultivation yield, Pandey and Vakil (2016) carried out a study. To achieve increased *B. subtilis* spore production, they devised a fed-batch method in a 2 L bioreactor. For the first five hours, a batch culture was maintained in a medium containing glucose. Fertilizer feeding started prior to glucose depletion, prolonging exponential growth and producing a high vegetative cell concentration (3.6×10^{10} cells/mL). The process of sporulation was initiated by the exhaustion of glucose, leading to an impressive rise in spore production that peaked at 7.4×10^9 spores/mL. The same procedure was used for *B. coagulans*, first achieving a dense cell population through batch culture and then undergoing fed-batch fermentation with sporadic additions of glucose. This method produced a biomass of 30 g/L (3.8×10^{11} cells/mL), 1.9×10^{11}/mL of notable spores, and about 81% sporulation efficiency. For significant biomass production to satisfy the organism's oxygen requirement, it was essential to maintain a dissolved oxygen concentration above a critical threshold of 20% dissolved oxygen (Popov & Algburi, 2021).

4.10.1 Two-Step Cultivation Method

In this method, *Bacillus* strains are initially cultured in a nutrient-rich medium for optimal vegetative cell growth, followed by transfer to a sporulation-inducing medium with limited nutrient availability. This method allows for the separation of vegetative growth and sporulation phases, ensuring a better control over sporulation efficiency. A two-stage strategy was implemented by Sen and Babu (2005), to cultivate and induce sporulation in *B. coagulans* RK-02 using a 20 L fermenter. The first stage focused on creating favourable conditions for biomass production, while the second stage maintained optimal conditions for sporulation during the stationary phase. The utilization of this technique led to a peak spore production of 9×10^{11} spores/g (Popov & Algburi, 2021). To enhance vitamin B12 yields from *B. megaterium*, a two-step optimization method was implemented in the fermentation process. In the initial approach, the fermentation was divided into three phases: aerobic growth to boost biomass, anaerobic growth to generate vitamin B12 precursors, and aeration to finalize vitamin B12 synthesis. The second strategy involved pinpointing and optimizing the optimal timing for each phase using the Box–Behnken design and RSM. This strategy aimed to determine the most suitable time for each phase, ensuring a high vitamin B12 yield (Mohammed et al., 2014).

Important growth nutrients are continuously supplied during an ongoing culture process, and part of the culture is continuously removed. A level controller is used to regulate the culture's volume. Continuous culture typically follows a fed-batch or batch process. A suitable amount of time after the start of continuous culture, a state of equilibrium can be reached if both the feed inflow and outflow rates stay constant and balanced (Cinar, 2003a).

The effect of lignocellulosic substrates on *Bacillus* spore development has not been thoroughly studied. Cassava root produced 8.32 and 1.35×10^8 spores/mL, respectively, during submerged fermentation, demonstrating superior spore formation for *Bacillus* KKU02 and KKU03 strains. Lactose and tapioca were combined to produce 5.92×10^8 spores/mL by *B. amyloliquefaciens* B-128. During submerged lignocellulose fermentation, our *B. amyloliquefaciens* B-1895 strain performed remarkably well, achieving spore counts ranging from 2.3 to 10.8×10^9 spores/mL (Khardziani et al., 2017). Experiments using *B. subtilis* revealed that batch cultivation resulted in 40.9% sporulation efficiency decreasing steadily from 100% to 17% as the glucose concentration increased from 3.5 to 20 g/L. This finding pertains to the impact of carbon sources on spore production and cultivation conditions. In the meantime, 20.5% sporulation efficiency was reported by the researchers in the fed-batch culture (Monteiro et al., 2005). When grown in a medium with a glucose concentration of 2.0 g/L, *Bacillus subtilis* EA-CB0575 exhibited the highest spore cell density (1.87×109) and sporulation efficiency (93.2%) (Posada-Uribe et al., 2015).

The fed-batch fermentation technique is extensively employed to achieve elevated cell density and spore production in *Bacillus* species. Various approaches for providing nutrients, such as pH-stat or DO-stat control, exponential feeding, and glucose-driven feeding, have been utilized to enhance productivity. This method has effectively been utilized for producing different bio-products using both natural and genetically modified strains. A study conducted by Cho and Kim et al. (2010) focused on nattokinase production, which is a fibrinolytic enzyme with potential for treating thrombosis. In this research, nattokinase production in *B. subtilis* was investigated through both batch and fed-batch cultures. The analysis of batch kinetics revealed a connection between nattokinase production and cell growth, underscoring the importance of harvesting the culture before the stationary phase to optimize production. To boost nattokinase activity, a fed-batch culture was performed using a pH-stat feeding method with an upper limit. The process initiated with an 8-hour batch phase succeeded by nutrient infusion. As nutrients were consumed, the pH level increased, prompting the automated feeding of a glucose and peptone blend when the pH surpassed 7.0. With the nutrient addition, the optical density (OD) continued to rise, reaching about 100 at 22 hours, marking a 2.5-fold surge compared to the batch culture. At the 19-hour mark, the nattokinase activity reached its peak at 7100 units/mL, indicating a 2.1-fold increase relative to the batch culture. Following feeding, the glucose concentrations in the fed-batch culture increased gradually and stayed constant at about 1 g/L. The batch culture's nattokinase activity per unit of optical density (unit/mL/OD) varied from 5 to 40 when 50 g/L of yeast extract was added. Increased specific activity was observed in batch and fed-batch cultures supplemented with 50 g/L of peptone, ranging from 15 to 100 units/mL/OD. To identify the precise peptone components that significantly contribute to the production of nattokinase, more investigation is required (Cho & Kim et al., 2010).

4.11 SCALED-UP *BACILLUS* SPP. PROBIOTICS PRODUCTION

Many industrial processes, including the synthesis of enzymes, antibiotics, and biofuels, use *Bacillus species*. Their rapid growth, no pathogenicity, and efficient protein secretion characteristics have positioned them as favourable organisms in the industrial sector (Shah et al., 2016). These *Bacillus* strains are frequently selected as probiotics for human use, harnessed in aquaculture to counteract harmful bacterial growth, and particularly employed to enhance water quality by reducing the buildup of organic substances (Wang et al., 2019; Yi et al., 2018). These bacteria, often classified as Gram-positive, exhibit a heightened ability to convert organic matter back into CO_2 compared to Gram-negative counterparts, which helps minimize the accumulation of particulate and dissolved organic carbon (Kuebutornye et al., 2019). Scaling up is the process of increasing *Bacillus* species production from laboratory to industrial levels. It entails the following common actions.

4.11.1 Strain Selection

The initial and pivotal step in producing probiotics is the choice of species. The strain you opt for will solely rely on your intended goals for the particular probiotic product and any potential health assertions you aim to convey. Whether your aim is to develop supplements promoting immune system strength, aiding in digestive health, or facilitating a resilient response to occasional stress, these factors will guide your selection. Each species boasts distinct characteristics and benefits, with certain species being particularly adept at reinforcing robust immune systems.

4.11.2 Media Planning

It is crucial to choose species that are resistant to acid and bile as well as an efficient formula for the right quantity of raw materials. It is a good idea to check their digestive capacity as well. After that, the chosen strains are fermented and stabilized.

In a bioprocessing lab, conduct research studies on the probiotic strain to determine what growth-promoting variables and nutrients can be changed. After a particular ratio of nutrients and process variables has been established, large-scale production starts.

4.11.3 Fermentation

With components gathered from all around the world, probiotics can be produced. Probiotics are produced to keep their purity and freshness, too. Up to 6 weeks are needed for bacterial cultivation. It is impossible to accelerate the process of cultivation as a culture develops. A material provider's unique strain ID codes are frequently used. Since nobody else can give you the same strain IDs, you must rely on the same vendor. Additionally, some raw materials might not be easily available at the contracted manufacturer in the necessary quantity, which would increase the turnaround time (Sanders et al., 2016). Fermentation sterilizes every ingredient and piece of machinery to avoid unintentional infection. The fermentation process involves adding a strain to nutrient-rich media in a large tank, which is then subjected to a warm bath to facilitate the growth of the microorganisms. This process is commonly used in the production of various products, including fermented foods and beverages, as well as in the manufacturing of pharmaceuticals and other industrial applications (Dahiya & Nigam, 2023) (Table 4.4).

TABLE 4.4
Comparison of Industrial- and Laboratory-Scale Fermentation Process for *Bacilli* Production

	Industrial Fermentation	Lab-Scale Fermentation
Aim and objective	Set the groundwork for industrial fermentation Investigate fundamental conditions and choose appropriate feedstock, to find the key factors	Final products obtained, quality control and stable production
Scale	Small scale, from 100 mL to 50 L	Large scale, from 100 L to 100 KL
Equipment	Bench-scale fermenters in shaker flasks, other lab equipment, such as a desk centrifuge	Centrifuge disc for fermentation vessels, numerous related pieces of gear
Process control	Simple, for instance, culture media, oxygen, pH and temperature	Complicated, different control method, stir, sterilization, ventilation, pressure, chilling, feeding, and scale-up controls, among others

Key factors in *Bacillus* fermentation:

- Growth medium and nitrogen ratio
- Nitrogen supply
- Air supply and foam control
- Sporulation proportion
- Suitable post-cultivation procedures

4.11.4 Centrifugation

After cultures are created, centrifugation separates probiotic organisms from metabolites. Another important aspect of the development of probiotics is probiotic stability, which needs specific consideration.

After being packaged, instant probiotic products start to lose their uniformity and freshness. Several strategies are employed for long-term care to guarantee supplement stability and efficacy. These actions are very important and have an impact on the survival and application of probiotic species.

- *Refrigeration*: The probiotic microorganisms are cryogenically preserved at extremely cold temperatures.
- *Preventing heat and moisture exposure*: This safeguards the bacteria from humidity by employing diverse drying techniques.
- *Cryodesiccation*: A slower yet milder approach.
- *Spray drying*: A quicker method involving moderately elevated temperatures that are non-lethal for bacterial viability. Following these steps, the probiotic is converted into a powdered form

4.11.5 Blending and Bottling

The multi-ethnic makeup along with additional probiotic powders produced a balanced mixture that was equally dispersed. To the probiotics, other crucial substances are added, including prebiotics, flavourings, and binders, to create different doses, probiotics that match the health goal, and many more. Final dosage forms of the mixture are created, including pills, capsules, and powder.

Probiotics require a variety of conditions to thrive, including temperature, humidity, and light. The manufacturing expiration date is impacted by these stress-specific factors. They must therefore, be securely packaged and transported as a result. Avoid harsh sunshine, extreme heat, and humidity.

4.12 CONCLUSION

In conclusion, this chapter delves into the intricate physiology of spore formation in *Bacillus*, with a specific focus on its significance for probiotic production. Throughout the chapter, we explored the multifaceted processes involved in spore development, highlighting the molecular mechanisms, regulatory factors, and environmental cues that govern this vital biological phenomenon. *Bacillus* spores play a crucial role in probiotic applications due to their inherent stability and resilience, allowing for enhanced shelf life and survivability during gastrointestinal transit. Understanding the underlying spore formation physiology is essential for optimizing probiotic production processes and maximizing their therapeutic potential. Key findings suggest spore transcriptional regulators, sporulation-specific sigma factors, and various signal transduction pathways. Various environmental factors such as nutrient availability, temperature, pH, carbon, and nitrogen also significantly influence sporulation efficiency, emphasizing the importance of precise cultivation conditions.

Moreover, insights into the spore germination process shed light on the development of efficient probiotic delivery systems, ensuring spore viability and probiotic functionality upon consumption. However, it is crucial to acknowledge that the physiology of spore formation in *Bacillus* is still an active area of research. Further studies are needed to elucidate undiscovered aspects and potential regulatory nodes that could be exploited to enhance probiotic production and functionality. Overall, this chapter provides a comprehensive overview of the physiology of spore formation in *Bacillus*, offering valuable insights into the development of robust and effective probiotic-based interventions for improving human health. By deciphering the intricacies of spore development, researchers and biotechnologists can unlock the full potential of *Bacillus* probiotics, paving the way for novel and innovative applications in the field of microbiome-based therapeutics. As we continue to delve deeper into this fascinating area of research, we can anticipate ground-breaking discoveries that will undoubtedly shape the future of probiotic production and its impact on human and animal well-being.

BIBLIOGRAPHY

Abo, B. O., Gao, M., Wang, Y., Wu, C., Ma, H., & Wang, Q. (2019). Lignocellulosic biomass for bioethanol: An overview on pretreatment, hydrolysis and fermentation processes. *Reviews on Environmental Health*, 34(1), pp.57–68.

Andersson, A., Rönner, U., & Granum, P. E. (1995). What problems does the food industry have with the spore-forming pathogens Bacillus cereus and Clostridium perfringens? *International Journal of Food Microbiology*, 28(2), pp.145–155. https://doi.org/10.1016/0168-1605(95)00053-4

Bandaiphet, C., & Prasertsan, P. (2006). Effect of aeration and agitation rates and scale-up on oxygen transfer coefficient, kLa in exopolysaccharide production from Enterobacter cloacae WD7. *Carbohydrate Polymers*, 66(2), pp.216–228. https://doi.org/10.1016/j.carbpol.2006.03.004

Berikashvili, V., Sokhadze, K., Kachlishvili, E., Elisashvili, V., & Chikindas, M. L. (2018). Bacillus amyloliquefaciens spore production under solid-state fermentation of lignocellulosic residues. *Probiotics and Antimicrobial Proteins*, 10(4), pp.755–761. https://doi.org/10.1007/s12602-017-9371-x

Bokossa, I., Adjoumani, M., Bassoum, O., Gnagni, D., Loukou, N. E., & Kouadio, K. M. (2014). Bacillus species isolated from fish ponds: Potential for controlling bacterial diseases of aquaculture importance in Côte d'Ivoire. *African Journal of Microbiology Research*, 8(37), pp.3461–3466.

Box, G. E. P., & Wilson, K. B. (1951). On the experimental attainment of optimum conditions. *Journal of the Royal Statistical Society: Series B* (Methodological), 13(1), pp.1–45.

Chen S., Hu Y. (2011) Use of *Bacillus subtilis* in purification of slightly-polluted water. Acta Sci Circumst 31, pp.1594–1601

Chen, Z.-M., Li, Q., Liu, H.-M., Yu, N., Xie, T.-J., Yang, M.-Y., Shen, P., & Chen, X.-D. (2010). Greater enhancement of *Bacillus subtilis* spore yields in submerged cultures by optimization of medium composition through statistical experimental designs, *Applied Microbiology and Biotechnology*, 85(5). https://doi.org/10.1007/s00253-009-2162-x

Chen, Z.-M., Wei, Q., Chen, J.-C., Hu, B., Yang, L.-Z., Chen, G.-Y., & Hu, G.-B. (2009b). Optimization of the composition of the sporulation medium for Bacillus thuringiensis by statistical experimental methods. *Journal of Applied Microbiology*, 107(6), pp.2031–2040.

Chen, Z.-M., Zhang, X.-M., Chen, J.-C., Chen, J.-L., Chen, Y.-J., Zhang, J.-Q., & Zhang, D.-J. (2009a). Optimization of fermentation conditions for the production of *Bacillus subtilis* WB800N as a feed additive by a central composite design. *Bioresource Technology*, 100(17), pp.4082–4086.

Chisti, Y. (2001). Computer simulation of biomass productivity in plant cell suspension cultures. *Biotechnology Advances*, 19(4), pp.291–295.

Cho, Y.-H., Song, J. Y., Kim, K. M., Kim, M. K., Lee, I. Y., Kim, S. B., Kim, H. S., Han, N. S., Lee, B. H., & Kim, B. S. (2010). Production of nattokinase by batch and fed-batch culture of Bacillus subtilis, *New Biotechnology*, 27(4), pp.341–346. https://doi.org/10.1016/j.nbt.2010.06.003

Cinar, A. (2003a). Batch-fed-batch-batch sequencing in bioprocessing: A review. *Journal of Scientific & Industrial Research*, 62(8), pp.725–730.

Cinar, A. (2003b). *Batch Fermentation: Modeling, Monitoring, and Control.* Marcel Dekker.

Cordero, H., Ngeles, M., & Cuest, A. (2014). Use of probiotic bacteria against bacterial and viral infections in shellfish and fish aquaculture. In *Sustainable Aquaculture Techniques*. InTech. https://doi.org/10.5772/57375

Cutting, S. M. (2011). Bacillus probiotics, *Food Microbiology*, 28(2), pp.214–220. https://doi.org/10.1016/j.fm.2010.03.007

Dahiya, D., & Nigam, P. S. (2023). Use of characterized microorganisms in fermentation of non-dairy-based substrates to produce probiotic food for gut-health and nutrition. *Fermentation*, 9(1), p.1. https://doi.org/10.3390/fermentation9010001

De Vries, Y. P., Atmadja, R. D., Hornstra, L. M., de Vos, W. M., & Abee, T. (2005). Influence of glutamate on growth, sporulation, and spore properties of Bacillus cereus ATCC 14579 in defined medium. *Applied and Environmental Microbiology*, 71, pp.3248–3254.

Elisashvili, V., Kachlishvili, E., & Chikindas, M. L. (2019). Recent advances in the physiology of spore formation for bacillus probiotic production. *Probiotics and Antimicrobial Proteins*, 11(3). https://doi.org/10.1007/s12602-018-9492-x

Elisashvili, V., Kachlishvili, E., Tsiklauri, N., Jobava, M., Metreveli, E., Chikindas, M., & Chanishvili, N. (2018a). The optimization of lactic acid bacteria strains with antifungal activity for food biopreservation. *Food Technology and Biotechnology*, 56(2), pp.245–254.

Elisashvili, V., Pataridze, I., Kachlishvili, N., Tsiklauri, G., Chilashvili, M., Petriashvili, M., Khidasheli, L., Elisashvili, T., Choiset, M., Schved, J. L., Mollé, C., Yvon, M., Chobert, J.-M., Gelbíčová, E., Kala, R., & Minchev, G. (2018b). Antibacterial metabolites produced by Bacillus spp. and Lactic Acid Bacteria against foodborne pathogens. *World Mycotoxin Journal*, 11(2), pp.263–277.

Fedorova, N. V., Kozlova, Y. N., & Zakharevich, N. V. (2019). Effect of culturing conditions on growth, spore formation, and metabolic activity of Bacillus subtilis 534. *Applied Biochemistry and Microbiology*, 55(6), pp.697–706.

Finney, L. A., & O'Halloran, T. V. (2003). Transition metal speciation in the cell: Insights from the chemistry of metal ion receptors. *Science*, 300(5621), pp.931–936. https://doi.org/10.1126/science.1085049

Fujita, M., González-Pastor, J. E., & Losick, R. (2005). High- and low-threshold genes in the Spo0A Regulon of *Bacillus subtilis*. *Journal of Bacteriology*, 187(4), pp.1357–1368. https://doi.org/10.1128/JB.187.4.1357-1368.2005

Giavasis, I., Harvey, L. M., & McNeil, B. (2006). The effect of agitation and aeration on the synthesis and molecular weight of gellan in batch cultures of *Sphingomonas paucimobilis*. *Enzyme and Microbial Technology*, 38(1–2), pp.101–108. https://doi.org/10.1016/j.enzmictec.2005.05.003

Henriques, A. O., & Moran Jr, C. P. (2007). Structure, assembly, and function of the spore surface layers. *Annual Review of Microbiology*, 61, pp.555–588.

Hlordzi, V., Kuebutornye, F. K. A., Afriyie, G., Abarike, E. D., Lu, Y., Chi, S., & Anokyewaa, M. A. (2020). The use of Bacillus species in maintenance of water quality in aquaculture: A review, *Aquaculture Reports*, 18, p.100503. https://doi.org/10.1016/j.aqrep.2020.100503

Hoffmann, T., Frankenberg, N., Marino, M., & Jahn, D. (1998). Ammonification in Bacillus subtilis utilizing dissimilatory nitrite reductase is dependent on resE and associated with anaerobic growth. *Journal of Bacteriology*, 180(4), pp.853–862.

Hölker, U., Höfer, M., & Lenz, J. (2004). Biotechnological advantages of laboratory-scale solid-state fermentation with fungi, *Applied Microbiology and Biotechnology*, 64(2), pp.175–186. https://doi.org/10.1007/s00253-003-1504-3

Hong, Y., Yu, M., Li, T., Zhao, J., Liu, Y., & Wei, C. (2005). Identification and characterization of a heterotrophic nitrifying and aerobic denitrifying bacterium strain A1010. *Journal of Environmental Sciences*, 17(4), pp.579–583.

James, G., Das, B. C., Jose, S., & Rejish Kumar, V. J. (2021b). Bacillus as an aquaculture friendly microbe. *Aquaculture International*, 29(1), pp.323–353. https://doi.org/10.1007/s10499-020-00630-0

James, G., Nithya, C., Shainabanu, K., & Sasikumar, S. (2021a). Evaluation of Bacillus cereus, Paenibacillus polymyxa, and Pseudomonas sp. as biocontrol agents against pathogenic vibrios in aquaculture. *Microbial Pathogenesis*, 160, p.105160.

James, G., Sahu, N. P., & Meena, D. K. (2021c). Impact of dietary supplementation of Bacillus subtilis on growth performance, nutrient utilization, gut health, and immunity in Nile tilapia (Oreochromis niloticus). *Fish Physiology and Biochemistry*, 47(1), pp.117–130.

Khardziani, T., Anisimova, I., Natsvlishvili, N., Pirtskhalava, M., Baratashvili, D., &Elisashvili, V. (2017c). Optimization of the cultivation conditions of Bacillus amyloliquefaciens B128, an antagonist of the causal agent of apple ring rot. *Journal of Basic Microbiology*, 57(9), pp.718–727.

Khardziani, T., Kachlishvili, E., Sokhadze, K., Elisashvili, V., Weeks, R., Chikindas, M. L., Chistyakov, V. (2017b). Elucidation of Bacillus subtilis KATMIRA 1933 potential for spore production in submerged fermentation of plant raw materials. *Probiotics and Antimicrobial Proteins*. https://doi.org/10.1007/s12602-017-9303-9

Khardziani, T., Sokhadze, K., Kachlishvili, E., Chistyakov, V., & Elisashvili, V. (2017a). Optimization of enhanced probiotic spores production in submerged cultivation of bacillus amyloliquefaciens B-1895. *Journal of Microbiology, Biotechnology and Food Sciences*, 7(2), pp.132–136. https://doi.org/10.15414/jmbfs.2017.7.2.132-136

Khardziani, T., Sokhadze, K., Kachlishvili, E., Chistyakov, V., & Elisashvili, V. (2021). Optimization of enhanced probiotic spores production in submerged cultivation of Bacillus amyloliquefaciens B-1895. *Journal of Microbiology, Biotechnology and Food Sciences*, 2021, pp.132–136.

Kuebutornye, F. K., Abarike, E. D., & Lu, Y. (2019). A review on the application of Bacillus as probiotics in aquaculture. Fish & shellfish immunology, 87, pp.820–828.

Lalloo, R., Ramjugernath, D., & Görgens, J. (2007). The effect of inoculum age on the production of poly(3-hydroxybutyrate) in a batch culture by Bacillus cereus. *Journal of Biotechnology*, 131(4), pp.359–363.

Li, B., Cai, D., & Chen, S. (2021). Metabolic engineering of central carbon metabolism of bacillus licheniformis for enhanced production of poly-γ-glutamic acid. *Applied Biochemistry and Biotechnology*. https://doi.org/10.1007/s12010-021-03619-4

Liu, Y., Du, J., Lai, Q., Zeng, R., Ye, D., Xu, J., & Shao, Z. (2017). Proposal of nine novel species of the Bacillus cereus group. *International Journal of Systematic and Evolutionary Microbiology*, 67(8), pp.2499–2508. https://doi.org/10.1099/ijsem.0.001821

Ma, Qingshan, He, Zengguo, (2020). Screening and characterization of nitrite-degrading bacterial isolates using a novel culture medium. *Journal of Ocean University of China*, 19. https://doi.org/10.1007/s11802-020-4093-1

Madigan, M. T., Martinko, J. M., Stahl, D. A., Clark, D. P. (2018). *Brock Biology of Microorganisms* (15th ed.). Pearson.

Mantzouridou, F., Roukas, T., & Kotzekidou, P. (2002). Effect of the aeration rate and agitation speed on β-carotene production and morphology of Blakeslea trispora in a stirred tank reactor: Mathematical modeling. *Biochemical Engineering Journal*, 10(2), pp.123–135. https://doi.org/10.1016/S1369-703X(01)00166-8

McKenney, P. T., Driks, A., & Eichenberger, P. (2013). The Bacillus subtilis endospore: Assembly and functions of the multilayered coat. *Nature Reviews Microbiology*, 11(1), pp.33–44. https://doi.org/10.1038/nrmicro2921

Meng, X.-H., Wang, Y.-Z., & Ge, J.-P. (2009). Isolation and characterization of Bacillus subtilis strain B-FS06 for biocontrol of the postharvest pathogenic fungi in apple fruit. *Letters in Applied Microbiology*, 49(2), pp.196–201.

Mohammed, Y., Lee, B., Kang, Z., Du, G. (2014). Development of a two-step cultivation strategy for the production of vitamin B12 by Bacillus megaterium. *Microbial Cell Factories*, 13. https://doi.org/10.1186/s12934-014-0102-7

Moir, A. (2006). How do spores germinate? *Journal of Applied Microbiology*, 101(3), pp.526–530. https://doi.org/10.1111/j.1365-2672.2006.02885.x

Mokoena, M. P. (2017). Lactic acid bacteria and their bacteriocins: Classification, biosynthesis and applications against uropathogens: A mini-review. Molecules, 22(8), p.1255.

Monteiro, C., Marques, P. I., Cavadas, B., Damião, I., Almeida, V., Barros, N., Barros, A., Carvalho, F., Gomes, S., & Seixas, S. (2018). Characterization of microbiota in male infertility cases uncovers differences in seminal hyperviscosity and oligoasthenoteratozoospermia possibly correlated with increased prevalence of infectious bacteria. *American Journal of Reproductive Immunology*, 79(6), p.e12838. https://doi.org/10.1111/aji.12838

Monteiro, S. M. S., Clemente, J. J., Carrondo, M. J. T., & Cunha, A. E. (2014). Enhanced spore production of Bacillus subtilis grown in a chemically defined medium. *Advances in Microbiology*, 4, pp.444–454.

Monteiro, S. M. Clemente, J. J., Henriques, A. O., Gomes, R. J., Carrondo, M. J., & Cunha, A. E. (2005). A procedure for high-yield spore production by *Bacillus subtilis*. *Biotechnology Progress*, 21(4). https://doi.org/10.1021/bp050062z

Nakano, M. M., Xia, L., & Zuber, P. (1998). Transcription initiation region of the srfA operon, which is controlled by the comP-comA signal transduction system in Bacillus subtilis. *Journal of Bacteriology*, 180(8), pp.1848–1855.

Nicholson, W. L., Munakata, N., Horneck, G., Melosh, H. J., &Setlow, P. (2000). Resistance of Bacillus endospores to extreme terrestrial and extraterrestrial environments. *Microbiology and Molecular Biology Reviews*, 64(3), pp.548–572.

Ochoa, T. J., & O'Ryan, M. (2017). Yersinia Species. *Principles and Practice of Pediatric Infectious Diseases*, pp.846–851. https://doi.org/10.1016/B978-0-323-40181-4.00148-1

Pandey, K. R., & Vakil, B. V. (2016). Development of bioprocess for high-density cultivation yield of the probiotic Bacillus coagulans and its spores. *Probiotics and Antimicrobial Proteins*, 8(3), pp.148–155.

Papagianni, M. (2012). Recent advances in engineering the central carbon metabolism of industrially important bacteria. *Microbial Cell Factories*, 11(50). https://doi.org/10.1186/1475-2859-11-50

Peng XL, Zhao WJ, Wang YS, Dai KL, Cen YK, Liu ZQ, Zheng YG. (2020) Enhancement of gibberellic acid production from *Fusarium fujikuroi* by mutation breeding and glycerol addition. 3 Biotech., 10(7): p.312. doi: 10.1007/s13205-020-02303-4.

Popov, I. V., & Algburi, A. (2021). Production of spores as probiotic formulations. In R. S. Singh, G. Arora, A. K. Srivastava, & P. K. Jha (Eds.), *Microbial Inoculants in Sustainable Agricultural Productivity* (pp.59–79). Springer Singapore.

Popov, I. V., Algburi, A., Prazdnova, E. V., Mazanko, M. S., Elisashvili, V., Bren, A. B., Chistyakov, V. A., Tkacheva, E. V., Trukhachev, V. I., Donnik, I. M., Ivanov, Y. A., Rudoy, D., Ermakov, A. M., Weeks, R. M., & Chikindas, M. L. (2021). A review of the effects and production of spore-forming probiotics for poultry. *Animals*, 11(7), p.1941. https://doi.org/10.3390/ani11071941

Posada-Uribe, L. F., Romero-Tabarez, M., Villegas-Escobar, V. (2015). Effect of medium components and culture conditions in *Bacillus subtilis* EA-CB0575 spore production. *Bioprocess and Biosystems Engineering*, 38(10), pp.1879–1888. http://doi.org/10.1007/s00449-015-1428-1

Poshadri, A., Deshpandeh, H. W., Khodke, U. M., & Katke, S. D. (2022). *Bacillus coagulans* and its spore as potential probiotics in the production of novel shelf-stable foods. *Current Research in Nutrition and Food Science*, 10(3), pp.858–870. http://doi.org/10.12944/CRNFSJ.10.3.4

Qiao, N., Yu, Y., Liu, J., Zhang, X., & Li, Z. (2020). Isolation and identification of Bacillus species and optimization of their culture conditions for nitrate removal in water. *Journal of Environmental Science and Health, Part A*, 55(9), pp.1121–1130.

Qingshan, M. A., & Zengguo, H. E. et al. (2019). Screening of high-efficiency denitrifying bacteria from marine environment and their application potential in aquaculture. *Marine Sciences*, 43(7), pp.30–38.

Rao, Y. K., Lee, M.-J., Chen, K., Wu, W.-S., & Tzeng, Y.-M. (2007a). Optimization of culture medium for production of microbial inulinase using statistically designed experiments. *Journal of Bioscience and Bioengineering*, 104(5), pp.373–378.

Rao, Y. K., Tsay, K.-J., Wu, W.-S., Tzeng, Y.-M. (2007b). Medium optimization of carbon and nitrogen sources for the production of spores from Bacillus amyloliquefaciens B128 using response surface methodology, *Process Biochemistry*, 42(4), pp.535–541. https://doi.org/10.1016/j.procbio.2006.10.007

Ren, H., Su, Y. T., & Guo, X. H. (2018). Rapid optimization of spore production from *Bacillus amyloliquefaciens* in submerged cultures based on dipicolinic acid fluorimetry assay. *AMB Express*, 8(1), p.21. https://doi.org/10.1186/s13568-018-0555-x

Rosenberg, A., Sinai, L., Smith, Y., & Ben-Yehuda, S. (2012). Dynamic expression of the translational machinery during *Bacillus subtilis* life cycle at a single cell level. *PLoS ONE*, 7(7), p.e41921. https://doi.org/10.1371/journal.pone.0041921

Sanders, M. E., Merenstein, D. J., Ouwehand, A. C., Reid, G., Salminen, S., Cabana, M. D., Paraskevakos, G., & Leyer, G. (2016). Probiotic use in at-risk populations. *Journal of the American Pharmacists Association*, 56(6), pp.680–686. https://doi.org/10.1016/j.japh.2016.07.001

Saritha, M., Arora, A., & Nain, L. (2012). Biological pretreatment of lignocellulosic substrates for enhanced delignification and enzymatic digestibility. *Indian Journal of Microbiology*, 52(2), pp.122–30. https://doi.org/10.1007/s12088-011-0199-x

Sen, R., & Babu, K. S. (2005). Modeling and optimization of the process conditions for biomass production and sporulation of a probiotic culture. Process Biochemistry, 40(7), pp.2531–2538.

Senok, A. C., Ismaeel, A. Y., & Botta, G. A. (2005). Probiotics: Facts and myths. *Clinical Microbiology and Infection*, 11(12), pp.958–966.

Shah, A. T., Favaro, L., Alibardi, L., Cagnin, L., Sandon, A., Cossu, R., Casella, S., & Basaglia, M. (2016). Bacillus sp. strains to produce bio-hydrogen from the organic fraction of municipal solid waste. *Applied Energy*, 176, pp.116–124. https://doi.org/10.1016/j.apenergy.2016.05.054

Shi, F., & Zhu, Y. (2007). Application of statistically-based experimental designs in medium optimization for spore production of Bacillus subtilis from distillery effluent. *BioControl*. 52, pp.845–853. https://doi.org/10.1007/s10526-006-9055-z

Soccol, C. R., Vandenberghe, L. P., Medeiros, A. B., Karp, S. G., Buckeridge, M., & Ramos, L. P. (2010). Bioethanol from lignocellulosic biomass: Current status and perspectives. In *Industrial biotechnology* (pp.51–88). Springer, Dordrecht.

Soccol, C. R., Vandenberghe, L. P., Spier, M. R., Medeiros, A. B., Yamaguishi, C. T., Lindner, J. D., ... Thomaz-Soccol, V. (2017). The potential of solid-state fermentation for production of enzymes: A review. *International Journal of Molecular Sciences*, 18(12), p.2562.

Sonenshein, A. L. (2000). Control of sporulation initiation in Bacillus subtilis. *Current Opinion in Microbiology*, 3(6), pp.561–566. https://doi.org/10.1016/S1369-5274(00)00141-7

Sooch, B. S., Singh, J., & Verma, D. (2023). 12 - Insights into metabolic engineering approaches for enhanced biobutanol production, Editor(s): Juan Gabriel Segovia-Hernandez, Shuvashish Behera, Eduardo Sanchez-Ramirez, In *Advances in Pollution Research, Advances and Developments in Biobutanol Production*, Woodhead Publishing, pp.329–361. https://doi.org/10.1016/B978-0-323-91178-8.00012-6

Sunitha, S., & Padmavathi, P. (2013). Role of Bacillus species in enhancement of phytoplankton production in wastewater. *International Journal of Current*.

Tian, Z., Hou, L., Hu, M., Gao, Y., Li, D., Fan, B., & Li, S. (2022). Optimization of Sporulation Conditions for Bacillus subtilis BSNK-5. *Processes*, 10(6), p.1133. https://doi.org/10.3390/pr10061133

Tzeng, Y.-M., Rao, Y. K., Tsay, K.-J., Wu, W.-S. (2008). Effect of cultivation conditions on spore production from Bacillus amyloliquefaciens B128 and its antagonism to Botrytis elliptica. *Journal of Applied Microbiology*, 104(5). https://doi.org/10.1111/j.1365-2672.2007.03683.x

Vakil, B. V. et al. (2016). Sporulation studies and preservation of Bacillus subtilis strains. *Journal of Pharmaceutical Negative Results*, 7(1), pp.1–7.

Veening, J.-W., Murray, H., & Errington, J. (2009). A mechanism for cell cycle regulation of sporulation initiation in Bacillus subtilis. *Genes & Development*, 23(16), pp.1959–1970. https://doi.org/10.1101/gad.528209

Wang, J. D., & Levin, P. A. (2009). Metabolism, cell growth and the bacterial cell cycle. *Nature Reviews Microbiology*, 7(11), pp.822–827. https://doi.org/10.1038/nrmicro2202

Wang, X., Li, Y., Wang, M., Wu, X., & Chen, X. (2005). Effects of Bacillus species on the growth, oxygen consumption, and ammonia-nitrogen excretion of Penaeus japonicus Bate. *Aquaculture*, 248(1–4), pp.223–231.

Wang, Y., Bi, L., Liao, Y., Lu, D., Zhang, H., Liao, X., Liang, J. B., & Wu, Y. (2019). Influence and characteristics of Bacillus stearothermophilus in ammonia reduction during layer manure composting. *Ecotoxicology and Environmental Safety*, 180, pp.80–87. https://doi.org/10.1016/j.ecoenv.2019.04.066

Wangka-Orm, C., Deeseenthum, S., & Leelavatcharamas, V. (2014). Low cost medium for spore production of Bacillus KKU02 and KKU03 and the effects of the produced spores on growth of giant freshwater prawn (Macrobrachiumrosenbergii de Man). *Pakistan Journal of Biological Sciences*, 17(8), pp.1015–22. https://doi.org/10.3923/pjbs.2014.1015.1022

Watanabe, M. M., Kato, Y., & Watanabe, M. M. (2008). Effects of nutrients and cell extracts from Bacillus subtilis on growth of three green algae and one diatom. *Fisheries Science*, 74(6), pp.1332–1338.

Xie, J.-B., Zhang, X.-S., Wang, Y., & Sun, X.-F. (2013). Enhancing the stability of an extremely versatile self-aggregating bacterial consortium. *Bioresource Technology*, 128, pp.477–483.

Yang, H., Xue, Y., Yu, X., & Liu, C. (2014). Colonization of Bacillus amyloliquefaciens CC09 in wheat leaf and its biocontrol effect on powdery mildew disease. Chinese Journal of Biological Control, 30(4), p.481.

Yi, Y., Zhang, Z., Zhao, F., Liu, H., Yu, L., Zha, J., & Wang, G. (2018). Probiotic potential of Bacillus velezensis JW: Antimicrobial activity against fish pathogenic bacteria and immune enhancement effects on Carassius auratus. *Fish & Shellfish Immunology*, 78, pp.322–330. https://doi.org/10.1016/j.fsi.2018.04.055

Ying, W., Zhu, R., Lu, W., & Gong, L. (2009b). A new strategy to apply Bacillus subtilis MA139 for the production of solid-state fermentation feed. *Letters in Applied Microbiology*, 49(2), pp.229–234. https://doi.org/10.1111/j.1472-765X.2009.02647.x

Ying, X., Liu, X., Shi, C., Lu, Z., & Yuan, Z. (2009a). Solid-state fermentation for production of a novel phytase from Bacillus subtilis Q7 using okara as substrate. *Bioresource Technology*, 100(19), pp.4580–4583.

Yuniarti, A., Arifin, N. B., Fakhri, M., & Hariati, A. M. (2019a, February). Effect of C: N ratio on the spore production of Bacillus sp. indigenous shrimp pond. In *IOP Conference Series: Earth and Environmental Science* (Vol. 236, No. 1, pp.012029). IOP Publishing. https://doi.org/10.1088/1755-1315/236/1/012029

Yuniarti, A. et al. (2019b). Effect of carbon and nitrogen sources and their ratio on spore production of *Bacillus* spp. SB 4. *Journal of Coastal Development*, 22(2), pp.337–346.

Zhang, P., Garner, W., Yi, X., Yu, J., Li, Y., & Setlow, P. (2010). Factors affecting variability in time between addition of nutrient germinants and rapid dipicolinic acid release during germination of spores of Bacillus species. *Journal of Bacteriology*, 192(14), pp.3608–3619. https://doi.org/10.1128/JB.00345-10

Zhang Y. R. (2013). Progress on probiotic research in China. *Food Research International*, 51(2), pp.507–514.

Zhang, Y.-R., Xiong, H.-R., Guo, X.-H. (2014). Enhanced viability of Lactobacillus reuteri for probiotics production in mixed solid-state fermentation in the presence of *Bacillus subtilis*, *Folia Microbiologica*, 59(1). https://doi.org/10.1007/s12223-013-0264-4

Zhao, L., Kong, J., & Wang, G. (2008). Comparison of submerged and solid-state fermentation for the production of lovastatin by Aspergillusterreus. *Process Biochemistry*, 43(12), pp.1462–1466.

Zhu, Y., Liu, M., & Lu, C.-P. (2007). Effect of culture conditions on endospore formation by Bacillus subtilis in submerged fermentation. *Biotechnology and Bioprocess Engineering*, 12(4), pp.373–379.

Zink, I. C., Amaral, V., & Oliveira, M. S. B. (2011). Use of probiotics in aquaculture. *ISRN Microbiology*, 2011, pp.1–13.

5 *Bacillus* in the Gastrointestinal (GI) Tract of Finfish and Shellfish

Einar Ringø
UiT The Arctic University of Norway, Tromsø, Norway

Koushik Ghosh
The University of Burdwan, Burdwan, India

Mehdi Soltani
Murdoch University, Perth, Australia
University of Tehran, Tehran, Iran

5.1 INTRODUCTION

Optimal gastrointestinal (GI) functionality is of importance for sustainable animal production, and three components namely diet, mucosa, and commensal microbiota are essential for gut health. Evaluation of the gut microbiota of finfish dates back to the late 1920s and early 1930s (Reed & Spence 1929; Gibbons 1933) and since then numerous studies have been published. The GI tract microbiota in fish is usually divided into the GI lumen microbiota (allochthonous), and those that adhere and colonize the mucosal surface, the autochthonous, although both microbiota communities can be modulated by several factors (e.g., Ringø et al. 2016). The intestinal microbiota contributes to many essential functions such as pathogen displacement, nutrient- and receptor competition, production of antimicrobial factors, structural factors (induction of IgA, tightening of tight junction, and immune system development), and several metabolic activities.

Since the first studies were published on salmonids by Trust and Sparrow (1974) and Trust (1975), numerous investigations have revealed *Bacillus* species in the GI tract of finfish and shellfish. Table 5.1 summarizes studies isolated and characterized as allochthonous and autochthonous *Bacillus* from the GI tract, and this is of importance to evaluate as they may contribute to nutrition, inhibit pathogen adherence and colonization, may affect the immune system, and may have potential as probiotics (e.g., Wang et al. 2008; Kuebutornye et al. 2019; Soltani et al. 2019; Ringø et al. 2020; James et al. 2021; Soto 2021; van Doan 2021; Shija et al. 2023).

More recently, anti-biofilm, anti-growth along with anti-quorum-sensing activities of the fish gut-associated bacilli have been demonstrated to be effective against bacterial infections in aquaculture and 52% of the isolates were noticed to effectively antagonize important fish pathogens including *Aeromonas hydrophila*, *A. salmonicida*, *A. bivalvium*, *A. veronii*, *Vibrio anguillarum*, *V. harveyi*, *V. parahaemolyticus*, *V. vulnificus*, *Photobacterium damselae*, *Tenacibaculum maritimum*, *Edwardsiella tarda*, and *Shigella sonnei* (Santos et al. 2021).

5.2 *BACILLUS* IN FINFISH

In numerous studies, counts of presumptive *Bacillus* spp. are reported, but without going into further identification (Table 5.1). These studies are not discussed further in the present chapter, even though they have investigated dietary effects (e.g., de Paula Silva et al. 2011; Green et al. 2013;

DOI: 10.1201/9781003503811-5

TABLE 5.1
***Bacillus* in the Gastrointestinal Tract of Finfish and Shellfish, Detected by Culture Based or Culture-Independent Methods (C-IM)**

Bacillus Isolated	Methodology	Isolated From	"Segments" of the GI Tract	Major Features	References
Finfish					
Bacillales	C-IM	Rainbow trout (*Oncorhynchus mykiss*)	EI content	ni	Goncalves and Gallardo-Escárate (2017)
Bacillus	Culture based	Freshwater salmonids	PC–MI–DI	ni	Trust and Sparrow (1974)
	Culture based	Chum salmon (*Oncorhynchus keta*)	EI content	ni	Trust (1975)
	Culture based	Rainbow trout	EI content	Adhesion and antagonism	Ramirez-Torrez et al. (2018)
	C-IM	Atlantic salmon (*Salmo salar*)	S/PC–MI–DI content	ni	Hovda et al. (2007)
	C-IM	Atlantic salmon	S/PC–MI–DI content	ni	Hovda et al. (2012)
	Culture based	Atlantic salmon	PI–DI–auto	Enzyme-producing and antagonism	Askarian et al. (2012)
	C-IM	Atlantic salmon	Feces	ni	Green et al. (2013)
	Culture based	Arctic charr (*Salvelinus alpinus*)	DI–auto	ni	Ringø et al. (2006)
	Cultivation	Rainbow trout	EI content	ni	Kim et al. (2007)
	C-IM	Rainbow trout	MI content	ni	Wong et al. (2013)
	C-IM	Rainbow trout	Feces	ni	Mente et al. (2018)
	C-IM	Rainbow trout	MI–DI content	ni	Ricaud et al. (2018)
	Culture based	Brown trout (*Salmo trutta*)	PI–auto	ni	Al-Hisnawi et al. (2015)
	Culture based	Common carp (*Cyprinus carpio*)	Intestine*	Enzyme producer	Bairagi et al. (2002)
	Culture based	Common carp	P–PI–MI–DI	ni	Jung-Schroers et al. (2016)
	Culture based	Grass carp (*Ctenopharyngodon idella*)	EI content	ni	Trust et al. (1979)
	Culture based	Grass carp	EI content	ni	Sugita et al. (1985)
	Culture based	Catla (*Catla catla*)	PI–DI–auto	Enzyme producer	Ray et al. (2010)
	Culture based	Catla	PI–auto	Enzyme producer	Ray et al. (2010)
	Culture based	Catla	PI–auto	Potential as probiotic	Nandi et al. (2017b)
	Culture based	Rohu (*Labeo rohita*)	EI content	Enzyme producer, fermenting grass pea (*Lathyrus sativus*) seed meal	Ramachandran et al. (2005)
	Culture based	Rohu	EI–auto	Enzyme producer, fermenting black gram seed meal	Ramachandran and Ray (2007)

Culture based	Rohu	MI–auto	ni	Ghosh et al. (2010)
Culture based	Rohu	PI–DI–auto	Enzyme producer	Ray et al. (2010)
Culture based	Murrel (*Channa punctatus*)	PI-–DI–auto	Enzyme producer	Banerjee et al. (2013)
C-IM	Northern snakehead (*Channa argus*)	Intestine*	ni	Miao et al. (2018)
Culture based	Stinging catfish (*Heteropneustes fossilis*)	PI–DI–auto	Enzyme-producer	Banerjee et al. (2013)
Culture based	Dwarf catfish (*Mystus vittatus*)	PI–DI–auto	Potential as probiotic	Nandi et al. (2017a)
Culture based	Goldfish (*Carassius auratus*)	Feces	ni	Sugita et al. (1988a)
Culture based	Goldfish	EI content	ni	Sugita et al. (1988b)
Culture based	Goldfish	Feces	Antagonism	Sugita et al. (1989)
C-IM	Goldfish	MI–auto	ni	Liu et al. (2018)
C-IM	Zebrafish (*Danio rerio*)	EI content	ni	Yang et al. (2017)
C-IM	Gilthead sea bream (*Sparus aurata*)	S and EI content	ni	De Paula Silva et al. (2011)
C-IM	Gilthead sea bream	EI content	ni	Castro et al. (2011)
Culture based	Striped bass (*Morone saxatilis*)	EI content	ni	Nedoluha and Westhoff (1995)
Culture based	Striped bass (*Morone saxatilis*)	EI content	ni	Nedoluha and Westhoff (1997)
C-IM	Sea bass (*Dicentrarchus labrax*) larvae	ni	ni	Gatesoupe et al. (2016)
Culture based	Dragonets (*Callionymus* sp.)	EI content	ni	Sugita et al. (1988c)
Culture based	Japanese whiting (*Sillago japonica*)	EI content	ni	Sugita et al. (1988c)
Culture based	Tilapia (*Sarotherodon niloticus*)	EI content	ni	Sugita et al. (1987)
Culture based	Japanese coastal fish	EI content	Antagonism	Sugita et al. (1998)
Culture based	Hybrid tilapia (*O. niloticus* x *O. aureus*)	EI content	ni	Al-Harbi and Uddin (2005)
C-IM	Nile tilapia (*Oreochromis niloticus*)	EI content	ni	Del'Duca et al. (2015)
Culture based	Five freshwater fish species	Ni	Amylase-producing	Sugita et al. (1997)
Culture based	Different fish species	Ni	ni	Izvekova et al. (2007)
C-IM	Totoaba (*Totoaba macdonaldi*)	IC–PC, PI, MI, DI	ni	Barreto-Curiel et al. (2018)
C-IM	Turbot (*Scophthalmus maximus*)	EI–auto	ni	Yang et al. (2018)
C-IM	Fine flounder (*Paralichthys adspersus*)	EI content	ni	Ramirez and Romero (2017)
C-IM	Atlantic cod (*Gadus morhua*) larvae	EI	ni	Vadstein et al. (2018)
Culture based	Atlantic cod juvenile	EI–auto	ni	Strøm and Olafsen (1990)
Culture based	Atlantic cod	DI–allo	ni	Olsen et al. (2008)

(Continued)

TABLE 5.1 (Continued)
***Bacillus* in the Gastrointestinal Tract of Finfish and Shellfish, Detected by Culture Based or Culture-Independent Methods (C-IM)**

Bacillus Isolated	Methodology	Isolated From	"Segments" of the GI Tract	Major Features	References
	C-IM	Atlantic cod	DI–auto	ni	Zhou et al. (2013)
	C-IM	Atlantic cod	DI content	ni	Ringø et al. (2014)
	Culture based	Siberian sturgeon (*Acipenser baerii*)	Gut content	Potential probiotic	Geraylou et al. (2014)
	C-IM	Cobia (*Rachycentron canadum*)	EI content	ni	Rasheeda et al. (2017)
	Culture based	Deep sea shark (*Centroscyllium fabricii*)	EI content	Antagonistic potential	Bindiya et al. (2015)
	Culture based	Lantern fish, Monterey Bay	EI content	ni	Sanchez et al. (2012)
	Culture based	Jaraqui (*Piaractus mesopotamicus*)	EI content	Antimicrobial peptide	Sirtori et al. 2006)
B. aerius	Culture based	Catla	DI–auto	Enzyme-producing and antagonism	Dutta et al. (2015)
B. aerophilus	Culture based	Indian major carp (*Cirrhinus mrigala*)	DI–auto	Potential probiotic	Mukherjee et al. (2016)
B. amyloliquefaciens	C-IM	Atlantic salmon	Feces	ni	Green et al. (2013)
	Culture based	Rohu	PI–DI–auto	Potential probiotic	Mukherjee et al. (2017)
	Culture based	Rohu	DI–auto	ni	Ghosh et al. (2010)
	Culture based	Catla	PI–auto	Potential as probiotic	Nandi et al. (2017b)
	Culture based	Mrigal	DI–auto	Enzyme-producing and antagonism	Dutta and Ghosh (2015)
	Culture based	Java barb (*Puntius javanicus*)	PI–auto	Potential probiotics	Nandi et al. (2017b)
	Culture based	Climbing perch (*Anabas testudineus*)	DI–auto	Enzyme-producing	Banerjee et al. (2015a)
	Culture based	Areolate grouper (*Epinephelus areolatus*)	Intestine*	Bacteriocin	An et al. (2015)
	C-IM	Southern flounder (*Paralichthys lethostigma*)	Intestine*	Antagonism	Chen et al. (2016a)
	C-IM	Southern flounder	Intestine*	Antagonism	Chen et al. (2016b)
	C-IM	*Indian major carp*	EI content	Enzyme-producing and antagonism	Kavitha et al. (2018)
B. altitudinis	Culture based	Mrigal	DI–auto	Enzyme-producing and antagonism	Banerjee et al. (2016)
B. aryabhattai	Culture based	Walking catfish (*Clarias batrachus*)	PI–auto	Enzyme-producing	Dey et al. (2016)
B. atrophaeus	Culture based	Stinging catfish	PI–auto	Phytase-producing	Khan and Ghosh (2012)

B. cereus	Culture based	Catla	PI–DI–auto	Enzyme-producing	Ray et al. (2010)
	Culture based	Mrigal	PI–DI–auto	Enzyme-producing	Ray et al. (2010)
	Culture based	Rohu	Intestine*	Enzyme-producing	Ghosh et al. (2002)
	Culture based	Rohu	PI–MI–DI–auto	ni	Ghosh et al. (2010)
	Culture based	Flat grey mullet (*Mugil cephalus*)	Intestine*	Protease-producing	Esakkiraj et al. (2009)
	Culture based	Channel catfish (*Ictalurus punctatus*)	EI content	Potential as probiotic	Ran et al. (2012)
	Culture based	Walking catfish	PI–auto	Enzyme-producing	Dey et al. (2016)
	Culture based	Wild olive flounder (*Paralichthys olivaceus*)	DI–auto	ni	Kim and Kim (2013)
	Culture based	Atlantic salmon	PI–DI–auto	Enzyme-producing and antagonism	Askarian et al. (2012)
	C-IM	*Indian major carp*	EI content	Enzyme-producing and antagonism	Kavitha et al. (2018)
B. circulans	Culture based	Rainbow trout	Intestine*	ni	Austin and Al-Zahrani (1988)
	Culture based	Rohu	Intestine*	Enzyme-producing	Ghosh et al. (2002)
	Culture based	Common carp (*Cyprinus carpio*)	Intestine*	Cellulase-producing	Ray et al. (2007)
	Culture based	Mossambicus tilapia (*Oreochromis mossambicus*)	Intestine*	Enzyme-producing	Saha et al. (2006)
B. coagulans	Culture based	Catla	PI–DI–auto	Enzyme-producing	Ray et al. (2010)
B. flexus	Culture based	Mrigal	PI–auto	Chitinase-producing	Banerjee et al. (2015b)
	Culture based	Walking catfish	PI–auto	Enzyme-producing	Dey et al. (2016)
B. licheniformis	Culture based	Atlantic salmon	PI–DI–auto	Did not reveal promising enzyme-producing activity and antagonism	Askarian et al. (2012)
	Culture based	Farmed olive flounder*	DI–auto	ni	Kim and Kim (2013)
	Culture based	Murrel	PI–DI–auto	Enzyme-producing	Banerjee et al. (2013)
	Culture based	Stinging catfish	PI–DI–auto	Enzyme-producing	Banerjee et al. (2013)
	Culture based	Walking catfish	DI–auto	Enzyme-producing	Banerjee et al. (2015a)
	Cultured based	Indian major carp	PI–auto	Potential as probiotic	Mukherjee et al. (2016)
	Culture based	Rohu	PI–auto	ni	Ghosh et al. (2010)
	Culture based	Rohu	PI–DI–auto	Inhibition of pathogen	Mukherjee et al. (2017)
	Culture based	Bata (*Labeo bata*)	PI–auto	Enzyme-producing	Mondal et al. (2010)

(*Continued*)

TABLE 5.1 (Continued)

***Bacillus* in the Gastrointestinal Tract of Finfish and Shellfish, Detected by Culture Based or Culture-Independent Methods (C-IM)**

Bacillus Isolated	Methodology	Isolated From	"Segments" of the GI Tract	Major Features	References
	Culture based	Long whiskers catfish (*Mystus gulio*)	MI–auto	Enzyme-producing	Das et al. (2014)
	Culture based	Nile tilapia (*Oreochromis niloticus*)	PI–auto	Potential as probiotic	Ghosh et al. (2017)
B. megaterium	Culture based	Rainbow trout	Intestine*	ni	Austin and Al-Zahrani (1988)
	C-IM	Atlantic salmon	Feces	ni	Green et al. (2013)
	Culture based	Wild olive flounder	DI–auto	ni	Kim and Kim (2013)
	Culture based	Mossambicus tilapia	Intestine*	Enzyme-producing	Saha et al. (2006)
	Culture based	Grass carp (*Ctenopharyngodon idella*)	PI–auto	Enzyme-producing and antagonism	Banerjee et al. (2016)
B. methylotrophicus	Culture based	Wild olive flounder	DI–auto	ni	Kim and Kim (2013)
	Culture based	Farmed olive flounder	DI–auto	ni	Kim and Kim (2013)
	Culture based	Channel catfish	EI content	Potential as probiotic	Ran et al. (2012)
	Culture based	Indian major carp (*Catla catla*)	PI–auto	Potential as probiotic	Mukherjee and Ghosh (2016)
	Culture based	Rohu	PI–DI–auto	Potential as probiotic	Mukherjee et al. (2017)
	Culture based	Rainbow trout	EI content	Adhesion and antagonism	Ramirez-Torrez et al. (2018)
B. nealsonii	Culture based	Rainbow trout	Intestine*	Non-pathogenic	Yilmaz et al. (2018)
B. pumilus	Culture based	Brown trout	DI–auto	ni	Al-Hisnawi et al. (2015)
	Culture based	Rainbow trout	DI–auto	ni	Ringø et al. (2016)
	Culture based	Rainbow trout	EI content	Adhesion	Ramirez-Torrez et al. (2018)
	Culture based	Rainbow trout	Intestine*	Non-pathogenic	Yilmaz et al. (2018)
	Culture based	Channel catfish	EI content	Potential as probiotic	Ran et al. (2012)
	Culture based	Silver carp (*Hypophthalmichthys molitrix*)	DI–auto	Chitinase-producing	Banerjee et al. (2015b)
	Culture based	Grouper	Intestine*	ni	Ma et al. (2010)
	Culture based	Mrigal	PI–auto	Enzyme-producing and antagonism	Banerjee et al. (2016)
	Culture based	Rohu	Intestine*	Enzyme-producing	Ghosh et al. (2002)
	Culture based	Rohu	PI–auto	Enzyme-producing and antagonism	Banerjee et al. (2016)

B. sonorensis	Culture based	Mrigal	DI–auto	Enzyme-producing and antagonism	Dutta and Ghosh (2015)
B. subtilis	Culture based	Atlantic salmon	PI–DI–auto	Enzyme-producing and antagonism	Askarian et al. (2012)
	C-IM	Atlantic salmon	Feces	ni	Green et al. (2013)
	Culture based	Rainbow trout	Intestine*	Non-pathogenic	Yilmaz et al. (2018)
	Culture based	Channel catfish	EI content	Potential as probiotic	Ran et al. (2012)
	Culture based	Stinging catfish	PI–DI–auto	Enzyme-producing	Banerjee et al. (2013)
	Culture based	Rohu	PI–MI–DI–auto	ni	Ghosh et al. (2010)
	Culture based	Rohu	EI content	Antagonism	Giri et al. (2011)
	Culture based	Rohu	PI–auto	Phytase-producing	Khan and Ghosh (2012)
	Culture based	Rohu	PI–auto	Solid-state fermentation	Khan and Ghosh (2013)
	Culture based	Mrigal	Intestine*	Used as synbiotic	Kumar et al. (2018)
	Culture based	Bata	PI–auto	Enzyme-producing	Mondal et al. (2010)
	Culture based	Jaraqui	MI content	Cellulase-producing	Peixoto et al. (2011)
	C-IM	Cyprinid species	DI content	ni	Li et al. (2015a, 2015b)
	Culture based	Nile tilapia	PI–DI–auto	Tannase-producing	Talukdar et al. (2016)
	Culture based	Mozambique tilapia	Intestine*	Cellulase-producing	Ray et al. (2007)
	Culture based	Mozambique tilapia	MI–auto	Enzyme-producing	Sarkar and Ghosh (2014)
	C-IM	Black carp (*Mylopharyngodon piceus*)	EI content	ni	He et al. (2013)
	C-IM	Gibel carp (*Carassius gibelio*)	EI content	ni	He et al. (2013)
	C-IM	Bluntnose black bream (*Megalobrama amblycephala*)	EI content	ni	He et al. (2013)
	C-IM	Indian major carp	EI content	Enzyme-producing and antagonism	Kavitha et al. (2018)
	C-IM	Indian major carp	Intestine*	Antagonism	Banerjee et al. (2017)
	C-IM	Southern flounder	Intestine*	Antagonism	Chen et al. (2016b)
B. subtilis subsp. *subtilis*	Culture based	Mrigal	MI–auto	Phytase-producing	Das and Ghosh (2013)
	Culture based	Mrigal	MI–auto	Phytase-producing, solid-state fermentation	Das and Ghosh (2015)
B. subtilis subsp. *spizizenii*	Culture based	Indian major carp	DI–auto	Potential as probiotic	Mukherjee and Ghosh (2016)

(*Continued*)

TABLE 5.1 (Continued)
***Bacillus* in the Gastrointestinal Tract of Finfish and Shellfish, Detected by Culture Based or Culture-Independent Methods (C-IM)**

***Bacillus* Isolated**	**Methodology**	**Isolated From**	**"Segments" of the GI Tract**	**Major Features**	**References**
B. tequilensis	Culture based	Silver carp	PI–auto	Enzyme-producing and antagonism	Banerjee et al. (2016)
B. thermoamylovorans	C-IM	Arctic charr	EI content	ni	Nyman et al. (2017)
B. thuringiensis	Culture based	Atlantic salmon	PI–DI–auto	Enzyme-producing and antagonism	Askarian et al. (2012)
	Culture based	Rainbow trout	Intestine*	Non- pathogenic	Yilmaz et al. (2018)
B. stratosphericus	Culture based	Indian major carp	PI–auto	Potential as probiotic	Mukherjee et al. (2016)
	Culture based	Piau-com-pinta (*Leporinus friderici*)	MI content	Cellulase-producing	Peixoto et al. (2011)
Solibacillus silvestris	Culture based	Indian major carp	DI–auto	Potential as probiotic	Mukherjee et al. (2016)
Shellfish					
Bacilli	C-IM	Pacific white shrimp (*Litopenaeus vannamei*)	Intestine*	ni	Zhu et al. (2016)
*Bacillus**	Culture based	Tiger shrimp (*Penaeus monodon*)	Intestine*	ni	Jasmin et al. (2016)
	C-IM	Pacific white shrimp	Intestinal content	Phytase	Gao et al. (2019)
	C-IM	Oriental river prawn (*Macrobrachium nipponense*)	Intestinal content	ni	Tzeng et al. (2015)
Bacillus cereus	Culture based	Tiger shrimp	Intestine*	Growth and immune enhancer	Chandran et al. (2014)
	C-IM	Pacific white shrimp	Intestine*	ni	Vargas-Albores et al. (2017)
B. thuringiensis	C-IM	Pacific white shrimp	Intestine*	ni	Vargas-Albores et al. (2017)
B. vireti	Culture based	Freshwater prawn (*Macrobrachium rosenbergii*)	Intestine*	Used as probiotic	Hindu et al. (2018)
B. weihenstephanensis	C-IM	Pacific white shrimp	Intestine*	ni	Vargas-Albores et al. (2017)

Source: Adapted from Soltani et al. (2019).

Note: * no further information was given, DI = distal intestine, PI = proximal intestine, MI = mid-intestine, EI = entire intestine, P = pseudogaster, PC = pyloric ceca, S= stomach, MI = mid intestine, PI = posterior intestine, Ni = no further information available, EA = early stage.

Ni = no further information available; P = pseudogaster; PC = pyloric ceca; S = stomach.

Barreto-Curriel et al. 2018; Nyman et al. 2017), effects of antibiotic supplementation (Austin & Al-Zahrani 1988; Sugita et al. 1988a, 1989), production of enzymes (e.g., Sugita et al. 1997; Bairagi et al. 2002; Ray et al. 2010; Askarian et al. 2012), putative probionts (e.g., Geraylou et al. 2014; Nandi et al. 2017a, 2017b), effect of prebiotics (Ringø et al. 2006), antimicrobial potential (Sugita et al. 1989, 1998), seasonal variations (Hovda et al. 2012), effect of stress (Olsen et al. 2008), and wild *vs.* farmed fish (Strøm & Olafsen 1990; Ramirez & Romero 2017). Readers with special interests in these studies can consult the original papers.

In the following sections, bacilli identified as *B. aerius, B. aerophilus, B. amyloliquefaciens, B. altitudinis, B. aryabhattai, B. atrophaeus, B. badius, B. brevis, B. cereus, B. circulans, B. clausii, B. coagulans, B. flexus, B. lentus, B. licheniformis, B. megaterium, B. methylotrophicus, B. pacificus, B. pasteurii, B. pumilus, B. sonorensis, B. sphaericus, B. subtilis, B. tequilensis, B. thermoamylovorans, B. thuringiensis, B. stratosphericus, B. vireti, B. weihenstephanensis*, and *Solibacillus silvestris* isolated from the finfish and shellfish GI tract are discussed.

Bacillus aerius. To our knowledge, only one study has isolated this bacterium from the GI tract of finfish, catla (*Catla catla*) (Dutta et al. 2015). The authors reported autochthonous *B. aerius* CCHIA in the distal intestine (DI) and probiotic characterizations were investigated. The strain revealed high extracellular enzyme activities of amylase, protease, lipase, cellulase, phytase, and xylanase; grows well at 8°C, but could not tolerate high temperatures; showed tolerance toward bile juice; and grows well in catla mucus but was potent to grow in gut mucus than skin mucus. In addition, the strain displayed antagonistic activity against *Aeromonas hydrophila* and *Aeromonas salmonicida.*

Bacillus aerophilus. In a study evaluating the autochthonous microbiota of Indian major carp (*Cirrhinus mrigala*) by cultivation, Mukherjee et al. (2016) reported *B. aerophilus*, a bacterium with probiotic potential based on good growth in intestinal mucus, resistance to diluted bile juice (2–20%), safe for the target fish, and production of bacteriocin.

Bacillus amyloliquefaciens. This bacterium has been isolated in studies of eight finfish species, and culture-based methods were used in seven of them (Table 5.1). Most strains featured probiotic potential, enzyme production, antagonism, and bacteriocin production. More recently, this bacilli species has been isolated from the intestines of southern flounder (*Paralichthys lethostigma*) (Chen et al. 2016a, 2016b), rohu (*Labeo rohita*) (Mukherjee et al. 2017), striped catfish (*Pangasianodon hypophthalmus*) (Thy et al. 2017), and Indian major carp (Kavitha et al. 2018) using culture-based methods, and these studies revealed promising features, produced different exo-enzymes (digestive and anti-nutritional), grows in mucus, and inhibited *in vitro* growth of several pathogens.

Bacillus altitudinis*, *Bacillus aryabhattai*, and *Bacillus atrophaeus. These *Bacillus* species, all autochthonous, have been detected by cultivation in the DI of mrigal (Banerjee et al. 2016), and the proximal intestine (PI) of walking catfish (*Clarias batrachus*) (Dey et al. 2016) and stinging catfish (*Heteropneustes fossilis*) (Khan & Ghosh 2012), respectively. The major feature of these bacilli was that all produce enzymes, and one species, *B. altitudinis*, revealed antibacterial potential. The strain *B. altitudinis* isolated by Dey et al. (2016) was later used as probiotic supplement in a study with walking catfish (Dey et al. 2018). In a study that characterized gut bacteria from amur minnow (*Rhynchocypris lagowskii*), Elsadek et al. (2023) isolated *Bacillus aryabhattai* showing probiotic potential.

Bacillus cereus. This bacterium, capable of producing amylase, cellulase and protease has been isolated from the PI and DI of rohu (Ghosh et al. 2002) and mrigal by Ray et al. (2010). Askarian et al. (2012) also revealed *B. cereus* to be part of the bacterial community of the DI of Atlantic salmon (*Salmo salar*) and the strain displayed protease-, cellulase, lipase and chitinase activity, and *in vitro* growth inhibition of *A. salmonicida*. In a later study evaluating the microbial diversity in the gut of wild and farmed olive flounder (*Paralichthys olivaceus*), Kim and Kim (2013) revealed culturable autochthonous *B. cereus* in the DI of wild fish, but no information was presented regarding enzyme producing and antagonistic features. Dey et al. (2016) documented exo-enzyme producing *B. cereus* (KR809412) from the DI of the walking catfish (*Clarias batrachus*). In a study using culture-based method, Kavitha et al. (2018) recovered *B. cereus* FC3 from the intestine of Indian major carp, a strain revealing probiotic features.

Bacillus circulans. Studies using cultivation have displayed *B. circulans* in the GI tract of rainbow trout (*Oncorhynchus mykiss*) fed different antibiotics (Austin & Al-Zahrani 1988), and in the intestinal microbiota of rohu (Ghosh et al. 2002), common carp (*Cyprinus carpio*) (Ray et al. 2007), and Mossambicus tilapia (*Oreochromis mossambicus*) (Saha et al. 2006) (Table 5.1). In the studies with rohu, common carp, and tilapia, the bacterial strains revealed enzyme-producing activities.

Bacillus clausii. Ma et al. (2010) recovered this *Bacillus* species from the gut of orange-spotted grouper (*Epinephelus coioides*) and demonstrated a remarkable tolerance to the GI environment and antagonistic activities to some potential pathogenic bacteria.

Bacillus coagulans. Ray et al. (2010) isolated amylase-, cellulase-, and protease-producing autochthonous bacteria in the PI and DI of three species of Indian major carps, catla, mrigal, and rohu, by cultivation and one of the strains from the PI of catla closely matched *B. coagulans*.

Bacillus flexus. To our knowledge, only two studies have revealed this bacterial species in the GI tract of mrigal (Banerjee et al. 2015b) and walking catfish (Dey et al. 2016). Banerjee et al. (2015b) investigated the optimal conditions for chitinase production by strain CMF2 in mrigal by varying fermentation parameters, pH, incubation temperature, incubation time, carbon sources, colloidal chitin, etc. Dey et al. (2016) investigated the extracellular activities, amylase, protease, cellulase, and lipase of several autochthonous strains isolated from the PI and DI of walking catfish and showed that strain FG43 had a high similarity to *B. flexus*.

Bacillus licheniformis. *B. licheniformis* has been isolated from several species of cold water and warm water finfish species with various degrees of pathogen inhibition and enzyme production (Table 5.1). Ghosh et al. (2010) and Mondal et al. (2010) isolated this *Bacillus* species from the PI of rohu and bata (*Labeo bata*) by culture-based methods. In an investigation by Askarian et al. (2012), *B. licheniformis* was identified in the PI and DI of Atlantic salmon by culture method, but the strain did not display promising exo-enzyme activities or *in vitro* growth inhibition toward the four pathogens tested. Later, Banerjee et al. (2013) revealed that *B. licheniformis* was a member of the autochthonous enzyme-producing bacteria isolated from PI and DI of two species of Indian air-breathing fish, murrel (*Channa punctatus*), and stinging catfish using conventional culture techniques. Kim and Kim (2013) displayed culturable autochthonous exo-enzyme-producing *B. licheniformis* in the DI of farmed olive flounder. A strain of *B. licheniformis* was isolated from the mid-intestine (MI) of long whiskers catfish (*Mystus gulio*) by Das et al. (2014) and displayed exo-enzyme activities. Three studies isolated *B. licheniformis* from the gut of mrigal (Mukherjee et al. 2016), intestine of rohu (Mukherjee et al. 2017), and the PI of Nile tilapia (*Oreochromis niloticus*) (Ghosh et al. 2017) and revealed pathogen inhibition and potential probiotic characteristics.

Bacillus megaterium. Strains of *B. megaterium* with some enzyme-producing and antagonistic features in the intestine of finfish have been reported by Austin and Al-Zahrani (1988) in rainbow trout, Green et al. (2013) in Atlantic salmon, and Kim and Kim (2013) in the DI of wild olive flounder, but these authors only focused on gut microbiota evaluations and did not report any features of the bacteria. In contrast, Saha et al. (2006) and Banerjee et al. (2016) demonstrated exo-enzyme-producing and antagonistic *B. megaterium* in the intestine of Mozambique tilapia and in PI of grass carp (*Ctenopharyngodon idella*).

Bacillus methylotrophicus. In a study by Kim and Kim (2013), the authors revealed that *B. methylotrophicus* was a part of the microbial diversity in the DI of wild and farmed olive flounder. *B. methylotrophicus* was also isolated from the entire intestine (EI) of channel catfish (*Ictalurus punctatus*) (Ran et al. 2012) and from the PI of catla (Mukherjee & Ghosh 2016). Later, this *Bacillus* was isolated as part of the autochthonous gut microbiota of rohu intestine with various antagonistic effects toward pathogenic *Aeromonas* spp. More recently, Mukherjee et al. (2017) and Ramirez-Torrez et al. (2018) isolated *B. methylotrophicus* with potential probiotic features from the PI of rohu and the EI content of rainbow trout, respectively.

Bacillus mojavensis. Elsadek et al. (2023) isolated *B. mojavensis* from intestine of amur minnow displaying probiotic potential and anti-pathogenic potential against fish pathogens.

Bacillus nealsonii. To our knowledge, only one study has revealed *B. nealsonii* in the intestine of finfish, rainbow trout, in a study that investigated commonly used antibiotics against non-pathogenic and important fish pathogens (Yilmaz et al. 2018).

Bacillus pacificus. Recently, the whole genomic sequence of *B. pacificus* isolated from rohu was published (Paul et al. 2023). The strain was originally isolated from rohu intestine published in an MS thesis by Khan (2019).

Bacillus pumilus. The first study, showing *B. pumilus* in the intestine of fish, rohu, was carried out by Ghosh et al. (2002). The isolate produced extracellular protease, amylase, and cellulase, and the authors suggested that the bacterial strain might play an important role in rohu fingerlings nutrition. Ma et al. (2010) recovered *B. pumilus* from the gut of orange-spotted grouper (*Epinephelus coioides*). An allochthonous *B. pumilus* strain with probiotic potential was isolated from the EI of channel catfish by Ran et al. (2012). The autochthonous gut microbiota of cultured brown trout (*Salmo trutta*) was studied by Al-Hisnawi et al. (2015) and *B. pumilus* was identified as a member of the identified bacterial community in the DI. In addition, autochthonous *B. pumilus* with a chitinase-producing character was recovered from the DI of silver carp (*Hypophthalmichthys molitrix*) by Banerjee et al. (2015b). In their review devoted to modulation of gut microbiota by dietary manipulations, Ringø et al. (2016) reported autochthonous *B. pumilus* in the DI of rainbow trout fed linseed oil. The bacterium was not, however, detected in the DI of fish fed sunflower-, rapeseed-, or marine oil. Strains of enzyme-producing *B. pumilus* showing an antagonist characteristic were isolated from the PI of mrigal and rohu (Banerjee et al. 2016). Thy et al. (2017) isolated *B. pumilus* from the intestine of striped catfish and used the strain in a probiotic study and concluded that a mixed supplementation with *B. amyloliquefaciens* can be used as supplement to striped catfish. Recently, Ramirez-Torrez et al. (2018) successfully recovered an isolate of *B. pumilus* with good adhesion features from the EI of rainbow trout.

Bacillus sonorensis. Strains of autochthonous enzyme-producing *B. sonorensis* with antagonistic feature were recovered from the DI of mrigal by Dutta and Ghosh (2015), and based on their results, the authors suggested that *in vivo* studies were merited to clarify the effect of these strains on growth performance and health of the fish. In another study by Dutta et al. (2015), they isolated *B. sonorensis* CCHIPh, autochthonous from DI of catla. Strain CCHIPh revealed high extracellular enzyme activities of amylase, protease, lipase, cellulase, phytase, and xylanase; tolerating high temperatures; tolerance against bile juice; could utilize several carbon sources; and grows well in catla mucus but was potent to grow in gut mucus than skin mucus. Furthermore, strain CCHIPh displayed antagonistic activity against *A. hydrophila*, *A. salmonicida*, and *Aeromonas veronii*.

Bacillus stratosphericus. Two culture-based studies revealed cellulase activity by a strain of *B. stratosphericus* isolated from the MI of Piau-com-pinta (*Leporinus friderici*) (Peixoto et al. 2011). Mukherjee et al. (2016) demonstrated the probiotic potential of an autochthonous *B. stratosphericus* strain isolated from the PI of Indian major carp.

Bacillus subtilis. This species has been identified in the GI tract of several finfish species (Table 5.1). In an early study, Newaj-Fyzul (2007) reported that a *B. subtilis* AB1 strain isolated from the digestive tract of rainbow trout was an effective supplement at controlling infection by a fish pathogenic *Aeromonas* sp. ABE1. Rajan and Revathi (2011) isolated *B. subtilis* from the intestine of the ornamental fish platy (*Xiphophorus maculatus*), and the bacteria revealed probiotic properties. Askarian et al. (2012) reported *B. subtilis* as autochthonous in the PI and DI of Atlantic salmon, and the strain displayed high lipase and chitinase activity and *in vitro* growth inhibition of *A. salmonicida*. Later studies were reported by Green et al. (2013). He et al. (2013) investigating the microbial communities in the intestinal content of black carp (*Mylopharyngodon piceus*), gibel carp (*Carassius gibelio*), and bluntnose black bream (*Megalobrama amblycephala*) revealed *B. subtilis*. A *B. subtilis* strain with a remarkable antimicrobial activity against *Edwardsiella ictaluri* was isolated from the channel catfish intestine (Ran et al. 2012). Banerjee et al. (2013) recovered *B. subtilis* from the PI and DI of Indian air-breathing fish, murrel, and stinging catfish using culture-base method. Furthermore, based on culture methods, *B. subtilis* has been recovered from the DI content of cyprinid species (Li et al. 2015a), the intestines of southern flounder (Chen et al. 2016a, 2016b), and the PI and DI of Nile tilapia (Talukdar et al. 2016) and Indian major carp (Kavitha et al. 2018). Das and Ghosh (2013) isolated autochthonous *B. subtilis* subsp. *subtilis* with high phytase activity from the MI of mrigal (Das & Ghosh 2013); this isolate was later used in a solid-state fermentation study (Das & Ghosh 2015). A potential probiotic, autochthonous *B. subtilis* subsp. *spizizenii* was isolated from the DI of Indian major carp (Mukherjee & Ghosh 2016), and the authors suggested the bacterium to be a bio-control agent, although further *in vivo* studies are required. A later study by Dutta et al. (2018) reported isolation of autochthonous *B. subtilis* subsp. *spizizenii* from DI of rohu and the isolate revealed probiotic features as it grew well in fish mucus, tolerated diluted bile juice, showed high enzymatic activities, and showed inhibition of *in vitro* growth of *A. hydrophila*, *Pseudomonas putida*, and *Pseudomonas fluorescens*. In a recent study, *B. subtilis* S17 isolated from Sardine (*Sardina pilchardus*) intestine exhibited a strong antibacterial activity against Vibriosis (*Vibrio harveyi* and *Vibrio anguillarum*) and was concluded as a potential probiotic candidate for aquaculture (Jlidi et al. 2022).

Bacillus tequilensis. This species is seldom isolated from finfish, as only one recent study has revealed its presence and then in PI of silver carp (Banerjee et al. 2016). The isolate, HMF6X, showed high cellulase and xylanase activities *vs*. the other strains isolated from the fish. In addition, strain HMF6X showed antagonistic activity toward *A. salmonicida*. Dutta et al. (2018) reported isolation of autochthonous *B. tequilensis* from PI of rohu and as the isolate showed putative probiotic features, grew well in fish mucus, tolerated diluted bile juice, showed high enzymatic activities, and showed inhibition of *in vitro* growth of *A. hydrophila*, *A. salmonicida*, and *A. veronii*, the authors suggested the isolate as a promising probiotic in aquaculture. In a later study, probiotic potential of this strain (*Bacillus tequilensis* KF623287) was demonstrated as both dietary supplement and water additive, and improved growth, nutrient utilization, as well as immunity were recorded in rohu fingerlings (Dutta & Ghosh 2021).

Bacillus thermoamylovorans. To our knowledge, *B. thermoamylovorans* has only been detected in one finfish study, by Nyman et al. (2017), investigating the bacterial community in the intestinal content of Arctic charr (*Salvelinus alpinus*). No further evaluation of the extracellular enzymes production and/or probiotic potential was, however, carried out; these topics merit further investigations.

Bacillus thuringiensis. In a study evaluating the culturable autochthonous gut bacteria in Atlantic salmon fed diets with or without chitin, Askarian et al. (2012) reported that *B. thuringiensis* isolated from the DI produced enzymes and inhibited *in vitro* growth of four fish pathogens tested. Later, Yilmaz et al. (2018) displayed *B. thuringiensis* in the intestine of rainbow trout. Another strain of *B. thuringiensis* SS4 isolated from the gut of freshwater fish (*Systomus sarana*) exhibited high protease production potential, recommending its prospective applications in food and pharmaceutical industries (Dhayalan et al. 2022).

Bacillus velezensis. To the best of our knowledge, this bacterium is only reported in one fish study, the intestine of amur minnow, and the strain showed probiotic potential and anti-pathogenic potential against fish pathogens (Elsadek et al. 2023).

Poly-β-hydroxybutyrate production by bacilli isolated from the digestive tract of various fish species. In the study of Kaynar and Beyatli (2009), the authors isolated poly-β-hydroxybutyrate production by 30 strains of *Bacillus* from the intestines of various fish species and the strains were identified as *B. pasteurii, B. badius, B. circulans, B. licheniformis, B. megaterium, B. thuringiensis, B. brevis, B. cereus, B. sphaericus, B. subtilis, B. coagulans, B. lentus*, and *B. pumilus*. Further information of the fish species investigated in the study is presented in the paper.

Solibacillus silvestris. Rheims et al. (1999) first isolated a bacterium they named *Bacillus silvestris* from forest soil, but Krishnamurthi et al. (2009) re-classified the bacterium as *Solibacillus silvestris*. To our knowledge, only one study has reported *S. silvestris* in the GI tract of fish, Indian major carp (Mukherjee et al. 2016).

***Brevibacillus parabrevis*.** Different strains of *Brevibacillus parabrevis* with diverse exo-enzyme-producing ability were recorded from the GI tracts of two brackish water fish species, *Scatophagus argus* and *Terapon jarbua*.

5.3 *BACILLUS* IN SHELLFISH

In a recent review, Soto (2021), evaluated the potential of bacilli as probiotics, effects on feed intake, modulation of the gut microbiota, and pathogens control in shrimp. As the GI tract is an important organ in probiotic studies, it is worth mentioning that the GI tract of shellfish consists of three main segments: foregut (stomach), midgut, and hindgut, and during the last 20 years, several studies have been published on the gut microbiota of shellfish (e.g., Daniels et al. 2010; Zhu et al. 2016; Li et al. 2018; Mongkol et al. 2018; Diwan et al. 2023). Less information is, however, available on bacilli in the GI tract of shellfish compared to that reported for finfish (Table 5.1). In shellfish studies, *B. cereus* is isolated from tiger shrimp (*Penaeus monodon*) and the bacilli revealed growth and immune enhancer features (Chandran et al. 2014). A study by Vargas-Albores et al. (2017) evaluating the gut microbiota of Pacific white shrimp (*Litopenaeus vannamei*) by a culture-independent method revealed *B. cereus*, *B. thuringiensis*, and *Bacillus weihenstephanensis* (Vargas-Albores et al. 2017).

BIBLIOGRAPHY

Al-Hisnawi A, Ringø E, Davies SJ, Waines P, Bradley G, Merrifield DL (2015) First report on the autochthonous gut microbiota of brown trout (*Salmo trutta*). *Aquaculture Research* **46**, pp.2962–2971.

Al-Harbi AH, Uddin MN (2005) Microbiological quality changes in the intestine of hybrid tilapia (*Oreochromis niloticus* x *Oreochromis aureus*) in fresh and frozen storage condition. *Letters in Applied Microbiology* **40**, pp.486–490.

An J, Zhu W, Ying L, Zhang X, Sun L, Hong P, et al. (2015) Purification and characterization of a novel bacteriocin CAMT2 produced by *Bacillus amyloliquefaciens* isolated from marine fish *Epinephelus areolatus*. *Food Control* **51**, pp.278–282.

Askarian F, Zhou Z, Olsen RE, Sperstad S, Ringø E (2012) Culturable autochthonous gut bacteria in Atlantic salmon (*Salmo salar* L.) fed diets with or without chitin. Characterization by 16S rRNA gene sequencing, ability to produce enzymes and *in vitro* growth inhibition of for fish pathogens. *Aquaculture* **326–329**, pp.1–8.

Austin B, Al-Zahrani AMJ (1988) The effect of antimicrobial compounds on the gastrointestinal microflora of rainbow trout, *Salmo gairdneri* Richardson. *Fish Biology* **33**, pp.1–14.

Bairagi A, Sarkar Ghosh K, Sen SK, Ray AK (2002) Duck weed (*Lemna polyrhiza*) leaf meal as a source of feedstuff in formulated diets for rohu (*Labeo rohita* Ham.) fingerlings after fermentation with a fish intestinal bacterium. *Bioresource Technology* **85**, pp.17–24.

Banerjee G, Dan SK, Nandi A, Ghosh P, Ray AK (2015a) Autochthonous gut bacteria in two Indian air-breathing fish, climbing perch (*Anabas testudineus*) and walking catfish (*Clarias batrachus*): Mode of association, identification and enzyme producing ability. *Polish Journal of Microbiology* **64**(4): pp.361–368.

Banerjee G, Ray AK, Askarian F, Ringø E (2013) Characterisation and identification of enzyme-producing autochthonous bacteria from the digestive tract of two Indian air-breathing fish. *Beneficial Microbes* **4**, pp.277–284.

Banerjee G, Nandi A, Ray AK (2017) Assessment of hemolytic activity, enzyme production and bacteriocin characterization of *Bacillus subtilis* LR1 isolated from the gastrointestinal tract of fish. *Archives of Microbiology* **199**, pp.115–124.

Banerjee S, Mukherjee A, Dutta D, Ghosh K (2015b) Evaluation of chitinolytic gut microbiota in some carps and optimization of culture conditions for chitinase production by the selected bacteria. *Journal of Microbiology, Biotechnology and Food Sciences* **5**,pp.12–19.

Banerjee S, Ghosh K (2016) Bio-processing of linseed oil-cake through solid state fermentation by non-starch polysaccharide degrading fish gut bacteria. *Fermentation Technology* **5**, pp.1–10.

Banerjee S, Mukherjee A, Dutta D, Ghosh K (2016) Non-starch polysaccharide degrading gut bacteria in Indian major carps and exotic carps. *Jordan Journal of Biological Sciences* **9**(1): pp.69–78.

Barreto-Curiel, F., Ramirez-Puebla, S. T., Ringø, E., Escobar-Zepeda, A., Godoy-Lozano, E. E., and Vazquez-Duhalt, R. (2018) Effects of extruded aquafeed on growth performance and gut microbiome of juvenile *Totoaba macdonaldi*. *Animal Feed Science and Technology*, **245**, pp.91–103. doi: 10.1016/j.anifeedsci.2018.09.002

Bindiya ES, Tina KJ, Raghul SS, Bhat SG (2015) Characterization of deed sea fish gut bacteria with antagonistic potential, from *Centroscyllium fabricii* (deep sea shark). *Probiotics and Antimicrobial Proteins* **7**, pp.157–163.

Castro C, Couto A, Diógenes AF, Corraze G, Panserat S, Serra CR, et al. (2011) Vegetable oil and carbohydrate-rich diets marginally affected intestine histology, digestive enzymes activities, and gut microbiota of gilthead sea bream juveniles. *Fish Physiology and Biochemistry* doi:10.1007/s10695-018-0559-9

Chandran MN, Iyapparaj P, Moovendhan Ramasubburayan R, Prakash S, Immanuel G, et al. (2014) Influence of probiotic bacterium *Bacillus cereus* isolated from the gut of wild shrimp *Penaeus monodon* in turn as a potent growth promoter and immune enhancer in *P. monodon*. *Fish & Shellfish Immunology* **36**, pp.38–45.

Chávez-Crooker P, Obreque-Contreras J (2010) Bioremediation of aquaculture wastes. *Current Opinion in Biotechnology* **21**, pp.313–317.

Chen Y, Li J, Xiao P, Zhu W, Mo ZL (2016a) The ability of marine *Bacillus* spp. isolated from fish gastrointestinal tract and culture pond sediment to inhibit growth of aquatic pathogenic bacteria. *Iranian Journal of Fisheries Sciences* **15**, pp.701–714.

Chen Y, Li J, Xiao P, Li GY, Yue S, Huang J, et al. (2016b) Isolation and characterization of *Bacillus* spp. M001 for potential application in turbot (*Scophthalmus maximus* L) against *Vibrio anguillarum*. *Aquaculture Nutrition* **22**, pp.374–381.

Daniels CL, Merrifield DL, Boothroyd DP, Davies SJ, Factor JP, Arnold KE (2010) Effect of dietary Bacillus spp. and mannan oligosaccharides (MOS) on European lobster (Homarus gammarus L.) larvae growth performance, gut morphology and gut microbiota. *Aquaculture*, **304**, pp.49–57.

Das P, Ghosh K (2013) Evaluation of phytase-producing ability by a fish gut bacterium, *Bacillus subtilis* subsp. *subtilis*. *Journal of Biological Sciences* **13**, pp.691–700.

Das P, Ghosh K (2015) Improvement of nutritive value of sesame oil cake in formulated diets for rohu, *Labeo rohita* (Hamilton) after bio-processing through solid state fermentation by a phytase-producing fish gut bacterium. *International Journal of Aquatic Biology* **3**, pp.89–101.

Das P, Mandal S, Khan A, Manna SK, Ghosh K (2014) Distribution of extracellular enzyme-producing bacteria in the digestive tracts of 4 brackish water fish species. *Turkish Journal of Zoology* **38**, pp.79–88.

Del'Duca A, Cesar DE, Abreu PC (2015b) Bacterial community of pond's water, sediment and in the guts of tilapia (*Oreochromis niloticus*) juveniles characterized by fluorescent *in situ* hybridization technique. *Aquaculture Research* **46**, pp.707–715.

Del'Duca A, Cesar DE, Diniz CG, Abreu PC (2015a) Evaluation of the presence and efficiency of potential probiotic bacteria in the gut of tilapia (*Oreochromis niloticus*) using the fluorescent *in situ* hybridization technique. *Aquaculture* **388–391**, pp.115–121.

De Paula Silva FC, Nicoli JR, Zambonino-Infante JL, Kaushik S, Gatesoupe F-J (2011). Influence of the diet on the microbial diversity of faecal and gastrointestinal contents of gilthead sea bream (*Sparus aurata*) and intestinal contents in goldfish (*Carassius auratus*). *FEMS Microbiology and Ecology* **78**, pp.285–296.

Deschamps AM, Mahoudeau G, Lebeault JM (1980) Fast degradation of kraft lignin by bacteria. *European Journal of Applied Microbiology and Biotechnology*, **9**, pp.45–51.

Dey A, Ghosh K, Hazra N (2016) Evaluation of extracellular enzyme-producing autochthonous gut bacteria in walking catfish, *Clarias batrachus* (L.). *Journal of Fisheries* **4**, pp.345–352.

Dey A, Ghosh K, Hazra N (2018) Effects of probiotics-encapsulated live feed on growth and survival of juvenile *Clarias batrachus* (Linnaeus, 1758) after differential exposure to pathogenic bacteria. *SAARC Journal of Agriculture* **16**, pp.105–113.

Dhayalan A, Velramar B, Govindasamy B, Ramalingam KR, Dilipkumar A, Pachiappan P (2022) Isolation a bacterial strain from the gut of the fish, *Systomus sarana*, identification of the isolated strain, optimized production of its protease, the enzyme purification, and partial structural characterization. *Journal of Genetic Engineering and Biotechnology* **20**, p.24.

Diwan AD, Harke SN, Panche AN (2023) Host-microbe interaction in fish and shellfish: An overview. *Fish and Shellfish Immunology Reports* **4**, p.100091.

Dutta D, Banerjee S, Mukherjee A, Ghosh K (2015) Selection and probiotic characterization of exoenzyme-producing bacteria isolated from the gut of *Catla catla* (Actinopterygii: Cypriniformes: Cyprinidae). *Acta Ichthyologica et Piscatoria* **45**, pp.373–384.

Dutta D, Banerjee S, Mukherjee A, Ghosh K (2018) Potential gut adherent probiotic bacteria isolated from rohu, *Labeo rohita* (Actinopterygii: Cypriniformes, Cyprinidae): Characterisation, exo-enzymes production, pathogen inhibition, cell surface hydrophobicity, and bio-film formation. *Acta Ichthyologica et Piscatoria* **48**, pp.221–233.

Dutta D, Ghosh K (2015) Screening of extracellular enzyme-producing and pathogen inhibitory gut bacteria as putative probiotics in mrigal, *Cirrhinus mrigala* (Hamilton, 1822). *International Journal of Fisheries and Aquatic Studies* **2**, pp.310–318.

Dutta D, Ghosh K (2021) Improvement of growth, nutrient utilization and haemato-immunological parameters in rohu, *Labeo rohita* (Hamilton) using *Bacillus tequilensis* (KF623287) Through diets or as water additive. *Aquaculture Nutrition* **27**, pp.29–47.

Ebnetorab SMA, Ahari H, Kokoolaki S (2020) Isolation, biochemical and molecular detection of *Bacillus subtilis* and *Bacillus licheniformis* from the digestive system of rainbow trout (*Oncorhynchus mykiss*) and its inhibitory effect on *Aeromonas hydrophila*. *Iranian Journal of Fisheries Sciences* **19**, pp.2824–2845.

Elsadek M, Wang S, Wu Z, Wang J, Wang X, Zhang Y, et al. (2023) Characterisation of *Bacillus* spp. isolated from the intestine of *Rhynchocypris lagowskii* as a potential probiotic and their effects on fish pathogens. *Microbial Pathogenesis* **180**, p.106163l.

Esakkiraj P, Immanuel G, Sowmya SM, Iyapparaj P, Palavesam A (2009) Evaluation of protease-producing ability of fish gut isolate *Bacillus cereus* for aqua feed. *Food and Bioprocess Technology* **2**, pp.383–390.

Gao S, Pan L, Huang F, Song M, Tian C, Zhang M (2019) Metagenomic insights into the structure and function of intestinal microbiota of the farmed Pacific white shrimp (*Litopenaeus vannamei*). *Aquaculture* **499**, pp.109–118.

Gatesoupe F-J, Huelvan C, Le Bayon N, Le Delliou H, Madec L, Mouchel O, et al. (2016) The highly variable microbiota associated to intestinal mucosa correlates with growth and hypoxia resistance of sea bass, *Dicentrarchus labrax*, submitted to different nutritional histories. *BMC Microbiology* **16**, p.266.

Geraylou Z, Vanhove MPM, Souffreau C, Rurangwa E, Buyse J, Ollevier F (2014) *In vitro* selection and characterization of putative probiotics isolated from the gut of *Acipenser baerii* (Brandt, 1869). *Aquaculture Research* **45**, pp.341–352.

Ghanei-Motlagh R, Mohammadian T, Gharbi D, Khosravi M, Mahmoudi E, Zarea M, et al. (2021) Quorum quenching probiotics modulated digestive enzymes activity, growth performance, gut microflora, haemato-biochemical parameters and resistance against *Vibrio harveyi* in Asian seabass (*Lates calcarifer*). *Aquaculture* **531**, p.735874.

Ghosh K, Banerjee S, Moon UM, Khan HA, Dutta D (2017) Evaluation of gut associated extracellular enzyme-producing and pathogen inhibitory microbial community as potential probiotics in Nile tilapia, *Oreochromis niloticus*. *International Journal of Aquaculture* **7**, pp.143–158.

Ghosh K, Roy M, Kar N, Ringø E (2010) Gastrointestinal bacteria in rohu, *Labeo rohita* (Actinopterygii: Cypriniformes: Cyprinidae): Scanning electron microscopy and bacteriological study. *Acta Ichthyologica et Piscatoria* **40**, pp.129–135.

Ghosh K, Sen SK, Ray AK (2002) Characterization of bacilli isolated from gut of rohu, *Labeo rohita*, fingerlings and its significance in digestion. *Journal of Applied Aquaculture* **12**, pp.33–42.

Ghosh K, Sen SK, Ray AK (2004) Growth and survival of rohu, *Labeo rohita* (Hamilton, 1822) spawn fed diets fermented with intestinal bacterium, *Bacillus circulans*. *Acta Ichthyologica Et Piscatoria* **34** (2): pp.155–165

Giri SS, Sukumaran V, Sen SS, Vinumonia J, Nazeema-Banu B, Jena PK (2011) Antagonistic activity of cellular components of potential probiotic bacteria, isolated from the gut of *Labeo rohita*, against *Aeromonas hydrophila*. *Probiotics and Antimicrobial Proteins* **3**, pp.214–222.

Gibbons NE The slime and intestinal flora of some marine fishes. *Contrib. Can. Biol. Fish.*, **8**, pp.275–290 (1933).

Goncalves AT, Gallardo-Escárate C (2017) Microbiome dynamic modulation through functional diets based on pre- and probiotics (mannan-oligosaccharides and *Saccharomyces cerevisiae*) in juvenile rainbow trout (*Oncorhynchus mykiss*). *Journal of Applied Microbiology* **122**, pp.1333–1347.

Green TJ, Smullen R, Barnes AC (2013) Dietary soybean protein concentrate-induced intestinal disorder in marine farmed Atlantic salmon, *Salmo salar* is associated with alterations in gut microbiota. *Veterinary Microbiology* **166**, pp.286–292.

He S, Wu Z, Liu Y, Wu N, Tao Y, Xu L, et al. (2013) Effects of dietary 60g kg^{-1} dried distiller's grains in least-cost practical diets on production and gut allochthonous bacterial composition of cage-cultured fish: Comparison among fish species with different natural food habits. *Aquaculture Nutrition* **19**, pp.765–772.

Hindu SV, Chandrasekaran N, Mukherjee A, Thomas J (2018) Effect of dietary supplementation of novel probiotic bacteria *Bacillus vireti* 01 on antioxidant defence system of freshwater prawn challenge with *Pseudomonas aeruginosa. Probiotics & Antimicrobial Proteins* **10**, pp.356–366.

Hovda BF, Fontanillas R, McGurk C, Obach A, Rosnes JT (2012) Seasonal variations in the intestinal microbiota of farmed Atlantic salmon (*Salmo salar* L.). *Aquaculture Research* **43**, pp.154–159.

Hovda BF, Lundestad BT, Fontanillas R, Rosnes JT (2007) Molecular characterisation of the intestinal microbiota of farmed Atlantic salmon (*Salmo salar* L.). *Aquaculture* **272**, pp.581–588.

Izvekova GI, Izvekov EI, Plotnikov AO (2007) Symbiotic microflora in fishes of different ecological groups. *Biology Bulletin* **34**, pp.610–618.

James G, Das BC, Jose S, Kumar R (2021) *Bacillus* as aquaculture friendly microbe. *Aquaculture International* **29**, pp.323–353.

Jasmin MY, Wagaman H, Yin TA, Ina-Salwany MY, Daud HM, Karim M (2016) Screening and evaluation of local bacteria isolated from shellfish as potential probiotics against pathogenic Vibrios. *Journal of Environmental Biology* **37**, pp.801–809.

Jlidi M, Akremi I, Ibrahim AH, Barbra W, Ben Ali M, Ben Ali M (2022) Probiotic properties of *Bacillus* strains isolated from the gastrointestinal tract against pathogenic Vibrios. *Frontiers of Marine Science* **9**, p.884244.

Jung-Schroers V, Adamek M, Jung A, Harris S, Doza O-S, Baumer A, et al. (2016) Feeding of β-1,3/1,6-glucan increases the diversity of the intestinal microflora of carp (*Cyprinus carpio*). *Aquaculture Nutrition* **22**, pp.1026–1039.

Kanwal S, Abbas K, Ahmed T, Abdullah S, Naz H, Zia MA, et al. (2020) Description of isolated bacterial pathogens from diseased *Cirrhinus mrigala. Aquaculture Research* **52**, pp.2130–2137.

Kavitha M, Raja M, Perumal P (2018) Evaluation of probiotic potential of *Bacillus* spp. isolated from the digestive tract of freshwater fish *Labeo calbasu* (Hamilton, 1822). *Aquaculture Reports* **11**, pp.59–69.

Kaynar P, Beyatli Y (2009) Determination of poly-β-hydroxybutyrate production by *Bacillus* spp. isolated from the intestines of various fish species. *Fishery Science* **75**, pp.439–443.

Khan SU (2019) Identification of native probiotic bacteria to enhance growth and prevent disease in *Labeo rohita.* MS thesis. Bangabandhu Sheikh Mujibur Rahman Agricultural University, Gazipur, Bangladesh.

Khan A, Ghosh K (2012) Characterization and identification of gut-associated phytase-producing bacteria in some freshwater fish cultured in ponds. *Acta Ichthyologica et Piscatoria* **42**, pp.37–45.

Khan A, Ghosh K (2013) Evaluation of phytase production by fish gut bacterium, *Bacillus subtilis* for processing of *Ipomea aquatica* leaves as probable aquafeed ingredient. *Journal of Aquatic Food Product Technology* **22**, pp.508–519.

Kim D-H, Kim D-Y (2013) Microbial diversity in the intestine of olive flounder (*Paralichthys olivaceus*). *Aquaculture* **414–415**, pp.103–108.

Kim D-H, Brunt J, Austin B (2007) Microbial diversity of intestinal contents and mucus in rainbow trout (*Oncorhynchus mykiss*). *Journal of Applied Microbiology* **102**, pp.1654–1664.

Kong Y, Li M, Chu G, Liu H, Shan X, Wang G, et al. (2021) The positive effects of single and conjoint administration of lactic acid bacteria on *Channa argus*: Digestive enzyme activity, antioxidant capacity, intestinal microbiota and morphology. *Aquaculture* **531**, p.735852.

Krishnamurthi, S., Chakrabarti, T., and Stackebrandt, E. (1999) Re-examination of the taxonomic position of Bacillus silvestris Rheims et al. 1999 and proposal to transfer it to Solibacillus gen. nov. as Solibacillus silvestris comb. *Nov. International Journal of Systematic and Evolutionary Microbiology*, **59**, pp.1054–1058.

Kuebutornye FKA, Abarike ED, Lu Y (2019) A review on the application of *Bacillus* as probiotics in aquaculture. *Fish & Shellfish Immunology* **87**, pp.820–828.

Kuebutornye FKA, Lu Y, Abarike ED, Wang Z, Li Y, Sakyi ME (2020b) *In vitro* assessment of probiotic characteristics of three *Bacillus* species from the gut of Nile tilapia, *Oreochromis niloticus. Probiotics and Antimicrobial Proteins* **12**, pp.412–424.

Kuebutornye FKA, Wang Z, Lu Y, Abarike ED, Sakyi ME, Li Y, et al. (2020a) Effects of three host-associated *Bacillus* species on mucosal immunity and gut health of Nile tilapia, Oreochromis niloticus and its resistance against *Aeromonas hydrophila* infection. *Fish & Shellfish Immunology* **97**, pp.83–95.

Kumar P, Jain KK, Sardar P (2018) Effects of dietary synbiotic on innate immunity, antioxidant activity and disease resistance of *Cirrhinus mrigala* juveniles. *Fish & Shellfish Immunology* **80**, pp.124–132.

Li, E., Xu, C., Wang, X., Wang, S., Zhao, Q., and Zhang, M. (2018) Gut microbiota and its modulation for healthy farming of Pacific white shrimp Litopenaeus vannamei. *Reviews in Fisheries Science & Aquaculture*, **26**, pp.381–399.

Li Y, Zhang T, Zhang C, Zhu Y, Ding J, Ma Y (2015a) Bacterial diversity in the intestine of young farmed puffer fish *Takifugu rubripes. Chinese Journal of Oceanology and Limnology* **33**, pp.913–818.

Li T, Long M, Gatesoupe FJ, Zhang Q, Li A, Gong X (2015b) Comparative analysis of the intestinal bacterial communities in different species of carp by pyrosequencing. *Microbial Ecology* **69**, pp.25–36.
Little SF, Ivins BE (1999) Molecular pathogenesis of *Bacillus anthracis* infection. *Microbial Infection*, **1**(2): pp.131–139.
Liu J, Pang J-J, Tu Z-C, Wang H, Sha X-M, Shao Y-H, et al. (2018) The accumulation, histopathology, and intestinal microorganisms effect of waterborne cadmium on *Carassius auratus gibelio*. *Fish Physiology and Biochemistry*. https://doi.org/10.1007/s10695-018-0557-2
Luis-Villaseñor IE, Campa-Córdova AI, Ascencio-Valle FJ (2012) Probiotics in larvae and juvenile white leg shrimp *Litopenaeus vannamei*, Chapter 27, pp. 601–622. InTech Open Access, https://doi.org/10.5772/50123
Ma RL, Yang HL, Sun YZ, Ye JD, Zhang CX, Lin WY (2010) Biological characteristics of two *Bacillus* strains isolated from the gut of grouper *Epinephelus coioides*. *Fisheries Science* **29**, pp.505–509.
Matias HB, Yusoff FM, Shariff M, Azhar O (2002) Effects of commercial microbial products on water quality in tropical shrimp culture ponds. *Asian Fisheries Science* **15**, pp.239–248.
Mente E, Nikouli E, Antonopoulou E, Martin SAM, Kormas KA (2018) Core versus diet-associated and post-prandial bacterial communities of the rainbow trout (*Oncorhynchus mykiss*) midgut and faeces. *Biology Open* **7**, p.bio034397.
Miao S, Zhao C, Zhu J, Hu J, Dong X, Sun L (2018) Dietary soybean meal affects intestinal homoeostasis by altering the microbiota, morphology and inflammatory cytokine gene expression in northern snakehead. *Scientific Reports* **8**, p.113.
Mongkol P, Bunphimpapha P, Rungrassamee W, Arayamethakorn S, Klinbunga S, Menasveta P, and Chaiyapechara S (2018) Bacterial community composition and distribution in different segments of the gastrointestinal tract of wild-caught adult Penaeus monodon. *Aquaculture Research*, **49**, pp.378–392.
Mondal S, Roy T, Ray AK (2010) Characterization and identification of enzyme-producing bacteria isolated from the digestive tract of bata, *Labeo bata*. *Journal of the World Aquaculture Society* **41**, pp.369–376.
Mukherjee A, Dutta D, Banerjee S, Ringø E, Breines EM, Hareide E, Ghosh K (2016) Potential probiotics from Indian major carp, *Cirrhinus mrigala*. Characterization, pathogen inhibitory activity, partial characterization of bacteriocin and production of exoenzymes. *Research in Veterinary Science* **108**, pp.76–84.
Mukherjee A, Dutta D, Banerjee S, Ringø E, Breines EM, Hareide E, Ghosh K (2017) Culturable autochthonous gut bacteria in rohu, *Labeo rohita*. *In vitro* growth inhibition against pathogenic *Aeromonas* spp., stability in gut, bio-safety and identification by 16S rRNA gene sequencing. *Symbiosis* **73**, pp.165–177.
Mukherjee A, Ghosh K (2016) Antagonism against fish pathogens by cellular components and verification of probiotic properties in autochthonous bacteria isolated from the gut of an Indian major carp, *Catla catla* (Hamilton). *Aquaculture Research* **47**, pp.2243–2255.
Nandi A, Banerjee G, Dan SK, Ghosh P, Ghosh K, Ray AK (2017b) Screening of autochthonous intestinal microbiota as candidate probiotics isolated from four freshwater teleosts. *Current Science* **113**, pp.767–773.
Nandi A, Dan SK, Banerjee G, Ghosh P, Ghosh K, Ringø E, Ray AK (2017a) Probiotic potential of autochthonous bacteria isolated from the gastrointestinal tract of four freshwater teleost. *Probiotics and Antimicrobial Proteins* **9**, pp.12–21.
Nedoluha PC, Westhoff D (1995) Microbial analysis of striped bass (*Morone saxatilis*) grown in flow-through tanks. *Journal of Food Protection* **58**, pp.1363–1368.
Nedoluha PC, Westhoff D (1997) Microbial analysis of striped bass (*Morone saxatilis*) grown in a recirculation system. *Journal of Food Protection* **60**, pp.948–953.
Newaj-Fyzul A, Adesiyun AA, Mutani A, Ramsubhag A, Brunt J, Austin B (2007) *Bacillus subtilis* AB1 controls *Aeromonas* infection in rainbow trout, (*Oncorhynchus mykiss*, Walbaum). *Journal of Applied Microbiology* **103**, pp.1699–1706.
Nyman A, Huyben D, Lundh T, Dicksved J (2017) Effects of microbe- and mussel-based diets on the gut microbiota in Arctic charr (*Salvelinus alpinus*). *Aquaculture Reports* **5**, pp.34–40.
Olsen RE, Sundell K, Ringø E, Myklebust R, Hemre G-I, Hansen T, Karlsen Ø (2008) The acute stress response in fed and food deprived Atlantic cod, *Gadus morhua* L. *Aquaculture* **280**, pp.232–241.
Paiva-Maia E, Alves-Modesto G, Otavio-Brito L, Olivera A, Vasconcelos-Gesteira TC (2013) Effect of a commercial probiotic on bacterial and phytoplankton concentration in intensive shrimp farming (*Litopenaeus vannamei*) recirculation systems. *Latin American Journal of Aquaculture Research*, **41**(1): pp.126–137.
Park Y, Lee S, Hong J, Kim D, Moniruzzaman M, Bai SC (2017) Use of probiotics to enhance growth, stimulate immunology and confer disease resistance to *Aeromonas salmonicida* in rainbow trout (*Oncorhynchus mykiss*). *Aquaculture Research* **48**, pp.2672–2682.

Paul SI, Khan SU, Sarkar MK, Foysal MJ, Salam MA, Rahman MM (2023) Whole-genome sequence of *Bacillus pacificus* CR121, a fish probiotic candidate. *Microbiology Resource Announcements* **12**. https://doi.org/10.1128/mra01206-22

Peixoto SB, Cladera-Olivera F, Daroit DJ, Brandelli A (2011) Cellulase-producing *Bacillus* strains isolated from the intestine of Amazon basin fish. *Aquaculture Research* **42**, pp.887–891.

Rajan MR, Revathi U (2011) Role of probiotics in ornamental fish platy *Xiphophorus maculatus*. *Journal of Pure and Applied Microbiology* **5**, pp.819–823.

Ramachandran S, Bairagi A, Ray AK (2005) Improvement of nutritive value of grass pea (*Lathyrus sativus*) seed meal in the formulated diets for rohu, *Labeo rohita* (Ham.) fingerlings after fermentation with a fish gut bacterium. *Bioresource Technology* **96**, pp.1465–1472.

Ramachandran S, Ray AK (2007) Nutritional evaluation of fermented black gram seed meal for rohu *Labeo rohita* (Hamilton) fingerlings. *Journal of Applied Ichthyology* **23**, pp.74–79.

Ramirez C, Romero J (2017) Fine flounder (*Paralichthys adspersus*) microbiome showed important differences between wild and reared specimens. *Frontiers in Microbiology* **8**, p.271.

Ramirez-Torrez JA, Monroy-Dosta MC, Hernández-Hernández LH, Castro-Mejia, Bustos-Martinez JA, Hamdan-Partida A (2018) Presumptive probiotic isolated from *Oncorhynchus mykiss* (Walbaum, 1792), cultivated in Mexico. *International Journal of Aquatic Sciences* **9**, pp.3–12.

Ran C, Carrias A, Williams MA, Capps N, Dan BCT, et al. (2012) Identification of *Bacillus* strains for biological control of catfish pathogens. *PLoS ONE* **7**, p.e45793.

Ray AK, Bairagi A, Sarkar Ghosh K, Sen SK (2007) Optimization of fermentation conditions for cellulase production by *Bacillus subtilis* CY5 and *Bacillus circulans* TP3 isolated from fish gut. *Acta Ichthyologica et Piscatoria* **37**, pp.47–53.

Ray AK, Roy T, Mondal S, Ringø E (2010) Identification of gut-associated amylase, cellulase and protease-producing bacteria in three species of Indian major carps. *Aquaculture Research* **41**, pp.1462–1469.

Reed, G. B., and Spence, C. M. (1929) The intestinal and slime flora of the haddock: A preliminary report. *Contrib. Can. Biol. Fish* **4**, pp.257–264.

Rheims, H., Fruhling, A., Schumann, P., Rohde, M., and Stackebrandt, E. (1999) Bacillus silvestris sp. nov., a new member of the genus Bacillus that contains lysine in its cell wall. *International Journal of Systematic and Evolutionary Microbiology* **49**, pp.795–802.

Ricaud K, Rey M, Plagnes-Juan E, Larroquet L, Even M, Quillet E, et al. (2018) Composition of intestinal microbiota in two lines of rainbow trout (*Oncorhynchus mykiss*) divergently selected for muscle fat content. *The Open Microbiology Journal* **12**, pp.308–320.

Ringø E, Song, SK (2016) Application of dietary supplements (synbiotics and probiotics in combination with plant products and β-glucans) in aquaculture. *Aquaculture Nutrition* **22**, pp.4–24.

Ringø E, Sperstad S, Myklebust R, Mayhew TM, Olsen RE (2006) The effect of dietary inulin on aerobic bacteria associated with hindgut of Arctic charr (*Salvelinus alpinus* L.). *Aquaculture Research* **37**, pp.891–897.

Ringø E, van Doan H, Lee SH, Soltani M, Hoseinifar SH, Ramasamy H, Song SK (2020) Probiotics, lactic acid bacteria and bacilli: Interesting supplementation for aquaculture. *Journal of Applied Microbiology* **129**, pp.116–136.

Ringø E, Zhou Z, Gonzalez Vecino JL, Wadsworth S, Romero J, Krogdahl Å, Olsen RE, Dimitroglou A, Foey A, Davies S, Owen M, Lauzon HL, Løvmo Martinsen L, De Schryver P, Bossier P, Sperstad S, Merrifield DL (2016) Effects of dietary components on the gut microbiota of aquatic animals: A never-ending story? *Aquaculture Nutrition* **22**, pp.219–282,

Ringø E, Zhou Z, He S, Olsen RE (2014) Effect of stress on intestinal microbiota of Arctic charr, Atlantic salmon, rainbow trout and Atlantic cod: A review. *African Journal of Microbiology Research* **8**, pp.609–618.

Saha S, Roy RN, Sen SK, Ray AK (2006) Characterization of cellulase-producing bacteria from the digestive tract of tilapia, *Oreochromis mossambica* (Peters) and grass carp, *Ctenopharyngodon idella* (Valenciennes). *Aquaculture Research* **37**, pp.380–388.

Santos RA, Oliva-Teles A, Pousao-Ferreira P, Jerusik R, Saavedra MJ, Enes P, et al. (2021) Isolation and characterization of fish-gut *Bacillus* spp. as source of natural antimicrobial compounds to fight aquaculture bacterial diseases. *Marine Biotechnology* **23**, pp.276–293.

Sarkar, B., and Ghosh, K. (2014) Gastrointestinal microbiota in Oreochromis mossambicus (Peters) and Oreochromis niloicus (Linnaeus): Scanning electron microscopy and microbiological study. *International Journal of Fisheries and Aquatic Studies* **2**, pp.78–88.

Shija VM, Amoah K, Cai J (2023) Effects of *Bacillus* probiotics on the immunological responses of Nile tilapia (*Oreochromis niloticus*): A review. *Fishes* **8**, p.366.

Sirtori LR, Cladera-Olivera F, Lorenzini DM, Tsai SM, Brandelli A (2006) Purification and partial characterization of an antimicrobial peptide produced by *Bacillus* sp. strain P45, a bacterium from the Amazon basin fish *Piaractus mesopotamicus*. *Journal of General and Applied Microbiology* **52**, pp.357–363.

Slepecky R, Hemphill E (2006) The genus *Bacillus*. Nonmedical. In: *The Prokaryotes*, Vol. 4 (Dworkin M, Falkow S, Rosenberg E, Schleifer K-H, Stackebrandt E, eds), pp.530–562. Springer, New York, USA.

Soltani M, Ghosh K, Hoseinifar SH, Kumar V, Lymbery A, Roy S, Ringø E (2019) Genus *Bacillus*, promising probiotics in aquaculture: Aquatic animal origin, bio-active components, bioremediation in fish and shellfish. *Reviews in Fisheries Science & Aquaculture* **27**, pp.331–379.

Soto HO (2021) Feed intake improvement, gut microbiota modulation and pathogens control by using *Bacillus* species in shrimp aquaculture. *World Journal of Microbiology and Biotechnology* **37**, p.28.

Strøm E, Olafsen JA (1990) The indigenous microflora of wild-captured juvenile cod in net-pen rearing. In: *Microbiology of Poecilotherms* (Lesel R, ed.), pp.181–184.

Sugita H, Fukumoto M, Koyama H, Deguchi Y (1985) The intestinal microflora of carp *Cyprinus carpio*, grass carp *Ctenopharyngodon idella*, and tilapia *Sarotherodon niloticus*. *Bulletin of the Japanese Society of Scientific Fisheries* **51**, pp.1325–1329.

Sugita H, Fukumoto M, Koyama H, Deguchi Y (1988a) Changes in the faecal microflora of goldfish *Carassius auratus* with the oral administration of oxytetracycline. *Nippon Suisan Gakkaishi* **54**, pp.2181–2187.

Sugita H, Hirose Y, Matsuo N, Deguchi Y (1998) Production of antibacterial substances by *Bacillus* sp. strain NM 12, an intestinal bacterium of Japanese coastal fish. *Aquaculture* **165**, pp.269–280.

Sugita H, Miyajima C, Fukumoto M, Koyama H, Deguchi Y (1989) Effect of oxolinic acid on faecal microflora of goldfish (*Carassius auratus*). *Aquaculture* **80**, pp.163–174.

Sugita H, Kawasaki J, Deguchi Y (1997) Production of amylase by the intestinal microflora in cultured freshwater fish. *Letters in Applied Microbiology* **824**, pp.105–108.

Sugita H, Miyajima C, Iwata J-C, Arai S, Kubo T, Igarashi S et al. (1988c) Intestinal microflora of Japanese coastal fishes. *Nippon Suisan Gakkaishi* **54**, pp.875–882.

Sugita H, Tsunohara M, Fukumoto M, Deguchi Y (1987) Comparison of microflora between intestinal contents and fecal pellets of freshwater fishes. *Nippon Suisan Gakkaishi* **53**, pp.287–290.

Sugita H, Tsunohara T, Ohkoshi T, Deguchi Y (1988b) The establishment of an intestinal microflora in developing goldfish (*Carassius auratus*) of culture ponds. *Microbial Ecology* **15**, pp.333–344.

Talukdar S, Ringø E, Ghosh K (2016) Extracellular tannase-producing bacteria detected in the digestive tracts of seven freshwater fishes. *Acta Ichthyologica et Piscatoria* **46**, pp.201–210.

Thy HTT, Tri NN, Quy OM, Fotedar R, Kannika K, Unajak S, Areechon N (2017) Effects of dietary supplementation of mixe probiotic spores of *Bacillus amyloliquefaciens* 54A, and *Bacillus pumilus* 47B on growth, innate immunity and stress responses of striped catfish (Pangasianodon hypohthalmus). *Fish & Shellfish Immunology* **60**, pp.391–399.

Trust TJ (1975) facultative anaerobic bacteria in the digestive tract of chum salmon (*Oncorhynchus keta*) maintained in fresh water under defined culture conditions. *Applied Microbiology* **29**, pp.663–668.

Trust TJ, Bull LM, Currie BR, Buckley JT (1979) Obligate anaerobic bacteria in the gastrointestinal microflora of the grass carp (*Ctenopharyngodon idella*), goldfish (*Carassius auratus*), and rainbow trout (*Salmo gairdneri*). *Journal of the Fisheries Research Board of Canada* **36**, pp.1174–1179.

Trust TJ, Sparrow RAH (1974) The bacterial flora in the alimentary tract of freshwater salmonid fishes. *Canadian Journal of Microbiology* **20**, pp.1219–1128.

Tzeng T-D, Pao Y-Y, Chen P-C, Weng FC-H, Jean WD, Wang D (2015) Effects of host phylogeny and habitats on gut microbiomes of oriental river prawn (*Macrobrachium nipponense*). *PLOS One* **10**, p.e0132860.

Vadstein O, Attramadal KJK, Bakke I, Forberg T, Olsen Y, Verdegem M, et al. (2018) Managing the microbial community of marine fish larvae: A holistic perspective for larviculture. *Frontiers in Microbiology* **9**, p.1820.

Van Doan H (2021) *Bacillus* spp. In aquaculture – mechanisms and application: A updated view. In: *Probiotic Bacteria and Postbiotic Metabolism: Role on Animal and Human Health.* (Mojgani N, Dadar M, eds), pp.1–58. Springer Nature, Singapore.

Vargas-Albores F, Porchas-Cornejo MA, Martinez-Porshas M, Villalpando E, Gollas-Galván T, Martinez-Córdova LR (2017) Bacterial biota of shrimp intestine is significantly modified by the use of a probiotic mixture: A high throughput sequencing approach. *Helgoland Marine Research* **71**, p.5.

Wang Y-B, Li J-R, Lin J (2008) Probiotics in aquaculture: Challenges and outlook. *Aquaculture* **281**, pp.1–4.

Wang Y-B, Xu Z-R, Xia M-S (2005) The effectiveness of commercial probiotics in northern white shrimp *Penaeus vannamei* ponds. *Fisheries Sciences* **71**, pp.1036–1041.

Wei Y, Liu J, Wang L, Duan M, Ma Q, Xu H, et al. (2023) Influence of fish protein hydrolysate on intestinal health and microbial communities in turbot Scophthalmus maximus. *Aquaculture* **576**, p.739827.

Wong S, Waldrop T, Summerfelt S, Davidson J, Barrows F, Kenney B, et al. (2013) Aquacultured rainbow trout (*Oncorhynchus mykiss*) possess a large core intestinal microbiota that is resistant to variation in diet and rearing density. *Applied and Environmental Microbiology* **79**, pp.4974–4984.

Yang H-T, Zou S-S, Zhai L-J, Wang Y, Zhang F-M, An L-G, et al. (2017) Pathogen invasion changes the intestinal microbiota composition and induces innate immune responses in the zebrafish intestine. *Fish & Shellfish Immunology* **71**, pp.35–42.

Yang P, Hu H, Liu Y, Li Y, Ai Q, Xu W, et al. (2018) Dietary stachyose altered the intestinal microbiota profile and improved intestinal barrier function of juvenile turbot, *Scophthalmus maximus*. *Aquaculture* **486**, pp.98–106.

Yilmaz S, Sova M, Ergün S (2018) Antimicrobial activity of trans-cinnamic acid and commonly used antibiotics against important fish pathogens and nonpathogenic isolates. *Journal of Applied Microbiology* **125**, pp.1714–1727.

Zhou Z, Karlsen Ø, He S, Olsen RE, Yao B, Ringø E (2013) The effect of dietary chitin on the autochthonous gut microbiota of Atlantic cod (*Gadus morhua* L.). *Aquaculture Research* **44**, pp.1889–1900.

Zhu J, Dai W, Qiu Q, Dong C, Zhang J, Xiong J (2016) Contrasting ecological processes and functional compositions between intestinal bacterial community in healthy and diseased shrimp. *Microbial Ecology* **72**, pp.975–985.

6 Products by Bacilli
Exoenzymes

Koushik Ghosh
The University of Burdwan, Burdwan, India

Einar Ringø
UiT The Arctic University of Norway, Tromsø, Norway

Mehdi Soltani
University of Tehran, Tehran, Iran
Murdoch University, Perth, Australia

6.1 INTRODUCTION

Discovering the core gut microbiome is considered to be crucial for understanding the ecology of microbial consortia and to define a stable and healthy bacterial community in animal intestines (Shade & Handelsman 2012). Previous studies pertaining to isolation of gut-associated microorganisms using conventional culture-dependent methods documented that *Bacilli* (the phylogenetic group *Firmicutes*) were dominant within the fish GI tract (Ray et al. 2012). High-throughput sequencing analyses of the gut microbial composition also demonstrated *Firmicutes* as one of the dominant bacterial phyla (core microbiota) in the teleosts (Li et al. 2015; Liu et al. 2016; Lyons et al. 2017; Dehler et al. 2017). Species of *Bacillus* are ubiquitous in the environment and are one of the most commonly used probiotics in aquaculture (Hong et al. 2005; Zokaeifar et al. 2012a, 2012b; Zokaeifar et al. 2014). *Bacillus* spp. are heterogenous, both phenotypically and genotypically (Slepecky & Hemphill 2006). In consequence, they exhibit quite diverse physiological properties, e.g., the ability to degrade different substrates derived from plant or animal sources including cellulose, starch, proteins, hydrocarbons (Lutz et al. 2006), and diverse anti-nutritional factors (Ghosh et al. 2019). Furthermore, *Bacillus* species can be heterotrophic nitrifiers, denitrifiers, nitrogen fixers, iron precipitators, selenium oxidizers, oxidizers and reducers of manganese, facultative chemolithotrophs, acidophiles, alkalophiles, psychrophiles, thermophiles, and others (Slepecky & Hemphill 2006; Abriouel et al. 2011).

Being metabolically active, the genus *Bacillus* produces a wide arsenal of useful enzymes and numerous antimicrobial compounds that may have antibacterial, anti-viral, or anti-fungal activity (Prieto et al. 2012). The probiotic attributes of the bacilli often lie with their metabolites that encompass an extensive range of substances with diverse biological functions. Generally, *Bacillus* spp. are well known as major producers of proteinaceous substances (Zokaeifar et al. 2012a), which include both enzymes and bacteriocins or bacteriocin-like inhibitory substances (BLIS). This chapter provides an overview on the exo-enzymes produced by bacilli and their beneficial attributes, with notes on the present status of knowledge and prospects in aquaculture. Although the enzyme-producing ability of bacilli from diverse sources is widely known, we focus here on the autochthonous enzyme-producing bacilli recorded from fish GI tracts. Autochthonous gut-adherent bacteria seem to be ideal for aquaculture application, since the use of microorganisms sourced from fish might ensure their colonization and enzyme supplementation within the intestine and would also be likely to minimize the risk of harmful effects by the microorganisms or their metabolites (Ghosh et al. 2019).

DOI: 10.1201/9781003503811-6

6.2 GUT-ASSOCIATED BACILLI AND THEIR EXO-ENZYMES

Unlike the ruminants and higher vertebrates, the probable contribution of the endosymbionts to the nutritional physiology of fish has been recognized only recently (Ray et al. 2012). Apart from the endogenous digestive enzymes, exogenous enzymes produced and supplemented by the autochthonous microbiota could be considered as one of the major secondary factors affecting nutrition and feed utilization in fish. In fact, a wide range of enzymes, viz., carbohydrases, phosphatases, esterases, lipases, and peptidases, produced by gut bacteria might contribute to the digestive processes in fish (Ghosh et al. 2019). Extensive studies on Indian major carps (e.g., Ray et al. 2012; Mandal & Ghosh 2013; Das & Ghosh 2014; Dutta et al. 2015; Dutta and Ghosh, 2015; Banerjee et al. 2016; Mukherjee & Ghosh 2016; Mukherjee et al. 2016; Mukherjee et al. 2017) and other teleosts (e.g., Cahil 1990; Ringø et al. 1995, 2010, 2016; Llewellyn et al. 2016; Al-Hisnawi et al. 2015; Hoseinifar et al. 2016; Ringø & Song 2016) have indicated the presence of autochthonous gut-associated microorganisms in fish and their beneficial attributes in nutrition. The enzymes of nutritional importance produced by the gut bacteria may be categorized into (1) digestive enzymes, e.g., protease, amylase, and lipase, and (2) degradation enzymes, e.g., non-starch polysaccharide (NSP) – degrading enzymes, phytase, tannase, and chitinase. The review of Ray et al. (2012) illustrated the contribution of the diverse exo-enzyme-producing gut bacteria to the nutrition and well-being of the host fish, with gut-associated bacilli being recognized as one of the major groups of bacteria within the fish gut. Ghosh et al. (2019) also highlighted the role of fish gut microbiota, particularly numerous strains of *Bacillus*, in enzymatic degradation of plant-derived anti-nutritional factors. In the following section, we will present an update of research in this field, concentrating on the occurrence and importance of exo-enzyme-producing, gut-associated bacilli in fish (Table 6.1).

6.3 EXO-ENZYME-PRODUCING BACILLI: DIGESTIVE ENZYMES

Endogenous digestive enzymes in fish hydrolyze organic macromolecules (mainly carbohydrate, protein, and lipid) into simpler compounds. In addition, as established for higher vertebrates, supplementation of digestive enzymes (viz., amylase, protease, and lipase) from microbial source could be of importance to improve nutrient utilization in fish. Although preliminary studies on microbial amylase activity within the fish gut noticed the presence of bacterial amylase, characterization and identification of the specific amylase-producing strains were rarely carried out (Lesel et al. 1986; Das & Tripathi 1991; Bairagi et al. 2002). On the other hand, microorganisms with efficient proteolytic activity are widespread in nature because of their rapid growth and *Bacillus* spp. are by far the most common among them (Ray et al. 2012). To the authors' knowledge, occurrence of proteolytic and amylolytic bacilli (*B. circulans*, *B. cereus*, and *B. pumilus*) was first reported in the gut of an Indian major carp, rohu, by Ghosh et al. (2002) who correlated it with the feeding habit, although enzyme activities were not quantified. Subsequently, Esakkiraj et al. (2009) documented extracellular protease production by *B. cereus* isolated from the gut of flathead grey mullet (*Mugil cephalus*) and found that bacterial protease could utilize different preparations of tuna-processing wastes (e.g., raw fish meat, defatted fish meat, and alkali or acid hydrolyzate) as a source of nitrogen. Mondal et al. (2010) detected both protease- and amylase-producing ability in *B. licheniformis* and *B. subtilis* from the gut of bata. Similarly, Ray et al. (2010) isolated various strains of bacilli from the gut of three Indian major carps, viz., catla (*B. coagulans* and *B. cereus*), mrigal (*Bacillus* sp. and *B. cereus*), and rohu (*Bacillus* sp.), and protein or starch hydrolyzing abilities were demonstrated. Askarian et al. (2012) recorded the presence of amylase- and protease-producing bacilli (*B. thuringiensis*, *B. cereus*, *B. subtilis*, and *Bacillus* sp.) in the gut of a marine teleost (Atlantic salmon) and Das et al. (2014) reported *Brevibacillus parabrevis* and *B. licheniformis* isolated from two brackish water fish species, crescent perch (*Terapon jarbua*) and long whiskers catfish, respectively, both with considerable amylase and protease activities *in vitro*. Since then, diverse *Bacillus* spp. capable of producing amylolytic and proteolytic enzymes have been documented from several fish species

TABLE 6.1
Overview of the Research Endeavors Depicting Occurrence of Enzyme-Producing Bacilli in Fish Gut

Source Fish	Strains of Bacilli*	Enzymes Produced	References
Mozambique tilapia (*Oreochromis mossambicus*)	*B. circulans* TM1	Cellulase	Saha et al. (2006)
Grass carp (*Ctenopharyngodon idella*)	*B. megaterium* CI3		
Bata (*Labeo bata*)	*Bacillus licheniformis* EF635428 *B. subtilis* EF032683	Amylase, protease, cellulase	Mondal et al. (2010)
Catla (*Catla catla*) Mrigal (*Cirrhinus mrigala*) Rohu (*Labeo rohita*)	*B. coagulans* FJ627946 *B. cereus* FJ188299 *Bacillus* sp. FJ613631	Amylase, protease, cellulase	Ray et al. (2010)
Pacu (*Piaractus mesopotamicus*) Piau-com-piñata (*Leporinus friderici*)	*B. subtilis* P6 *B. velezensis* P11	Cellulase	Peixoto et al. (2011)
Atlantic salmon (*Salmo salar*)	*B. subtilis* 1015 *B. thuringiensis* 1114 *B. cereus* 1113, 1115 *Bacillus* sp. 1112	Protease, lipase, phytase, chitinase	Askarian et al. (2012)
Murrel (*Channa punctatus*) Stinging catfish (*Heteropneustes fossilis*)	*Bacillus* sp. CPF3, HFH4 *B. licheniformis* CPF4, CPH6, CPH7, HFF1, HFF3 *B. subtilis* HFH7	Amylase, protease, cellulase	Banerjee et al. (2013)
Brackish water fish species; *Terapon jarbua* and *Mystus gulio*	*Brevibacillus parabrevis* KF377322, KF377324 *B. licheniformis* KF377323	Amylase, protease, lipase, cellulase	Das et al. (2014)
Walking catfish (*Clarias batrachus*), Climbing perch (*Anabas testudineus*)	*B. amyloliquefaciens* ATH1 *B. licheniformis* CBF2, CBF4, CBH5, CBH6, CBH7	Amylase, protease, cellulase	Banerjee et al. (2015a)
Mrigal Silver carp (*Hypophthalmichthys molitrix*)	*B. pumilus* KF454036 *B. flexus* KF454035	Chitinase	Banerjee et al. (2015b)
Mrigal	*B. amyloliquefaciens* KF623290 *B. sonorensis* KF623291	Amylase, protease, lipase, cellulase, phytase, xylanase	Dutta and Ghosh (2015)
Catla	*B. aerius* KF623288 *B. sonorensis* KF623289	Amylase, protease, lipase, cellulase, phytase, xylanase	Dutta et al. (2015)
Rohu Mrigal Silver carp Grass carp	*B. pumilus* KF640221, KF640223 *B. tequilensis* KF640219 *B. megaterium* KF640220 *B. altitudinis* KF640222	Cellulase, xylanase	Banerjee et al. (2016)
Mrigal	*B. stratosphericus* KM277362 *B. aerophilus* KM277363 *B. licheniformis* KM277364 *Solibacillus silvestris* KM277365	Amylase, protease, lipase, cellulase, phytase, xylanase	Mukherjee et al. (2016)
Catla	*B. methylotrophicus* KF559344, KF559345 *B. subtilis* subsp. *spizizenii* KF559346	Amylase, protease, lipase, cellulase, phytase, xylanase	Mukherjee and Ghosh (2016)

(*Continued*)

TABLE 6.1 (Continued)
Overview of the Research Endeavors Depicting Occurrence of Enzyme-Producing Bacilli in Fish Gut

Source Fish	Strains of Bacilli*	Enzymes Produced	References
Rohu, catla, mrigal, silver carp, grass carp, common carp, Nile tilapia (*O. niloticus*)	*B. subtilis* KP765736 *Brevibacillus agri* KP765734	Tannase	Talukdar et al. (2016)
Striped dwarf catfish (*Mystus vittatus*)	*Bacillus* sp. KP256503, KP256506	Amylase, protease, lipase, cellulase	Nandi et al. (2017a)
Catla Java barb (*Puntius javanicus*)	*Bacillus* sp. KP256501.1 *Bacillus amyloliquefaciens* KP256502.1, KT719406.1	Amylase, protease, lipase, cellulase	Nandi et al. (2017b)
Walking catfish (*C. batrachus*)	*B. aryabhattai* KP784311 *B. flexus* KR809411 *B. cereus* KR809412	Amylase, protease, lipase, cellulase	Dey et al. (2016)
Rohu	*B. methylotrophicus* KU556164 *B. amyloliquefaciens* KU556165 *B. licheniformis* KU556167	Amylase, protease, lipase, cellulase, phytase, xylanase	Mukherjee et al. (2017)
Nile tilapia	*B. licheniformis* KT362744	Amylase, protease, lipase, cellulase, phytase, xylanase	Ghosh et al. (2017)
Climbing perch	*Paenibacillus polymyxa* MF457398.1	Amylase, protease, lipase, cellulase	Midhun et al. (2017)
Freshwater carp, *Labeo calbasu*	*B. subtilis* (KX756706) *B. cereus* (KX756707) *B. amyloliquefaciens* (KX775224)	Amylase, protease, lipase	Kavitha et al. (2018)

* Species names are followed by either GenBank Accession Numbers (when available) or strain numbers.

(Table 6.1), including Indian major carps (Dutta & Ghosh 2015; Dutta et al. 2015; Mukherjee & Ghosh 2016; Mukherjee et al. 2016; Mukherjee et al. 2017; *Kavitha* et al. 2018); climbing perch (*Anabas testudineus*) (Banerjee et al. 2015a; Midhun et al. 2017); walking catfish (Banerjee et al. 2015a; Dey et al. 2016), grass carp, and rohu (Guo et al. 2016; Banerjee et al. 2017); and striped dwarf catfish (*Mystus vittatus*) (Nandi et al. 2017a).

Gut bacteria might induce lipolysis either by enzymatic breakdown of triglyceride through direct bacterial action or by altering pancreatic lipase activity with bacterial proteases (Ringø et al. 1995). Although reports on specific lipase-producing bacilli from fish gut are scarce, some of the studies describing amylase-, protease-, or cellulase-producing bacilli within fish gut also addressed lipolytic activity. Thus, lipase-producing bacilli were detected in the guts of Indian major carps (Dutta & Ghosh 2015; Dutta et al. 2015; Mukherjee & Ghosh 2016; Mukherjee et al. 2016, Mukherjee et al. 2017); Atlantic salmon (Askarian et al. 2012); brackish water fishes, crescent perch and wishers catfish (Das et al. 2014); catfishes (Dey et al. 2016; Nandi et al. 2017a); and Nile tilapia (Ghosh et al. 2017).

6.4 EXO-ENZYME-PRODUCING BACILLI: DEGRADATION ENZYMES

Cellulose and hemicelluloses (e.g., xylans) are the major NSPs in plant feedstuffs commonly encountered by the fish under culture condition, either through natural food (algae, phytoplankton, detritus, and aquatic macrophytes) or through formulated diets as there is a thrust to replace animal sources in fish feed with plant ingredients such as rice bran, wheat husks, and different oil cakes. The principal endogenous polysaccharide-digesting enzymes in fish specifically hydrolyze the α-glycosic linkages of starch and yield glucose. Cellulose remains mostly indigestible in monogastric animals due to the presence of β-(1→4) glycosidic linkages and the lack of the endogenous cellulase. Likewise, β-glucanases and β-xylanases capable of digesting other NSPs are also either rare or not present in fish (Kuźmina 1996). Symbiotic gut microorganisms are likely to be involved in the fermentative degradation of cellulose and hemicelluloses for the host fish (Clements 1997). Thus, among the degradation enzymes, emphasis has been given on the ability of the gut microbiota to produce cellulase as the major NSP-degrading enzyme (for review, see Ray et al. 2012). Although the presence of microbial cellulase within the fish gut was first indicated in 1971 in the common carp (Shcherbina & Kazlauskiene 1971), involvement of gut-associated bacilli in cellulase production was detected much later (Ghosh et al. 2002). Protease- and amylase-producing bacilli (*B. circulans*, *B. cereus*, and *B. pumilus*) isolated from rohu were also found to be efficient in producing cellulase, although none of the enzyme-producing ability was quantified (Ghosh et al. 2002). Importantly, the presence of diverse exo-enzyme-producing bacilli was correlated with the omnivorous feeding attitude of the concerned carp species. Subsequently, a large number of cellulose-degrading bacilli have been recorded in the guts of grass carp and tilapia (Saha et al. 2006); rohu, catla, and mrigal (Ray et al. 2010); bata (Mondal et al. 2010); pacu (*Piaractus mesopotamicus*) and piau-com-pinta (Peixoto et al. 2011); walking catfish (Dey et al. 2016) and striped dwarf catfish (Nandi et al. 2017a). While studying the *in vitro* cellulase-producing ability of the gut bacilli, Ray et al. (2007) noticed that cellulase production by *B. subtilis* CY5 and *B. circulans* TP3, isolated from the gut of common carp and Mozambique tilapia, respectively, was enhanced under optimized condition through solid-state fermentation (SSF). Further, Peixoto et al. (2011) noticed the cellulolytic potential of *B. subtilis* P6 and *Bacillus velesensis* P11 maximum residual cellulase activity at pH 7.0–9.0. These observations were instrumental in terms of the future utilization of the cellulolytic bacilli in bio-processing of plant feedstuffs *in vitro*.

Compared to cellulase production, reports on xylanase-producing gut microorganisms in fish are meager (German & Bittong 2009; Banerjee & Ghosh 2014; Banerjee et al. 2016; see Table 6.1). Banerjee et al. (2016) carried out screening of cellulose- and xylan-degrading autochthonous gut bacteria from six freshwater carps. In their study, the strains *B. pumilus* LRF1X, *B. pumilus* CMF1C, *B. tequilensis* HMF6X, *B. megaterium* CtIF1C, and *B. altitudinis* CMH8X revealed both xylan- and cellulose-degrading ability. Banerjee and Ghosh (2016) also demonstrated the degradation of cellulose and xylan under SSF by *B. pumilus* and *B. tequilensis* isolated from the proximal intestines of rohu and silver carp, respectively. Furthermore, cellulase- and xylanase-producing ability of autochthonous bacilli isolated from the gut of rohu, catla, and mrigal has been reported (Dutta et al. 2015; Dutta & Ghosh 2015; Mukherjee et al. 2016; Mukherjee & Ghosh 2016; Mukherjee et al. 2017; see Table 6.1).

Phytase is another important degradation enzyme, supplementation of which might improve the availability of phosphorus and other minerals bound to phytic acid by hydrolysis of the phytate compounds (Oatway et al. 2001). Protein-rich oil cakes used in aquafeed formulation are the major source of phytate compounds. Only a few reports have considered exogenous phytase activity represented by gut bacteria in fish. The first study to report phytase-producing bacilli within fish gut was accomplished by Roy et al. (2009), who depicted two phytase-producing strains of *B. licheniformis* from rohu. Subsequently, Askarian et al. (2012) demonstrated phytase activity by autochthonous *B. subtilis*, *B. thuringiensis*, *B. cereus* and an unknown *Bacillus* sp. isolated from the gut of Atlantic salmon fed with or without chitin supplementation in the diet, although phytase-producing ability was not quantified. In a comprehensive investigation of phytase-producing bacteria in freshwater teleosts, Khan and Ghosh (2012) documented *B. subtilis* LB1.4 and *B. atrophaeus* GC1.2 isolated from the digestive tracts of minor carps, bata and Indian river shad (*Gudusia chapra*), respectively, as efficient phytase-producing strains. Further, the phytase-producing capacity of *B. subtilis* was evaluated *in vitro* in a later study under SSF and the phytate-degrading ability of this strain was established (Khan & Ghosh 2012). More recent studies have reported the phytase-producing ability of autochthonous exo-enzyme-producing bacilli isolated from the Indian major carps (Dutta et al. 2015; Dutta & Ghosh 2015; Mukherjee et al. 2016; Mukherjee & Ghosh 2016; Mukherjee et al. 2017).

Tannins are widespread in nature and are considered the most common among the plant-derived antinutritional factors. The tannin-degrading ability of *Bacillus* and some other genera has been recorded by Deschamps et al. (1980). Tannase-producing microbiota have been detected in the digestive tract of ruminants feeding on tannin-rich forage (Goel et al. 2005), but information on tannase-producing bacteria from fish gut is scanty as studies carried out on this topic are inadequate. To our knowledge, only one study has found tannase-producing bacilli in fish gut; Talukdar et al. (2016) reported tannase activity in *B. subtilis* KP765736 and *Brevibacillus agri* KP765734 isolated from Nile tilapia. The authors hypothesized that the tannase-producing bacteria established a symbiotic relationship with the host fish and have adapted to the neutral or alkaline pH of the fish gut (Talukdar et al. 2016).

Chitin is considered as the second most abundant biological material in the world after cellulose that forms a major constituent of many fish food organisms, viz., protozoans, coelenterates, crustaceans, mollusks, fungi, and green algae (Ray et al. 2012). Although the first report of bacterial chitin destruction involved *Bacillus chitinovorus* isolated from an aquatic source (Benecke 1905), chytinolytic bacilli from fish gut were documented much later. Askarian et al. (2012) recorded chitinase-producing ability of bacilli in the digestive tract of Atlantic salmon fed with (*B. subtilis*) or without (*B. thuringiensis*, *B. cereus*, and *Bacillus* sp.) chitin-supplemented diets. Further, Banerjee et al. (2015b) detected potent chitinolytic activity of *B. pumilus* KF454036 and *B. flexus* KF454035 isolated from the digestive tracts of silver carp and mrigal, respectively, and suggested the possibility of using chitinolytic bacilli from fish gut for chitinase production or as probiotics to improve feed efficiency in fish.

6.5 CONCLUSION

Thus, diverse strains of bacilli with enzymatic potential have been detected in the GI tract of carnivorous, herbivorous, or omnivorous fish species (for review, see Ray et al. 2012; Ghosh et al. 2019; Soltani et al. 2019). It has been hypothesized that symbiotic microorganisms might aid in the digestion of food and supply of energy to the host (Ray et al. 2012). Being the major colonizers within the gut, bacilli are expected to play a pivotal role in this process. Surprisingly, research endeavors on the endosymbiotic bacteria and its function in digestion of the dominant aquatic vertebrate – the fish – have been initiated of late. The studies pertaining to enzyme-producing ability of the gut-associated bacilli or other bacteria are mostly based on *in vitro* investigations. Very few of the previous studies made attempt to realize the contribution of the bacterial enzymes *in vivo*. For example, cellulase activity sharply decreased when the fish were fed diets containing antibiotics, suggesting that the gut cellulase activity is largely contributed by the gut microbiota (Saha & Ray 1998; Roy & Ghosh 2008). However, actual contribution of the endosymbiotic bacilli and its enzymes in digestion and metabolism of fish is not very well understood and needs to be worked out. The herbivorous and omnivorous fish with long and coiled intestine are likely to have more bacteria (and thus, more bacilli) colonized in their GI tract in comparison to the carnivorous one with much shorter intestine. Starch-, cellulose-, or other NSP-degrading bacilli have been frequently detected in the GI tract of herbivorous/omnivorous fish species (as compared to the carnivorous fish), which could be indicative of fermentative degradation of complex polysaccharides within their gut. Despite simpler form of nutrients, the main fermentation products of gut microbial metabolism are the short-chain fatty acids (SCFAs), e.g., acetate, propionate, and butyrate (Hoseinifar et al. 2017a). It has been hypothesized that SCFAs and their salts could modulate the immune responses in finfish (Hoseinifar et al. 2017a, 2017b, 2017c). However, SCFA-producing bacilli in the fish gut have not been authentically reported and should be addressed in forthcoming studies. Apart from digestive and degradation enzymes, *Bacillus* species can also produce antioxidant enzymes, e.g., superoxidase dismutase (SOD) and glutathione, to eliminate the free radicals effectively (Li et al. 2012; Hindu et al. 2018). However, specific function of exo-enzyme producer *Bacillus* spp. to modulate antioxidant activities in the host fish might be an important area of research. Further, some strains of bacilli (e.g., *Bacillus cereus*, *Lactobacillus*, and *Bifidobacterium*) are known to degrade the signaling molecules of pathogenic bacteria (quorum sensing) by enzymatic secretion/production of auto-inducer antagonists (Brown 2011). Therefore, likely application of the extracellular quorum-quenching molecules (*molecules that cause disruption* of *quorum* sensing) produced by the bacilli could be an alternative to the antibiotics used in aquaculture.

Structure and composition of fish gut microbiota are believed to play an important ecological function, which is strongly influenced by multiple factors that might include the host genetics, living environment, diet, and phylogeny (Liu et al. 2016). Whether host genetic background determines the host trophic level and influences the individual gut microbiome diversity along with occurrence of exo-enzyme producer bacilli needs to be investigated. In this context, whether enzyme-producing bacilli coordinate or shape the host gut microbiota needs to be clarified.

BIBLIOGRAPHY

Abriouel H, Farzan CMAP, Omar NB, Galvez A (2011) Diversity and applications of *Bacillus* bacteriocins. *FEMS Microbiol Reviews* **35**, pp.201–232.

Al-Hisnawi A, Ringø E, Davies SJ, Waines P, Bradley G, Merrifield DL (2015) First report on the autochthonous gut microbiota of brown trout (*Salmo trutta*). *Aquaculture Research* **46**, pp.2962–2971.

Askarian F, Zhou Z, Olsen RE, Sperstad S, Ringø E (2012) Culturable autochthonous gut bacteria in Atlantic salmon (*Salmo salar* L.) fed diets with or without chitin. Characterization by 16S rRNA gene sequencing, ability to produce enzymes and *in vitro* growth inhibition of for fish pathogens. *Aquaculture* **326–329**, pp.1–8.

Bairagi A, Sarkar Ghosh K, Sen SK, Ray AK (2002) Duck weed (*Lemna polyrhiza*) leaf meal as a source of feedstuff in formulated diets for rohu (*Labeo rohita* Ham.) fingerlings after fermentation with a fish intestinal bacterium. *Bioresource Technology* **85**, pp.17–24.

Banerjee G, Dan SK, Nandi A, Ghosh P, Ray AK (2015a) Autochthonous gut bacteria in two Indian air-breathing fish, climbing perch (*Anabas testudineus*) and walking catfish (*Clariasbatrachus*): Mode of association, identification and enzyme producing ability. *Polish Journal of Microbiology* **64(4)**: pp.361–368.

Banerjee G, Nandi A, Ray AK (2017) Assessment of hemolytic activity, enzyme production and bacteriocin characterization of *Bacillus subtilis* LR1 isolated from the gastrointestinal tract of fish. *Archives of Microbiology* **199**, pp.115–124.

Banerjee G, Ray AK, Askarian F, Ringø E (2013) Characterisation and identification of enzyme-producing autochthonous bacteria from the digestive tract of two Indian air-breathing fish. *Beneficial Microbes* **4**, pp.277–284.

Banerjee S, Ghosh K (2014) Enumeration of gut associated extracellular enzyme-producing yeasts in some freshwater fishes. *Journal of Applied Ichthyology* **30**, pp.986–993.

Banerjee S, Ghosh K (2016) Bio-processing of linseed oil-cake through solid state fermentation by non-starch polysaccharide degrading fish gut bacteria. *Fermentation Technology* **5**, pp.1–10.

Banerjee S, Mukherjee A, Dutta D, Ghosh K (2015b) Evaluation of chitinolytic gut microbiota in some carps and optimization of culture conditions for chitinase production by the selected bacteria. *Journal of Microbiology, Biotechnology and Food Sciences* **5**,pp.12–19.

Banerjee S, Mukherjee A, Dutta D, Ghosh K (2016) Non-starch polysaccharide degrading gut bacteria in Indian major carps and exotic carps. *Jordan Journal of Biological Sciences* **9(1)**: pp.69–78.

Benecke W (1905) Über *Bacillus chitinovorous*, Einen chitinzersetzenden Spaltpilz. *Botanische Zeitung* **63**, pp.227–242.

Brown M (2011) Modes of action of probiotics: Recent developments. *Journal of Animal and Veterinary Advances* **10**, pp.1895–1900.

Cahil MM (1990) Bacterial flora of fishes: A review. *Microbial Ecology* **19**, pp.21–41.

Clements KD (1997) Fermentation and gastrointestinal microorganisms in fishes. In: *Gastrointestinal Microbiology Vol I. Gastrointestinal Ecosystems and Fermentations* (Mackie, R.I. & White, B.A. eds), pp.156–198. Chapman & Hall, New York.

Das KM, Tripathi SD (1991) Studies on the digestive enzymes of grass carp, *Ctenopharyngodon idella* (Val.). *Aquaculture* **92**, pp.21–32.

Das P, Ghosh K (2014) The presence of phytase in yeasts isolated from the gastrointestinal tract of four major carps [*Labeo rohita* (Hamilton, 1822), *Catla catla*(Hamilton, 1822), *Cirrhinus mrigala* (Hamilton, 1822), *Hypophthalmichthys molitrix* (Valenciennes, 1844)], climbing perch [*Anabas testudineus* (Bloch, 1792)] and Mozambique tilapia [*Oreochromis mossambicus* (Linnaeus, 1758)]. *Journal of Applied Ichthyology* **30**, pp.403–407.

Das P, Mandal S, Khan A, Manna SK, Ghosh K (2014) Distribution of extracellular enzyme-producing bacteria in the digestive tracts of 4 brackish water fish species. *Turkish Journal of Zoology* **38**, pp.79–88.

Dehler CE, Secombes CJ, Martin SAM (2017) Environmental and physiological factors shape the gut microbiota of Atlantic salmon parr (*Salmo salar* L.). *Aquaculture* **467**, pp.149–157.

Deschamps AM, Mahoudeau G, and Lebeault JM (1980) Fast degradation of Kraft Lignin by Bacteria. *European Journal of Applied Microbiology and Biotechnology*, **9**, pp.45–51.

Dey A, Ghosh K, Hazra N (2016) Evaluation of extracellular enzyme-producing autochthonous gut bacteria in walking catfish, *Clarias batrachus* (L.). *Journal of Fisheries* **4**, pp.345–352.

Dutta D, Banerjee S, Mukherjee A, Ghosh K (2015) Selection and probiotic characterization of exoenzyme-producing bacteria isolated from the gut of *Catla catla* (Actinopterygii: Cypriniformes: Cyprinidae). *Acta Ichthyologica et Piscatoria* **45**, pp.373–384.

Dutta D, Ghosh K (2015) Screening of extracellular enzyme-producing and pathogen inhibitory gut bacteria as putative probiotics in mrigal, *Cirrhinus mrigala* (Hamilton, 1822). *International Journal of Fisheries and Aquatic Studies* **2**, pp.310–318.

Esakkiraj P, Immanuel G, Sowmya SM, Iyapparaj P, Palavesam A (2009) Evaluation of protease-producing ability of fish gut isolate *Bacillus cereus* for aqua feed. *Food and Bioprocess Technology* **2**, pp.383–390.

German DP, Bittong RA (2009) Digestive enzyme activities and gastrointestinal fermentation in wood-eating catfishes. *Journal of Comparative Physiology B, Biochemical, Systemic, and Environmental Physiology* **179**, pp.1025–1042.

Ghosh K, Banerjee S, Moon UM, Khan HA, Dutta D (2017) Evaluation of gut associated extracellular enzyme-producing and pathogen inhibitory microbial community as potential probiotics in Nile tilapia, *Oreochromis niloticus*. *International Journal of Aquaculture* **7**, pp.143–158.

Ghosh K, Ray AK, Ringø E (2019) Applications of plant ingredients for tropical and sub-tropical finfish: Possibilities and challenges. *Reviews in Aquaculture* **11**, pp.793–815.

Ghosh K, Sen SK, Ray AK (2002) Characterization of bacilli isolated from gut of rohu, *Labeo rohita*, fingerlings and its significance in digestion. *Journal of Applied Aquaculture* **12**, pp.33–42.

Goel G, Puniya AK, Singh K (2005) Interaction of gut microflora with tannins in feeds. *Naturwissenschaften* **92**, pp.497–503.

Guo X, Chen D-D, Peng K-S, Cui Z-W, Zhang X-J, Li S, Zhang Y-A (2016) Identification and characterization of *Bacillus subtilis* from grass carp (*Ctenopharynodon idellus*) for use as probiotic additives in aquatic feed. *Fish and Shellfish Immunology* **52**, pp.74–84.

Hindu SV, Thanigaivel S, Vijayakumar S, Chandrasekaran N, Mukherjee A, Thomas J (2018) Effect of microencapsulated probiotic *Bacillus vireti* 01-polysaccharide extract of *Gracilaria folifera* with alginate-chitosan on immunity, antioxidant activity and disease resistance of *Macrobrachium rosenbergii* against *Aeromonas hydrophila* infection, *Fish and Shellfish Immunology* **73**, pp.112–120.

Hong HA, Duc LH, Cutting SM (2005) The use of bacterial spore formers as probiotics. *FEMS Microbiology Reviews* **29**, pp.813–835.

Hoseinifar SH, Dadar M, Ringø E (2017b) Modulation of nutrient digestibility and digestive enzyme activities in aquatic animals: The functional feed additives scenario. *Aquaculture research*, **48**, pp.3987–4000.

Hoseinifar SH, Ringø E, Shenavar Masouleh A, Esteban MÁ (2016) Probiotic, prebiotic and synbiotic supplements in sturgeon aquaculture: A review. *Reviews in Aquaculture* **8**, pp.89–102.

Hoseinifar SH, Sun Y-Z, Caipang CM (2017c) Short chain fatty acids as feed supplements for sustainable aquaculture: An updated view. *Aquaculture Research* **48**, pp.1380–1391.

Hoseinifar SH, Zou HK, Miandare HK, Doan HV, Romano N, Dadar M (2017a) Enrichment of common carp (*Cyprinus carpio*) diet with medlar (*Mespilus germanica*) leaf extract: Effects on skin mucosal immunity and growth performance. *Fish & Shellfish Immunology* **67**, pp.346–352.

Kavitha M, Raja M, Perumal P (2018) Evaluation of probiotic potential of *Bacillus* spp. isolated from the digestive tract of freshwater fish *Labeo calbasu* (Hamilton, 1822). *Aquaculture Reports* **11**, pp.59–69.

Khan A, Ghosh K (2012) Characterization and identification of gut-associated phytase-producing bacteria in some freshwater fish cultured in ponds. *Acta Ichthyologica et Piscatoria* **42**, pp.37–45.

Kuźmina VV (1996) Influence of age on digestive enzyme activity in some freshwater teleost. *Aquaculture* **148**, pp.25–37.

Lesel R, Fromageot C, Lesel M (1986) Cellulase digestibility in grass carp, *Ctenopharyngodonidella* and in gold fish, *Carassius auratus*. *Aquaculture* **54**, pp.11–17.

Li T, Long M, Gatesoupe FJ, Zhang Q, Li A, Gong X (2015) Comparative analysis of the intestinal bacterial communities in different species of carp by pyrosequencing. *Microbial Ecology* **69**, pp.25–36.

Li WF, Deng B, Cui ZW, Fu LQ, Chen NN, Zhou XX, Shen WY, Yu DY (2012) Several indicators of immunity and antioxidant activities improved in grass carp given a diet containing bacillus additive. *Journal of Animal and Veterinary Advances* **11**, pp.2392–2397.

Liu H, Guo X, Gooneratne R, Ruifang L, Cong Z, Fanbin Z, Weimin W. (2016). The gut microbiome and degradation enzyme activity of wild freshwater fishes influenced by their trophic levels. *Scientific Report* **6**, p.24340. doi: 10.1038/srep24340.

Llewellyn MS, McGinnity P, Dionne M, Letourneau J, Thonier F, Carvalho GR, Derome N (2016) The biogeography of the Atlantic salmon (*Salmo salar*) gut microbiome. The *ISME Journal* **10**, pp.1280–1284.

Lutz G, Chavarrìa M, Arias ML, Mata-Segreda JF (2006) Microbial degradation of palm (*Elaeis guineensis*) biodiesel. *Revista Biologia Tropìcal* **54**, pp.59–63.

Lyons PP, Turnbull JF, Dawson KA, Crumlish M. (2017) Exploring the microbial diversity of the distal intestinal lumen and mucosa of farmed rainbow trout *Oncorhynchus mykiss* (Walbaum) using next generation sequencing (NGS). *Aquaculture Research* **48**, pp.77–91.

Mandal S, Ghosh K (2013) Isolation of tannase-producing microbiota from the gastrointestinal tracts of some freshwater fish. *Journal of Applied Ichthyology* **29**, pp.145–153.

Midhun SJ, Neethu S, Vysakh A, Arun D, Radhakrishnan EK, Jyothis M (2017) Antibacterial activity and probiotic characterization of autochthonous *Paenibacillus polymyxa* isolated from *Anabas testudineus* (Bloch, 1792). *Microbial Pathogenesis* **113**, pp.403–411.

Mondal S, Roy T, Ray AK (2010) Characterization and identification of enzyme-producing bacteria isolated from the digestive tract of bata, *Labeo bata. Journal of the World Aquaculture Society* **41**, pp.369–376.

Mukherjee A, Dutta D, Banerjee S, Ringø E, Breines EM, Hareide E, Chandra G, Ghosh K (2016) Potential probiotics from Indian major carp, *Cirrhinusmrigala*. Characterization, pathogen inhibitory activity, partial characterization of bacteriocin and production of exoenzymes. *Research in Veterinary Science* **108**, pp.76–84.

Mukherjee A, Dutta D, Banerjee S, Ringø E, Breines EM, Hareide E, Chandra G, Ghosh K (2017) Culturable autochthonous gut bacteria in rohu, *Labeo rohita. In vitro* growth inhibition against pathogenic *Aeromonas* spp., stability in gut, bio-safety and identification by 16S rRNA gene sequencing. *Symbiosis* **73**, pp.165–177.

Mukherjee A, Ghosh K (2016) Antagonism against fish pathogens by cellular components and verification of probiotic properties in autochthonous bacteria isolated from the gut of an Indian major carp, *Catla catla* (Hamilton). *Aquaculture Research* **47**, pp.2243–2255.

Nandi A, Dan SK, Banerjee G, Ghosh P, Ghosh K, Ringø E, Ray AK (2017a) Probiotic potential of autochthonous bacteria isolated from the gastrointestinal tract of four freshwater teleost. *Probiotics and Antimicrobial Proteins* **9**, pp.12–21.

Nandi A, Banerjee G, Dan SK, Ghosh P, Ghosh K, Ray AK (2017b) Screening of autochthonous intestinal microbiota as candidate probiotics isolated from four freshwater teleosts. *Current Science* **113**, pp.767–773.

Oatway L, Vasanthan T, Helm JH (2001) Phytic acid. *Food Reviews International* **17**, pp.419–431.

Peixoto SB, Cladera-Olivera F, Daroit DJ, Brandelli A (2011) Cellulase-producing *Bacillus* strains isolated from the intestine of Amazon basin fish. *Aquaculture Research* **42**, pp.887–891.

Prieto ML, O'Sullivan L, Tan SP, McLoughlin P, Hughes H, O'Connor PM et al. (2012) Assessment of the bacteriocinogenic potential of marine bacteria reveals lichenicidin production by seaweed derived *Bacillus* spp. *Marine Drugs* **10**, pp.2280–2299.

Ray AK, Bairagi A, Sarkar Ghosh K, Sen SK (2007) Optimization of fermentation conditions for cellulase production by *Bacillus subtilis* CY5 and *Bacillus circulans*TP3 isolated from fish gut. *Acta Ichthyologica et Piscatoria* **37**, pp.47–53.

Ray AK, Ghosh K, Ringø E (2012) Enzyme-producing bacteria isolated from fish gut: A review. *Aquaculture Nutrition* **18**, pp.465–492.

Ray AK, Roy T, Mondal S, Ringø E (2010) Identification of gut-associated amylase, cellulase and protease-producing bacteria in three species of Indian major carps. *Aquaculture Research* **41**, pp.1462–1469.

Ringø E, Olsen RE, Gifstad TØ, Dalmo RA, Amlund H, Hemre GI, Bakke AM (2010) Prebiotics in aquaculture: A review. *Aquaculture Nutrition* **16**, pp.117–136.

Ringø E, Song, SK (2016) Application of dietary supplements (synbiotics and probiotics in combination with plant products and β-glucans) in aquaculture. *Aquaculture Nutrition* **22**, pp.4–24.

Ringø E, Strøm E, Tabachek JA (1995) Intestinal microflora of salmonids: A review. *Aquaculture Research* **26**, pp.773–789.

Ringø E, Zhou Z, Vecino JLG, Wadsworth S, Romero J, Krogdahl A et al. (2016) Effect of dietary components on the gut microbiota of aquatic animals. A never-ending story? *Aquaculture Nutrition* **22**, pp.219–282.

Roy M, Ghosh K (2008) Antibiotic sensitivity and effect of tetracycline on gut bacterial flora and enzyme activity in *Labeo rohita* (Hamilton) fingerlings. *Journal of the Inland Fisheries Society of India.* **40 (2)**: pp.21–26.

Roy T, Mondal S, Ray AK (2009) Phytase-producing bacteria in the digestive tracts of some freshwater fish. *Aquaculture Research* **40**, pp.344–353.

Saha AK, Ray AK (1998) Cellulase activity in rohu fingerlings. *Aquaculture International* **6**, pp.281–291.

Saha S, Roy RN, Sen SK, Ray AK (2006) Characterization of cellulase-producing bacteria from the digestive tract of tilapia, *Oreochromis mossambica* (Peters) and grass carp, *Ctenopharyngodon idella* (Valenciennes). *Aquaculture Research* **37**, pp.380–388.

Shade A, Handelsman J (2012) Beyond the Venn diagram: The hunt for a core microbiome. *Environmental Microbiology* **14**, pp.4–12.

Shcherbina MA, Kazlauskiene OP (1971) The reaction of the medium and the rate of absorption of nutrients in the intestine of carp. *Journal of Ichthyology* **11**, pp.81–85.

Slepecky R, Hemphill E (2006) The genus *Bacillus*. Nonmedical. The Prokaryotes, Vol. 4 (Dworkin M, Falkow S, Rosenberg E, Schleifer K-H &Stackebrandt E, eds), pp. 530–562. Springer, New York, USA.

Soltani M, Ghosh K, Hoseinifar S, Kumar V, Lymbery AJ, Roy S, Ringø E. (2019) Genus *Bacillus*, promising probiotics in aquaculture: Aquatic animal origin, bio-active components, bioremediation and efficacy in fish and shellfish. *Reviews in Fisheries Science & Aquaculture*. **27**, pp.331–379.

Talukdar S, Ringø E, Ghosh K (2016) Extracellular tannase-producing bacteria detected in the digestive tracts of seven freshwater fishes. *Acta Ichthyologica et Piscatoria* **46**, pp.201–210.

Zokaeifar H, Balcázar JL, Kamarudin MS, Sijam K, Arshad A, Saad CR (2012a) Selection and identification of non-pathogenic bacteria isolated from fermented pickles with antagonistic properties against two shrimp pathogens. *Journal of Antibiotics* **65**, pp.289–294.

Zokaeifar H, Balcázar JL, Saad CR, Kamarudin MS, Sijam K, Arshad A, Nejat N (2012b) Effects of *Bacillussubtilis* on the growth performance, digestive enzymes, immune gene expression and disease resistance of white shrimp, *Litopenaeus vannamei*. *Fish and Shellfish Immunology* **33**, pp.683–689.

Zokaeifar H, Babaei N, Saad CR, Kamarudin MS, Sijam K, Balcazar JL (2014) Administration of *Bacillus subtilis* strains in the rearing water enhances the water quality, growth performance, immune response, and resistance against *Vibrio harveyi* infection in juvenile white shrimp, *Litopenaeus vannamei*. *Fish and Shellfish Immunology* **36**, pp.68–74.

7 Bacteriocins Produced by *Bacillus* and Their Antibacterial Activity

Koushik Ghosh
The University of Burdwan, Burdwan, West Bengal, India

Einar Ringø
UiT The Arctic University of Norway, Tromsø, Norway

Mehdi Soltani
University of Tehran, Tehran, Iran
Murdoch University, Perth, Australia

7.1 INTRODUCTION

The antimicrobial compounds produced by bacteria are usually divided into two major groups: (1) non-ribosomal secondary metabolites, such as peptide or lipopeptide antibiotics; and (2) ribosomally synthesized proteins/peptides, such as bacteriocins (Abriouel et al. 2011; Lee & Kim 2011). Bacteriocins are ribosomally synthesized antimicrobial peptides produced by bacteria that often present bactericidal effects against other closely related species (Cotter et al. 2005) and are the most abundant and diverse of the bacterially produced antimicrobials (Riley 2009). These heterogeneous substances portray variable biochemical properties, inhibitory spectra, molecular weights, and mechanisms of action (O'Sullivan et al. 2002).

Antibiotics have long had a place in the fish health manager's toolbox for the treatment of bacterial diseases. The massive use of broad-spectrum antibiotics and antimicrobial drugs increases, however, the selective pressure for development resistance (Verschuere et al. 2000). The development of antibiotic resistance among the microorganisms associated with fish diseases has become a global concern in recent years (Kolndadacha et al. 2011). Thus, much interest has been paid to the search for novel antibacterial compounds, preferably proteins with prophylactic and/or curative potential, for which the pathogens may not develop resistance (Patil et al. 2001). In this regard, bacteriocins from natural sources have been suggested to be an alternative to control bacterial diseases in aquaculture (Kim et al. 2014; Sahoo et al. 2016).

Bacteriocins produced by lactic acid bacteria (LAB) are the most widely studied (Nes et al. 2007; Ringø et al. 2018), while the *Bacillus* spp. have been less studied in this respect (Abriouel et al. 2011). The genus *Bacillus* includes an assortment of industrially important species and has a history of safe use in both food and industry (Paik et al. 1997). Moreover, *Bacillus* is an interesting genus to investigate antimicrobial potential, because *Bacillus* spp. produce a large number of bacteriocins or bacteriocin-like inhibitory substances (BLISs) composed of several different chemical structures (Von Döhren 1995; Abriouel et al. 2011). The production of bacteriocins or BLIS has been described for many *Bacillus* species, including *B. thuringiensis* (Paik et al. 1997), *B. subtilis* (Zheng et al. 1999), and *B. cereus* (Bizani & Brandelli 2002). Like the LAB, the genus *Bacillus* also includes representatives that are generally recognized as safe, such as *B. subtilis* and *B. licheniformis* (Smitha & Bhat 2013) and, hence, can find application in aquaculture as probiotics for the

DOI: 10.1201/9781003503811-7

prevention of bacterial diseases such motile *Aeromonas* septicemia and vibriosis. Bacteriocins from *Bacillus* spp. are of increasing interest owing to their broader spectrum of inhibition that may include Gram-negative and Gram-positive bacteria of genera of *Aeromonas*, *Edwardsiella*, *Streptococcus*, *Pseudomonas*, and *Vibrio* (Dutta & Ghosh 2015; Chen et al. 2016,b; Sumathi et al. 2017).

The gut microbiota in some finfish species such as Indian carp has been reported to be fairly dominated by the genus *Bacillus* through both culture-dependent and culture-independent methods (Ringø et al. 2006; Ghosh et al. 2010; Ray et al. 2010; Sarkar & Ghosh 2014; Li et al. 2015, b; Das & Ghosh 2015; Mukherjee et al. 2016; Liu et al. 2016; Yilmaz et al. 2018). Further, fish gut-associated bacilli are known to play an important role in the prevention of infections in aquaculture through the production of antibacterial substances (e.g., Dimitroglou et al. 2011; Mukherjee et al. 2017; Nandi et al. 2017; Nandi et al. 2017; Ghosh et al. 2017). Although several studies on bacteriocins from *Bacillus* have exhibited their important aspects of food safety (Gautam & Sharma 2009; Abriouel et al. 2011; Nath et al. 2015), very few have addressed their potential to be used against bacterial diseases in aquaculture (Ran et al. 2012; Kim et al. 2014; Luo et al. 2014; Guo et al. 2016; Sahoo et al. 2016). Moreover, there have been few studies on the bacteriocinogenic potential of fish gut-associated bacilli (Sirtori et al. 2006; Giri et al. 2011; Mukherjee et al. 2017; Dutta et al. 2018). There is therefore an urgent need for screening and characterization of bacteriocinogenic bacilli from the fish gut and their antibacterial compounds. This chapter provides an overview of the diverse classes of bacteriocins produced by bacilli, an update on the efficacy of fish gut-associated *Bacillus* spp. against fish pathogens, and their prospective future applications in aquaculture.

7.2 CLASSES OF BACTERIOCINS PRODUCED BY *BACILLUS*

Species in the genus *Bacillus* are known to produce a wide variety of antimicrobial substances that include peptide or lipopeptide antibiotics, bacteriocins, and bacteriocin-like inhibitors (Stein 2005; Abriouel et al. 2011; Sumi et al. 2015). Antimicrobial peptides produced by bacteria through ribosomal synthesis are generally referred to as bacteriocins, which are a heterologous group of proteinaceous antimicrobial substances and are known to be produced by bacteria from every major group (Riley & Wertz 2002a, b). Many other antimicrobial substances that are not well characterized or in which the peptide nature of the compound has not been confirmed are referred to as BLISs (Abriouel et al. 2011). The *Bacillus* group of bacteria often produces lipopeptide antibiotics by non-ribosomal synthesis (e.g., iturins). This section will present an overview on the classification of bacteriocins or BLIS produced by the bacilli excluding the non-ribosomally synthesized peptides.

The classification scheme of LAB bacteriocins was primarily developed by Klaenhammer (1993), who grouped bacteriocins into four distinct classes: Class I or lantibiotics (post-translationally modified, thermostable peptides, containing lanthionine or derivatives, <5 kDa); Class II (unmodified small heat-stable peptides, <10 kDa); Class III (unmodified large heat-labile peptides, >30 kDa); and Class IV (large complex proteins, containing carbohydrates or lipid moieties). This grouping has formed the basis of all subsequent classification schemes for bacteriocins produced by Gram-positive bacteria. Readers with special interest are referred to the reviews of Mokoena (2017) and Ringø et al. (2018) where updated classification schemes of bacteriocins produced by LAB have been presented.

The genus *Bacillus* may be considered the second most important group for the production of bacteriocins and BLIS after the LAB. In view of the increasing number of bacteriocins described within different bacterial groups, it is very difficult to provide a combined classification scheme of bacteriocins (Nes et al. 2007). Therefore, the *Bacillus* bacteriocin classification system may be adopted independently even though some compounds produced by both *Bacillus* and LAB have very similar characteristics. Consequently, a simple classification scheme for the bacteriocins/BLIS produced by *Bacillus* spp. has been proposed by Abriouel et al. (2011). This scheme holds three classes of bacteriocins (Table 7.1): Class I (antimicrobial peptides that undergo post-translational modifications); Class II (small nonmodified and linear peptides, heat and pH stable, 0.77–10 kDa); and Class III (large proteins with phospholipase activity, >30 kDa).

TABLE 7.1
Diverse Classes of Bacteriocins Produced by *Bacillus* spp.

Classes		Characteristic Features	Bacteriocins Produced	Typical Producer Organism	References
Class I		Post-translationally modified peptides			
	Subclass I.1.	Single-peptide, elongated lantibiotics	Subtilin, ericin S, ericin A	*B. subtilis*	Parisot et al. (2008)
	Subclass I.2.	Other single-peptide lantibiotics	Sublancin 168, mersacidin, paenibacillin	*B. subtilis*	Dubois et al. (2009)
	Subclass I.3.	Two-peptide lantibiotics	Haloduracin, lichenicidin	*B. halodurans*	Lawton et al. (2007)
	Subclass I.4.	Other post-translationally modified peptides	Subtilosin A	*B. amyloliquifaciens*	Sutyak et al. (2008)
Class II		Non-modified peptides			
	Subclass II.1.	Pediocin-like peptides	Coagulin, SRCAM 37, SRCAM 602, SRCAM 1580	*B. coagulans*	Le Marrec et al. (2000)
	Subclass II.2.	Thuricin-like peptides	Thurincin H, thuricin S, thuricin 17, bacthuricin F4, cerein MRX1	*B. thuringiensis*	Gray et al. (2006)
	Subclass II.3.	Other linear peptides	Cerein 7A, cerein 7B, lichenin, thuricin 439	*B. cereus*	Oscáriz et al. (2000)
Class III		Large proteins	Megacin A-216, megacin A-19213	*B. megaterium*	Kiss et al. (2008)

Source: Adapted and modified from Abriouel et al. (2011).

The bacteriocins produced by *Bacillus* spp. display a broad range of molecular mass; the smallest bacteriocin-like compound (800 Da) is produced by *B. licheniformis* (Teixeira et al. 2009), whereas *B. thuringiensis* produces the largest bacteriocins (950 kDa), such as thuricin (Cherif et al. 2001). Many *Bacillus* bacteriocins belong to the lantibiotics, post-translationally modified peptides commonly dispersed among different bacterial groups. Lantibiotics are the best-described antimicrobial peptides with regard to biosynthesis mechanisms, genetic determinants, and peptide structure. Members of the genus *Bacillus* also produce many other unmodified bacteriocins within Class II of LAB bacteriocins (Klaenhammer 1993; Drider et al. 2006; Nes et al. 2007), including the pediocin-like bacteriocins (Class IIa) and the two-peptide bacteriocins (Class IIb), while others show completely novel peptide sequences (Abriouel et al. 2011). Some of the well-described bacteriocins produced by *Bacillus* spp. are subtilin by *B. subtilis* (Banerjee & Hansen 1988), cerein by *B. cereus* (Oscariz et al. 1999), bacillocin 490 by *B. licheniformis* (Martirani et al. 2002), haloduracin by *B. halodurans* (Lawton et al. 2007), thuricin by *B. thuringiensis* (Gray et al. 2006), subtilosin by *B. amyloliquifaciens* (Sutyak et al. 2008), and megacin by *B. megaterium* (Kiss et al., 2008), all of which are mostly active against Gram-positive organisms *such as Listeria monocytogenes, Gardnerella vaginalis, Streptococcus agalactiae*, *Staphylococcus aureus*, and *Leuconostoc mesenteroides*.

7.3 PATHOGEN INHIBITION BY BACILLI AND THEIR BACTERIOCINOGENIC ACTIVITY

Bacterial antagonism is a regular incident in nature that might play a vital role in maintaining the balance between potentially beneficial and pathogenic microorganisms associated with fish (Pandiyan et al. 2013). Growth inhibition of pathogenic bacteria by beneficial bacteria could be due to the individual or combined production of antibacterial metabolites (e.g., bacteriocins, siderophores, lysozymes, and proteases), competition for essential nutrients, alteration of pH by organic acid production, and competitive exclusion (Verschuere et al. 2000; De Vrese & Schrezenmeir 2008; Lalloo et al. 2010; Mukherjee & Ghosh 2016). Among these, antimicrobial peptides or bacteriocins have received major attention as an alternative bio-control agent-limiting colonization of pathogenic bacteria in the GI tract of fish (Ghanbari et al. 2013).

In general, the majority of bacteriocins of aquatic origin with antagonistic activities against bacterial pathogens have been reported from marine fish (Sahoo et al. 2016). Numerous studies have revealed that species of *Bacillus* are part of the autochthonous bacterial community of fish (Ray et al. 2012; Askarian et al. 2012; Ghosh et al. 2017). The use of autochthonous, rather than allochthonous, bacilli in aquaculture might be preferred so as to avoid harmful effects on the host fish and the normal microbiota, as the bacterial strains, target pathogens and hosts, all share the same ecological niche (Mukherjee et al. 2016, 2017). For the application of bacteriocinogenic bacilli, research endeavors in aquaculture may be grouped into three categories: (1) bacteriocin production has not been confirmed, but assumed to be the reason behind pathogen inhibition; (2) the proteinaceous nature of the BLIS has been confirmed with partial characterization; and (3) specific bacteriocins produced by bacilli have been characterized and identified. Most reports are associated with the first two categories. In contrast, studies on bacteriocinogenic bacilli from aquatic sources (e.g., water, soil, sediment, seaweed, and fish) and identification of the bacteriocins produced by aquatic animals/environments are scarce (Table 7.2).

TABLE 7.2
Bacteriocins from *Bacillus* spp. Characterized and Identified from Aquatic Resources

Bacteriocinogenic *Bacillus* spp.	Source	Bacteriocin/ BLIS (Molecular Weight)	Inhibited Strains	Reference
B. licheniformis P40	Amazon basin fish and *Leporinus* sp.	BLS P40	*Listeria monocytogenes*, *Bacillus cereus*, and clinical isolates of *Streptococcus* spp.	Cladera-Olivera et al. (2004)
Bacillus sp.	Amazon basin	BLIS (<14 kDa)	*L. monocytogenes* and *B. cereus*	Motta et al. (2004)
Bacillus sp. P45	Amazon basin fish (*Piaractus mesopotamicus*)	BLIS P45 (1.45 kDa)	*L. monocytogenes*, *B. cereus*, and clinical isolates of *Streptococcus* spp.	Sirtori et al. (2006)
B. licheniformis	Irish seaweeds (*Ulva* sp., *U. lactuca*)	Lichenicidin	*L. monocytogenes*, *Enterococcus faecalis*, *Salmonella typhimurium*, *Escherichia coli*, and *Cronobacter sakazakii*	Prieto et al. (2012)
B. pumilus R2	Sea soil	Pumiviticin (3.9 kDa)	*S. typhimurium*, *Proteus vulgaris*, *Micrococcus luteus*, and a wide range of lactic acid bacteria	Rajesh et al. (2012)

(Continued)

TABLE 7.2 (Continued)
Bacteriocins from *Bacillus* spp. Characterized and Identified from Aquatic Resources

Bacteriocinogenic *Bacillus* spp.	Source	Bacteriocin/ BLIS (Molecular Weight)	Inhibited Strains	Reference
B. licheniformis BTHT8	Coastal sediment and water samples	BL8 (1–4 kDa)	*Bacillus circulans*, *Staphylococcus aureus*, *Bacillus coagulans*, *B. cereus*, *Clostridium perfringens*, and *Bacillus pumilus*	Smitha and Bhat (2013)
B. sonorensis MT93	Marine soil	Sonorensin (6.27 kDa)	*L. monocytogenes* and *Staph. Aureus*	Chopra (2014)
B. amyloliquefaciens	Marine fish and *Epinephelus areolatus*	CAMT2 (20.0 kDa)	*L. monocytogenes, Staph. aureus, E. coli*, and *Vibrio parahaemolyticus*	An et al. (2015)

During the last two decades, information on antagonism by fish gut-associated bacilli against many species of pathogenic bacteria has become available (Table 7.3). The first study indicating antibacterial substances produced by bacilli isolated from fish GI tract was reported in a Japanese coastal fish (*Callionymus* sp.) (Sugita et al. 1998). Antibacterial compounds produced by *B. licheniformis* P40 isolated from an Amazon basin fish (*Leporinus* sp.) were found to be bactericidal and bacteriolytic to *Listeria monocytogenes* (Cladera-Olivera et al. 2004). Antimicrobial compounds produced by *B. subtilis* VSG1 isolated from an Indian major carp (rohu, *L. rohita*) were antagonistic to *Aeromonas hydrophila* (Giri et al. 2011). In addition, *B. subtilis* SG4 (Ghosh et al. 2007), *B. aerius* CCH1A, and *B. sonorensis* CCH1Ph (Dutta et al. 2015) isolated from mrigal; *B. methylotrophicus*

TABLE 7.3
Research Efforts Depicting Fish Gut-Associated Bacilli Antagonistic against Pathogenic Strains

Source Fish	Species of Bacilli	Antagonistic to Pathogens	References
Mrigal (*C. mrigala*)	*B. subtilis*	*A. hydrophila*	Ghosh et al. (2007)
Channel catfish (*Ictalurus punctatus*)	*B. subtilis* and *B. methylotrophicus*	*A. hydrophila* and *Edwardsiella ictaluri*	Ran et al. (2012)
Siberian sturgeon (*A. baerii*)	*B. cereus* and *B. circulans*	*A. hydrophila*	Geraylou et al. (2014)
Labeo rohita	*B. subtilis*	*A. hydrophila*	Giri et al. (2011)
Channel catfish (*I. punctatus*)	*B. subtilis*	*A. hydrophila*, *A. sobria*, and *A. caviae*	Luo et al. (2014)
Mrigal (*C. mrigala*)	*B. amyloliquefaciens* and *B. sonorensis*	*A. hydrophila*, *A. veronii*, *A. salmonicida*, *P. putida*, *P. fluorescens*, *B. mycoides*, and *Pseudomonas* sp.	Dutta and Ghosh (2015)
Catla (*C. catla*)	*B. aerius* and *B. sonorensis*	*A. hydrophila*, *A. veronii*, and *A. salmonicida*	Dutta et al. (2015)
Mrigal (*C. mrigala*)	*B. stratosphericus*, *B. aerophilus*, *B. licheniformis*, and *Solibacillus silvestris*	*A. hydrophila*, *A. salmonicida*, *P. fluorescens*, *P. putida*, and *B. mycoides*	Mukherjee et al. (2016)

(Continued)

TABLE 7.3 (Continued)
Research Efforts Depicting Fish Gut-Associated Bacilli Antagonistic against Pathogenic Strains

Source Fish	Species of Bacilli	Antagonistic to Pathogens	References
Catla (*C. catla*)	*B. methylotrophicus* and *B. subtilis* subsp. *spizizenii*	*A. hydrophila*, *A. salmonicida*, *P. fluorescens*, *P. putida*, and *B. mycoides*	Mukherjee & Ghosh (2016)
Rohu (*Labeo rohita*)	*B. subtilis*	*A. hydrophila*, *A. salmonicida*, *P. fluorescens*, and *B. mycoides*	Banerjee et al. (2016)
Striped dwarf catfish (*Mystus vittatus*)	*Bacillus* sp.	*A. hydrophila*, *A. salmonicida*, *A. sobria*, and *P. fluorescens*	Nandi et al. (2017)
Catla, Punti (*Puntius javanicus*)	*Bacillus* sp. and *B. amyloliquefaciens*	*A. hydrophila*, *A. salmonicida*, *A. sobria*, and *P. fluorescens*	Nandi et al. (2017)
Rohu	*B. methylotropicus*, *B. amyloliquefaciens*, and *B. licheniformis*	*A. hydrophila*, *A. salmonicida*, and *A. sobria*	Mukherjee et al. (2017)
Rohu	*B. megaterium*	*Streptococcus mutans*	Sumathi et al. (2017)
Grass carp	*B. subtilis*	*A. hydrophila*, *A. punctata*, *E. ictaluri*, *A. punctata f. intestinali*, *V. flurialis*, and *S. agalactiae*	Guo et al. (2016)
Flounder (*P. olivacues*)	*B. amyloliquefaciens*	*V. anguillarum*, *V. campbellii*, *V. vulnificus*, *V. parahamolyticus*, *E. tarda*, and *Streptococcus* sp.	Chen et al. (2016)
Nile tilapia (*O. niloticus*)	*B. licheniformis*	*A. salmonicida*, *Pseudomonas* sp., *Bacillus mycoides*, *P. putida*, and *P. fluorescens*	Ghosh et al. (2017)
Rainbow trout (O. *mykiss*)	*Bacillus* sp. and *B. pumilus*	*A. hydrophila* and *Enterobacter sakazakii*	Ramirez-Torrez et al. (2018)

isolated from channel catfish (Ran et al. 2012) and catla (Mukherjee & Ghosh 2016); and *B. cereus* and *B. circulans* obtained from the GI tract of other fish species (Lalloo et al. 2010; Geraylou et al. 2014) were found to be antagonistic against different strains of *A. hydrophila* pathogenic to fish. Strains of *B. methylotrophicus* isolated from soil or channel catfish intestine inhibited fish pathogens causing enteric septicemia (*E. ictaluri*) and motile aeromonad septicemia (*A. hydrophila*) (Ran et al. 2012). Another strain, *B. subtilis* BHI344, isolated from the GI tract of channel catfish was shown to depress the growth of pathogenic *A. hydrophila*, *A. sobria*, and *A. caviae in vitro* (Luo et al. 2014), while strain *B. sonorensis* CM2H3L isolated from the gut of mrigal inhibited *in vitro* growth of *A. salmonicida* (Dutta & Ghosh 2015). Mukherjee et al. (2016) found that *B. stratosphericus* KM277362, *B. aerophilus* KM277363, *B. licheniformis* KM277364, and *S. silvestris* KM277365 isolated from the GI tract of mrigal inhibited the *in vitro* growth of *A. hydrophila*, *A. salmonicida*, *P. fluorescens*, and *P. putida*. Inhibition of pathogenic aeromonads was also found in *B. methylotrophicus* (NR116240), *B. amyloliquefaciens* (NR117946), and *B. licheniformis* (NR118996) isolated from rohu (Mukherjee et al. 2017).

Among three bacilli species, *B. subtilis*, *B. cereus*, and *B. amyloliquefaciens*, recovered from the gut of Indian major carp, only *B. amyloliquefaciens* demonstrated antagonistic activity against three fish pathogens: *A. hydrophila*, *Acinetobacter* sp. and *Acinetobacter tandoii* (Kavitha et al., 2018). *B. subtilis* (ATCC 6633) inhibited *in vitro* growth of *A. hydrophila* and *P. fluorescens* (Aly et al. 2008), while *B. subtilis* strains isolated from grass carp intestine exhibited inhibitory activities against fish pathogenic bacteria, including *A. hydrophila*, *A. punctata*, *E. ictaluri*, *A. punctata f. intestinali*, *Vibrio flurialis*, and *Str. agalactiae*, but the inhibitory effect varied with the greatest effect against *A. hydrophila* and *A. punctata* (Guo et al., 2016). Among eight *Bacillus* strains isolated from the intestine of turbot (*Scophthalmus maximus*), olive flounder, grouper, and spotted sardine (*Clupanodon punctatus*), *B. amyloliquefaciens* M001 had the greatest antagonistic activity against multiple aquatic bacterial pathogens including *V. anguillarum, V. campbellii, V. vulnificus, V. parahamolyticus, Streptococcus* sp., and *Edwardsiella tarda* (Chen et al. 2016).

Characterization and identification of the antibacterial compounds produced by these gut bacteria were not evaluated in the above-mentioned studies, and inhibition was assumed to be due to bacteriocins or BLIS. Some of these observations made an attempt for partial purification and characterization of the inhibitory substances, and the proteinaceous nature of the BLIS was confirmed (Giri et al. 2011; Mukherjee et al. 2017). Several reports (e.g. Aly et al. 2008; Chen et al., 2016, b; Banerjee et al. 2017) have suggested that inhibitory activity by fish gut-associated bacilli against fish pathogens could be due to bacteriocins and there are some well-characterized bacteriocins produced by fish gut-associated *Bacillus* spp., for example, *Bacillus* sp. P45 isolated from Amazon basin fish (*Piaractus mesopotamicus*) (Sirtori et al., 2006) and *B. amyloliquefaciens* isolated from the marine fish yellow spotted rockcod (*Epinephelus areolatus*) (An et al., 2015). In a study by Banerjee et al. (2017), *B. subtilis* LR1 isolated from the GI tract of rohu exhibited a bacteriocin (~50 kDa) with inhibitory activity against four fish pathogens: *B. mycoides*, *A. salmonicida*, *P. fluorescens* and *A. hydrophila*. *B. subtilis* and *B. amyloliquefaciens* recovered from the GI tract of southern flounder exhibited antagonistic activity against *V. anguillarum, V. harveyi, V. vulnificus, Streptococcus* sp., and *Staphylococcus aureus* (Chen et al., 2016). Purification and characterization of the specific bio-active compounds and mechanisms behind pathogen inhibition are not, however, understood in most cases.

The use of purified bacteriocins in aquaculture is a controversial issue (Rather et al., 2017) and application of the bacteriocinogenic strains as probiotics has been suggested as more rational and economically feasible approach (Gatesoupe 2008; Karthikeyan & Santhosh, 2009; Issazadeh et al. 2012). The ability of bacteriocinogenic bacilli to sporulate would enable them to establish within the GI tract of their hosts and in the aquatic environments (Rather et al. 2017). Another benefit of using live *Bacillus* spp. is that they are not generally involved in the processes of horizontal gene transfer with Gram-negative bacteria (e.g., *Vibrio* and *Aeromonas* spp.) and, thus, are unlikely to acquire genes of antibiotic resistance or virulence from these species (Moriarty, 1999). The production of antimicrobial substances and sporulation capacity confer *Bacillus* strains with a double advantage in terms of their survival in different habitats. Furthermore, the proteinaceous nature of the bacteriocins or BLIS implies putative degradation of the anti-microbial substance within the GI tract of fish, suggesting their use as prophylactic or therapeutic feed supplements (Ringø et al. 2018). Further studies on purification, characterization, and identification of the bacteriocins from bacilli are, however, necessary to realize the application of the vast array of compounds produced by *Bacillus* species and also to develop a sustainable strategy to explore potential bacteriocinogenic bacilli for disease resistance and pathogen prevention in aquaculture.

BIBLIOGRAPHY

Abriouel H, Farzan CMAP, Omar NB, Galvez A (2011) Diversity and applications of *Bacillus* bacteriocins. *FEMS Microbiol Reviews* **35**, pp.201–232.

An J, Zhu W, Liu Y, Zhang X, Sun L, Hong P, Wang Y, Xu C, Xu D, Liu H. (2015) Purification and characterization of a novel bacteriocin CAMT2 produced by *Bacillus amyloliquefaciens* isolated from marine fish *Epinephelus areolatus*. *Food Control*, **51**, pp.278–282.

Aly SM, Ahmed YSG, Ghareeb AAA, Mohamed MF. (2008) Studies on *Bacillus subtilis* and *Lactobacillus acidophilus*, as potential probiotics, on the immune response and resistance of Tilapia nilotica (*Oreochromis niloticus*) to challenge infections. *Fish and Shellfish Immunology* 25,pp.128–136. https://doi.org/10.1016/j.fsi.2008.03.013

Askarian F, Zhou Z, Olsen RE, Sperstad S, Ringø E (2012) Culturable autochthonous gut bacteria in Atlantic salmon (*Salmo salar* L.) fed diets with or without chitin. Characterization by 16S rRNA gene sequencing, ability to produce enzymes and *in vitro* growth inhibition of for fish pathogens. *Aquaculture* **326–329**, pp.1–8.

Banerjee S, Hansen JN (1988) Structure and expression of a gene encoding the precursor of subtilin, a small protein antibiotic. *The Journal of Biological Chemistry* **263**, pp.9508–9514.

Banerjee S, Mukherjee A, Dutta D, Ghosh K (2016) Non-starch polysaccharide degrading gut bacteria in Indian major carps and exotic carps. *Jordan Journal of Biological Sciences* **9**(1): pp.69–78.

Banerjee G, Nandi A, Ray AK (2017) Assessment of hemolytic activity, enzyme production and bacteriocin characterization of *Bacillus subtilis* LR1 isolated from the gastrointestinal tract of fish. *Archives of Microbiology* **199**, pp.115–124.

Bizani D, Brandelli A. (2002) Characterization of a bacteriocin produced by a newly isolated *Bacillus* sp. *Journal of Applied Microbiology* **93**, pp.512–519.

Chen Y, Li J, Xiao P, Zhu W, Mo ZL (2016) The ability of marine *Bacillus* spp. isolated from fish gastrointestinal tract and culture pond sediment to inhibit growth of aquatic pathogenic bacteria. *Iranian Journal of Fisheries Sciences* **15**, pp.701–714.

Chen Y, Li J, Xiao P, Li GY, Yue S, Huang J, et al. (2016) Isolation and characterization of *Bacillus* spp. M001 for potential application in turbot (*Scophthalmus maximus* L) against *Vibrio anguillarum*. *Aquaculture Nutrition* **22**, pp.374–381.

Cherif A, Ouzari H, Daffonchio D, Cherif H, Ben Slama K, Hassen A, Jaoua S, Boudabous A (2001) Thuricin 7: A novel bacteriocin produced by *Bacillus thuringiensis* BMG1.7, a new strain isolated from soil. *Letters in Applied Microbiology* **32**, pp.243–247.

Chopra L, Singh G, Choudhary V, Sahoo DK. (2014) Sonorensin: An antimicrobial peptide, belonging to the heterocycloanthracin subfamily of bacteriocins, from a new marine isolate, Bacillus sonorensis MT93. *Applied and Environmental Microbiology* **80**, pp.2981–2990.

Cladera-Olivera F, Caron GR, Brandelli A (2004) Bacteriocin-like substance production by *Bacillus licheniformis* strain P40. *Letters in Applied Microbiology* **38**, pp.251–256.

Cotter PD, Hill C, Ross RP (2005) Bacteriocins: Developing innate immunity for food. *Nature Reviews Microbiology* **3**, pp.777–788.

Das P, Ghosh K (2015) Improvement of nutritive value of sesame oil cake in formulated diets for rohu, *Labeo rohita* (Hamilton) after bio-processing through solid state fermentation by a phytase-producing fish gut bacterium. *International Journal of Aquatic Biology* **3**, pp.89–101.

De Vrese D, Schrezenmeir M (2008) A comprehensive review that critically evaluates modes of actions and medicinal applications of prebiotics and synbiotics. *Advances in Biochemical Engineering/Biotechnology* **111**, 1–66.

Dimitroglou A, Merrifield DL, Carnevali O, Picchietti S, Avella M, Daniels C, et al. (2011) Microbial manipulations to improve fish health and production-A Mediterranean perspective. *Fish and Shellfish Immunology* **30**, pp.1–16.

Drider D, Fimland G, Hechard Y, Mcmullen LM, Prevost H (2006) The continuing story of class IIa bacteriocins. *Microbiology and Molecular Biology Reviews*, 70, pp.564–582.

Dubois JYF, Kouwen TRHM, Schurich A. KC, Reis CR, Ensing HT, Trip EN, Zweers, JC, van Dijl JM. (2009) Immunity to the bacteriocin sub lancin168 is determined by the SunI (YolF) protein of *Bacillus subtilis*. *Antimicrobial Agents and Chemotherapy* **53**, pp.651–661

Dutta D, Banerjee S, Mukherjee A, Ghosh K (2015) Selection and probiotic characterization of exoenzyme-producing bacteria isolated from the gut of *Catla catla* (Actinopterygii: Cypriniformes: Cyprinidae). *Acta Ichthyologica et Piscatoria* **45**, pp.373–384.

Dutta D, Ghosh K (2015) Screening of extracellular enzyme-producing and pathogen inhibitory gut bacteria as putative probiotics in mrigal, *Cirrhinus mrigala* (Hamilton, 1822). *International Journal of Fisheries and Aquatic Studies* **2**, pp.310–318.

Dutta D, Banerjee S, Mukherjee A, Ghosh K (2018) Potential gut adherent probiotic bacteria isolated from rohu, *Labeo rohita* (Actinopterygii: Cypriniformes: Cyprinidae): Characterisation, exo-enzyme production, pathogen inhibition, cell surface hydrophobicity, and bio-film formation. *Acta Ichthyologica et Piscatoria* **48**(3): pp.221–233.

Gatesoupe FJ (2008) Updating the importance of lactic acid bacteria in fish farming: Natural occurrence and probiotic treatments. *Journal of Molecular Microbiology and Biotechnology* **14**, pp.107–114.

Gautam N, Sharma N (2009) Bacteriocin: Safest approach to preserve food products *Indian Journal of Microbiology* **49**(3): pp.204–11. https://doi.org/10.1007/s12088-009-0048-3

Ghanbari M, Jami M, Kneifel W, Domig KJ (2013) Antimicrobial activity and partial characterization of bacteriocins produced by lactobacilli isolated from sturgeon fish. *Food Control* **32**, pp.379–385.

Ghosh S, Sinha A, Sahu C (2007) Isolation of putative probionts from the intestines of Indian major carps. *Israeli Journal of Aquaculture- Bamidgeh* **59**, pp.127–132.

Ghosh K, Banerjee S, Moon UM, Khan HA, Dutta D (2017) Evaluation of gut associated extracellular enzyme-producing and pathogen inhibitory microbial community as potential probiotics in Nile tilapia, *Oreochromis niloticus. International Journal of Aquaculture* **7**, pp.143–158.

Ghosh K, Roy M, Kar N, Ringø E (2010) Gastrointestinal bacteria in rohu, *Labeo rohita* (Actinopterygii: Cypriniformes: Cyprinidae): Scanning electron microscopy and bacteriological study. *Acta Ichthyologica et Piscatoria* **40**, pp.129–135.

Ghosh K, Sen SK, Ray AK (2002) Characterization of bacilli isolated from gut of rohu, *Labeo rohita*, fingerlings and its significance in digestion. *Journal of Applied Aquaculture* **12**, pp.33–42.

Ghosh K, Sen SK, Ray AK (2004) Growth and survival of rohu, *Labeo rohita* (Hamilton, 1822) spawn fed diets fermented with intestinal bacterium, *Bacillus circulans. Acta Ichthyologica Et Piscatoria* **34**(2): pp.155–165

Giri SS, Sukumaran V, Sen SS, Vinumonia J, Nazeema-Banu B, Jena PK (2011) Antagonistic activity of cellular components of potential probiotic bacteria, isolated from the gut of *Labeo rohita*, against *Aeromonas hydrophila. Probiotics and Antimicrobial Proteins* **3**, pp.214–222.

Geraylou Z, Vanhove MPM, Souffreau C, Rurangwa E, Buyse J, Ollevier F. (2014) *In vitro* selection and characterization of putative probiotics isolated from the gut of *Acipenser baerii* (Brandt, 1869). *Aquaculture Research* **45**, pp.341–352.

Gray EJ, Di Falco M, Souleimanov A, Smith DL (2006) Proteomic analysis of the bacteriocin thuricin 17 produced by *Bacillus thuringiensis* NEB17. *FEMS Microbiol Letters* **255**, pp.27–32.

Guo X, Chen D-D, Peng K-S, Cui Z-W, Zhang X-J, Li S, Zhang Y-A (2016) Identification and characterization of *Bacillus subtilis* from grass carp (*Ctenopharynodon idellus*) for use as probiotic additives in aquatic feed. *Fish and Shellfish Immunology*, **52**, pp.74–84, https://doi.org/10.1016/j.fsi.2016.03.017

Issazadeh K, Reza Majid-Khoshkhol Pahlaviani M, Massiha A (2012) Isolation of *Lactobacillus* species from sediments of Caspian Sea for bacteriocin production. In: *2nd International Conference on Biomedical Engineering and Technology IPCBEE* **34**: pp.79–84. IACSIT Press, Singapore.

Kavitha M, Raja M, Perumal P (2018) Evaluation of probiotic potential of *Bacillus* spp. isolated from the digestive tract of freshwater fish *Labeo calbasu* (Hamilton, 1822).

Karthikeyan V, Santhosh SW (2009) Study of bacteriocin as a food preservative and the *L.*

Kim D-H, Kim D-Y (2013) Microbial diversity in the intestine of olive flounder (*Paralichthys olivaceus*). *Aquaculture* **414–415**, pp.103–108.

Kim D-H, Brunt J, Austin B (2007) Microbial diversity of intestinal contents and mucus in rainbow trout (*Oncorhynchus mykiss*). *Journal of Applied Microbiology* **102**, pp.1654–1664.

Kim YK, Park IS, Kim DJ, Nam BH, Kim DG, Jee YJ, et al. (2014) Identification and characterization of a bacteriocin produced by an isolated *Bacillus* sp. SW1-1 that exhibits antibacterial activity against fish pathogens. *Journal of the Korean Society for Applied Biological Chemistry* **57**, pp.605–612.

Kiss A, Balikó G, Csorba A, Chuluunbaatar T, Medzihradszky KF, Alföldi L (2008) Cloning and characterization of the DNA ¨region responsible for megacin A-216 production in *Bacillus megaterium* 216. *Journal of Bacteriology* **190**, pp.6448–6457.

Klaenhammer TR (1993) Genetics of bacteriocins produced by lactic acid bacteria. *FEMS Microbiology Letters* **12**, pp.224–227.

Kolndadacha OD, Adikwu IA, Okaeme AN, Atiribom RY, Mohammed A, Musa YM (2011) The role of probiotics in aquaculture in Nigeria – A review. *Continental Journal of Fisheries Aquatic Science* **5**, pp.8–15.

Lalloo R, Ramchuran S, Ramduth D, Gorgens J, Gardiner N (2007) Isolation and selection of *Bacillus* spp. as potential biological agents for enhancement of water quality in culture of ornamental fish. *Journal of Applied Microbiology* **103**(5): pp.1471–1479.

Lalloo R, Moonsamy G, Ramchuran S, Görgens J, Gardiner N (2010) Competitive exclusion as a mode of action of a novel *Bacillus cereus* aquaculture biological agent. *Letters in Applied Microbiology* **50**, pp.563–570.

Lawton EM, Cotter PD, Hill C, Ross RP (2007) Identification of a novel two-peptide lantibiotic, haloduracin, produced by the alkaliphile *Bacillus halodurans* C-125. *FEMS Microbiology Letters* **267**, pp.64–71.

Lee H, Kim HY (2011) Lantibiotics, class I bacteriocins from the genus *Bacillus*. *Journal of Microbiology and Biotechnology* **21**, pp.229–235.

Le Marrec C, Hyronimus B, Bressollier P, Verneuil B, Urdaci MC. (2000) Biochemical and genetic characterization of coagulin, a new antilisterial bacteriocin in the pediocin family of bacteriocins, produced by *Bacillus coagulans* I4. *Applied and Environmental Microbiology* **66**, pp.5213–5220.

Li T, Long M, Gatesoupe FJ, Zhang Q, Li A, Gong X (2015) Comparative analysis of the intestinal bacterial communities in different species of carp by pyrosequencing. *Microbial Ecology* **69**, pp.25–36.

Li Z, Xu L, Liu W, Liu Y, Ringø E, Du Z, Zhou Z (2015) Protein replacement in practical diets altered gut allochthonous bacteria of cultured cyprinid species with different food habits. *Aquaculture International* **23**, pp.913–928.

Liu H, Guo X, Gooneratne R, Ruifang L, Cong Z, Fanbin Z, Weimin W. (2016). The gut microbiome and degradation enzyme activity of wild freshwater fishes influenced by their trophic levels. *Scientific Report*, **6**, pp.24340; doi:https://doi.org/10.1038/srep24340

Luo Z, Bai XH, Chen CF (2014) Integrated application of two different screening strategies to select potential probiotics from the gut of channel catfish *Ictalurus punctatus*. *Fisheries Science* **80**, pp.1269–1275.

Martirani L, Varcamonti M, Naclerio G, De Felice M (2002) Purification and partial characterization of bacillocin 490, a novel bacteriocin produced by a thermophilic strain of *Bacillus licheniformis*. *Microbial Cell Factories* **1**, pp.1–5.

Mokoena MP (2017) Lactic acid bacteria and their bacteriocins: Classification, biosynthesis and applications against uropathogens: A mini-review. *Molecules* **22**, p.1255.

Moriarty DJW (1999) Disease control in shrimp aquaculture with probiotic bacteria. In: *Proceedings of the Eighth International Symposium on Microbial Ecology* (ed. Bell CR, Brylinsky M, Johnson-Green P) pp.237–244. Halifax, NS, Canada: Atlantic Canada Society for Microbial Ecology.

Motta AS, Cladera-Olivera F, Brandelli A. (2004) Screening for antimicrobial activity among bacteria isolated from the Amazon basin. *Brazilian Journal of Microbiology* **35**,pp.307–310.

Mukherjee A, Ghosh K (2016) Antagonism against fish pathogens by cellular components and verification of probiotic properties in autochthonous bacteria isolated from the gut of an Indian major carp, *Catla catla* (Hamilton). *Aquaculture Research* **47**, pp.2243–2255.

Mukherjee A, Dutta D, Banerjee S, Ringø E, Breines EM, Hareide E, et al. (2016) Potential probiotics from Indian major carp, *Cirrhinus mrigala*. Characterization, pathogen inhibitory activity, partial characterization of bacteriocin and production of exoenzymes. *Research in Veterinary Science* **108**, pp.76–84.

Mukherjee A, Dutta D, Banerjee S, Ringø E, Breines EM, Hareide E, et al. (2017) Culturable autochthonous gut bacteria in rohu, *Labeo rohita*. *In vitro* growth inhibition against pathogenic *Aeromonas* spp., stability in gut, bio-safety and identification by 16S rRNA gene sequencing. *Symbiosis* **73**, pp.165–177.

Nandi A, Banerjee G, Dan SK, Ghosh P, Ghosh K, Ray AK (2017) Screening of autochthonous intestinal microbiota as candidate probiotics isolated from four freshwater teleosts. *Current Science* **113**, pp.767–773.

Nandi A, Dan SK, Banerjee G, Ghosh P, Ghosh K, Ringø E, Ray AK (2017) Probiotic potential of autochthonous bacteria isolated from the gastrointestinal tract of four freshwater teleost. *Probiotics and Antimicrobial Proteins* **9**, pp.12–21.

Nath S, Chowdhury S, Dora KC (2015) Application of *Bacillus* sp. as a biopreservative for food preservation. *International Journal of Engineering Research and Applications* **5**(4), (Part-2): pp.85–95.

Nes IF, Yoon SS, Diep DB (2007) Ribosomally synthesized antimicrobial peptides (bacteriocins) in lactic acid bacteria: A review. *Food Science and Biotechnology* **16**, pp.675–690.

Oscariz JC, Lasa I, Pisabarro AG (1999) Detection and characterization of cerein 7, a new bacteriocin produced by *Bacillus cereus* with a broad spectrum of activity. *FEMS Microbiol Letters* **178**, pp.337–341.

Oscariz JC, Pisabarro AG (2000) Characterization and mechanism of action of cerein 7, a bacteriocin produced by *Bacillus cereus* Bc7. *Journal of Applied Microbiology* **89**, pp.1–10.

O'Sullivan L, Ross RP, Hill C (2002) Potential of bacteriocin producing lactic acid bacteria for improvements in food safety and quality. *Biochimie* **84**, pp.593–604.

Paik HD, Bae SS, Pan JG (1997) Identification and partial characterization of tochicin, a bacteriocin produced by *Bacillus thuringiensis* subsp. *tochigiensis*. *Journal of Industrial Microbiology and Biotechnology* **19**, pp.294–298.

Pandiyan P, Balaraman D, Thirunavukkarasu R, George EGT, Subaramaniyan K, Manikkam S, et al. (2013). Probiotics in aquaculture. *Drug Invention Today* **5**, pp.55–59.

Parisot J, Carey S, Breukink E, Chan WC, Narbad A, Bonev B (2008). Molecular mechanism of target recognition by subtilin, a Class I lanthionine antibiotic. *Antimicrobial Agents and Chemotherapy* **52**, pp.612–618.

Patil R, Jeyasekaran G, Shanmugam SA, Shakila JR (2001) Control of bacterial pathogens, associated with fish diseases, by antagonistic marine actinomycetes isolated from marine sediments. *Indian Journal of Marine Science* **30**, pp.264–267.

Prieto ML, O'Sullivan L, Tan SP, McLoughlin P, Hughes H, O'Connor PM, Cotter PD, Lawlor PG, Gardiner GE. (2012) Assessment of the bacterioci-nogenic potential of marine bacteria reveals lichenicidin production by seaweed derived Bacillus spp. *Marine Drugs* **10**, pp.2280–2299.

Rajesh D, Karthikeyan S, Jayaraman G. (2012) Isolation and partial characterization of a new bacteriocin from *Bacillus pumilus* DR2 isolated from seawater. *CIBTech Journal of Microbiology* **1**, pp.33–41.

Ramirez-Torrez JA, Monroy-Dosta MC, Hernandez-Hernandez LH, Castro-Mejia Bustos-Martinez JA, Hamdan-Partida A. (2018) Presumptive probiotic isolated from *Oncorhynchus mykiss* (Walbaum, 1792), cultivated in Mexico. *The International Journal of Aquatic Science* **9**, pp.3–12.

Ran C, Carrias A, Williams MA, Capps N, Dan BCT, et al. (2012) Identification of *Bacillus* strains for biological control of catfish pathogens. *PLoS ONE* **7**, p.e45793.

Rather IA, Galope R, Bajpai VK, Lim J, Paek WK, Park Y (2017) Diversity of marine bacteria and their bacteriocins: Applications in aquaculture. *Reviews in Fisheries Science & Aquaculture* **25**(4): pp.257–269.

Ray AK, Bairagi A, Sarkar Ghosh K, Sen SK (2007) Optimization of fermentation conditions for cellulase production by *Bacillus subtilis* CY5 and *Bacillus circulans* TP3 isolated from fish gut. *Acta Ichthyologica et Piscatoria* **37**, pp.47–53.

Ray AK, Ghosh K, Ringø E (2012) Enzyme-producing bacteria isolated from fish gut: A review. *Aquaculture Nutrition* **18**, pp.465–492.

Ray AK, Roy T, Mondal S, Ringø E (2010) Identification of gut-associated amylase, cellulase and protease-producing bacteria in three species of Indian major carps. *Aquaculture Research* **41**, pp.1462–1469.

Riley M (2009) Bacteriocins, biology, ecology, and evolution. In: *Encyclopedia of Microbiology*, Schaechter M, Ed. 3rd ed. pp.32–44. Academic Press: Oxford, UK.

Riley MA, Wertz JE (2002a) Bacteriocin diversity: Ecological and evolutionary perspectives. *Biochimie* **84**, pp.357–364.

Riley MA, Wertz JE (2002b) Bacteriocins: Evolution, ecology, and application. *Annual Review of Microbiology* **56**, pp.117–137.

Ringø E, Hoseinifar SH, Ghosh K, Doan HV, Beck BR, Song SK (2018) Lactic acid bacteria in finfish-An update. *Frontiers in Microbiology* **9**, pp.1818

Ringø E, Sperstad S, Myklebust R, Mayhew TM, Olsen RE (2006) The effect of dietary inulin on aerobic bacteria associated with hindgut of Arctic charr (*Salvelinus alpinus* L.). *Aquaculture Research* **37**, pp.891–897.

Sahoo TK, Jena PK, Patel AK, Seshadri S (2016) Bacteriocins and their applications for the treatment of bacterial diseases in aquaculture: A review. *Aquaculture Research* **47**, pp.1013–1027.

Sarkar B, Ghosh K. (2014) Gastrointestinal microbiota in *Oreochromis mossambicus* (Peters) and *Oreochromis niloticus* (Linnaeus): Scanning electron microscopy and microbiological study. *International Journal of Fisheries and Aquatic Studies* 2, pp.78–88.

Sirtori LR, Cladera-Olivera F, Lorenzini DM, Tsai SM, Brandelli A (2006) Purification and partial characterization of an antimicrobial peptide produced by *Bacillus* sp. strain P45, a bacterium from the Amazon basin fish *Piaractus mesopotamicus*. *Journal of General and Applied Microbiology* **52**, pp.357–363.

Smitha S, Bhat SG (2013) Thermostable bacteriocin BL8 from *Bacillus licheniformis* isolated from marine sediment. *Journal of Applied Microbiology* **114**, pp.688–694.

Stein T (2005) *Bacillus subtilis* antibiotics: Structures, syntheses and specific functions. *Molecular Microbiology* **56**, pp.845–857.

Sugita H, Hirose Y, Matsuo N, Deguchi Y (1998) Production of antibacterial substances by *Bacillus* sp. strain NM 12, an intestinal bacterium of Japanese coastal fish. *Aquaculture* **165**, pp.269–280.

Sumathi C, Nandhini A, Padmanaban J (2017) Antagonistic activity of probiotic *Bacillus megaterium* against *Streptococcus mutans*. *International Journal of Pharmacological and Biological Sciences* **8**, pp.270–274.

Sumi CD, Yang BW, Yeo IC, Hahm YT (2015) Antimicrobial peptides of the genus *Bacillus*: A new era for antibiotics. *Canadian Journal of Microbiology* **61**(2): pp.93–103. 10.1139/cjm-2014-0613

Sutyak KE, Wirawan RE, Aroutcheva AA, Chikindas ML (2008) Isolation of the *Bacillus subtilis* antimicrobial peptide subtilosin from the dairy product-derived *Bacillus amyloliquefaciens*. *Journal of Applied Microbiology* **104**, pp.1067–1074.

Teixeira ML, Cladera OF, dos Santos J, Brandelli A (2009) Purification and characterization of a peptide from *Bacillus licheniformis* showing dual antimicrobial and emulsifying activities. *Food Research International* **42**, pp.63–68.

Verschuere L, Rombaut G, Sorgeloos P, Verstraete W (2000) Probiotic bacteria as biological control agents in aquaculture. *Microbiology and Molecular Biology Reviews* **64**, pp.655–671.

Von Döhren H (1995) Peptides: In: *Genetics and Biochemistry of antibiotic production* (Ed. Vining LC, Stuttard C). pp.129–171. Newton MA. Butterworth-Heinemann.

Yilmaz S, Sova M, Ergün S (2018) Antimicrobial activity of trans-cinnamic acid and commonly used antibiotics against important fish pathogens and nonpathogenic isolates. *Journal of Applied Microbiology*. https://doi.org/10.1111/jam.14097

Zheng G, Yan LZ, Vederas JC, Zuber P (1999) Genes of the bulb locus of *Bacillus subtilis* are required for production of the antilisterial bacteriocin subtilosin. *Journal of Bacteriology* **181**, pp.7346–7355.

8 *Bacillus* Probiotics as the Bioremediatory Tools in Aquaculture

Mehdi Soltani
Murdoch University, Perth, Australia
University of Tehran, Tehran, Iran

Felix Kofi Agbeko Kuebutornye
Institute of Aquaculture and Protection of Waters, České Budějovice, Czech Republic

Vivian Hlordzi
University College Cork, Cork, Republic of Ireland

Emmanuel Delwin Abarike
University for Development Studies, Tamale, Ghana

Yishan Lu
Guangdong Ocean University, Zhanjiang, China

Jan Mraz
Institute of Aquaculture and Protection of Waters, České Budějovice, Czech Republic

8.1 INTRODUCTION

Bioremediation is a process in which living organisms, for instance, beneficial microbial agents, such as *Bacillus* bacteria, are used to remove the pollutants from water and soil. With the rapid development of aquaculture activities worldwide, it is well known that aquatic ecosystems are greatly affected by pollutions. Water pollution can damage both the aquatic organisms and the terrestrial animals and birds. Especially, the contaminated water abolishes the aquatic animal life cycle and decreases their reproductions. Seeding of contaminated wastewater or rearing water of aquatic organisms with capable probiotics such as *Bacillus* spp. can simply and rapidly degrade the harmful substances such as the toxic gases. From the practical aquaculture point of view, this method is usually applicable in most water and soil treatments. The application of *Bacillus* probiotics may be natural or can be prepared in the laboratory to attack the target waste substances.

It is notable that the role of probiotics in bioremediation processes of rearing water and soil in aquaculture is great, as microbial bioremediation can be cost-effective and quickly destroy or immobilize contaminants in a way that finally increases the aquaculture production and protects human health and the environment (Gadd, 2000; Gheewala & Annachhatre, 1997; Heitzer & Sayler, 1993). However, the optimization of bioremediation still requires more research works in regard to the facilities such as exogenous, specialized microbes or genetically engineered microbes (Brim et al.,

DOI: 10.1201/9781003503811-8

2003). Thus, a fruitful cost-effective microbial bioremediation plan is dependent on hydro geologic conditions, the contaminant, microbial ecology, and other temporal parameters that are varied. In the bioremediation processes, the application of probiotics is to use the contaminants as the sources of nutrients or energy (Tang et al., 2007). Also, the bioremediation action via probiotics can be enhanced by adding some nutrients such as nitrogen and phosphorus, oxygen, and chemical substrates, e.g., methane, phenol, and toluene (Baldwin et al., 2008; Ma et al., 2007). From available literature members of bacteria genera of *Bacillus, Pseudomonas, Vibrio, Micrococcus, Rhodococcus, Sphingomonas, Acromobacter, Alcaligenes, Arthrobacter, Cinetobacter, Corneybacterium*, and *Flavobacterium* are commonly used in the bioremediation processes (Gupta et al., 2001, Kim et al., 2007). However, some *Bacillus* species, e.g., *B. subtilis* and *B. licheniformis*, and lactic acid bacteria, e.g., *Lactobacillus plantarum, L. casei*, and *Streptococcus lacti*, as well as photosynthetic bacteria, e.g., *Rhodopseudomonas palustris* and *Rhodobacter spaeroides*, are the main microorganisms for treatment of wastewaters (Narmadha & Kavitha, 2012). Regarding the treatment of the rearing water of aquatic animal species, the used probiotic must be non-pathogenic and safe for the environment and public health, and application of *Bacillus* probiotics has become more attractive and effective.

It has been demonstrated that the application of *Bacillus sp.* in the vicinity to pond aerators can reduce chemical oxygen demand and increase shrimp harvest (Porubcan, 1991). Over 15 species of *Bacillus* bacteria have been introduced as the commercial probiotic (bioremediators) products for pond aquaculture. Although the bioremediation agents such as *Bacillus sp.* are known as a significant management tool in fish and shrimp culture, their efficacy depends on various measures including understanding the nature of competition between species or strains of bacteria and their concentration in the water column. Therefore, bioremediation with some strains of *Bacillus sp.* may have a prefect effect on the growth and survival of the aquatic animal, while some other strains may cause an adverse effect. It has been shown that some members of the genus *Bacillus* including B. *subtilis, B. licheniformis, B. cereus*, and *B. coagulans* and species of *Phenibacillus polymyxa* (genus *Phenibacillus*) are decent examples of bacteria appropriate for bioremediation of organic detritus in the rearing water and soil/sediment of aquaculture ponds. It has also been demonstrated that some *Bacillus* probiotics such as *B. subtilis* and *B. licheniformis* in adequate concentrations are good competitors with the bacterial flora naturally existing in the water for the available organic matter, e.g., excess feed and shrimp faeces with consequent reduction of toxic gases, i.e., NH_3 and NO_2 (Sharma, 1999; Singh, 2002). This chapter is focused on the bioremediation functions of *Bacillus* probiotics for the treatment of rearing water of aquaculture species.

8.2 *BACILLUS* SPP. AS BIOREMEDIATORS OF AQUACULTURE WATER QUALITY

Aquaculture activity can be affected by various sources of pollutions, both internal and external; hence, bioremediation by probiotic bacteria has a great potential for removing contaminated rearing waters and soil. An efficacious bioremediation of rearing water by probiotics can happen through different ways including optimizing nitrification rates to maintain a low ammonia concentration; optimizing denitrification rates to remove excess nitrogen from the rearing ponds as nitrogen gas; increasing sulphide oxidation to reduce accumulated hydrogen sulphide; improving carbon mineralization to carbon dioxide to minimize accumulation of sludge; increasing primary pond productivity to improve the production of target aquatic organism and secondary crops; and sustaining a diverse and stable pond community to evade undesirable species becoming dominant (Balcázar et al., 2006; Devaraja et al., 2013; Divya et al., 2015). Effect of probiotics is, though, related with a good understanding of the nature of competition between species or strains of bacteria.

Bacillus spp. in the form of probiotics are particularly related with enhanced water quality parameters including nitrate, nitrite, and ammonia (Divya et al., 2015; Xie et al., 2013). For instance, addition of commercial *Bacillus* spp. probiotics on bacterial population and phytoplankton population in an intensive *L. vannamei* culture with a recirculation system enhanced water quality and improved total heterotrophic bacteria population in the sediment and significantly elevated density of Pyrrophyta algae (de Paiva-Maia et al., 2013).

The basis is that Gram-positive bacteria are more effective in converting organic matter into CO_2 than Gram-negative bacteria (Kumar et al., 2016). The accumulation of dissolved and particulate organic substances is quite a common phenomenon seen during shrimp culture processes, though high load of Gram-positive probiotic species such as *Bacillus* sp. can be administered to minimalize the loads of organic carbon in the aquaculture system (Divya et al., 2015). In addition, *Bacillus* bacteria are well-known microorganisms as potential bioremediators for organic wastes (Panigrahi et al., 1992; Thomas & Ward, 1992). These probiotic spore-forming bacteria contribute to the population of beneficial microbiome in aquatic ecosystems and control harmful microorganisms, subsequently improving the decomposition of unwanted organic substances in water and sediments (Chávez-Crooker & Obreque-Contreras, 2010).

Probiotic *Bacillus* can compete with natural microflora for available organic substances including aquatic animal faeces and uneaten feed. The result of such competition is determined by the capacity of the bacilli strain for enzyme production and water quality conditions including dissolved oxygen, water temperature, pH, and organic and inorganic substances. As an example, Yu et al. (2012) revealed that from nine strains of bacteria obtained from wastewater, *B. amyloliquefaciens* showed the highest ammonia nitrogen degrading activity at 35°C, pH 7.0, and aeration at 200 r/min. This degrading activity was gradually decreased when water temperature and pH exceeded 35°C and 7.0, respectively. Verstraete and Focht (1977) suggested that temperature is a serious factor in affecting bacterial growth, absorption, and utilization of substances, and this is associated with the level of bacterial enzyme activity. Additionally, the degradation capacity of ammonia nitrogen by bacilli bacteria is associated with environmental alkaline/acidic condition (Verstraete & Focht, 1977; Wilks et al., 2009). Further, degradation level is decreased with an enhancing carbon (glucose) rate, probably due to glucose excess inhibiting degradation of ammonia nitrogen, suggesting that ammonia nitrogen consumed in the medium is used to proliferate new bacterial cells. The statement that maximum ammonia nitrogen degradation was seen at a shaking speed of 200 r/min and gradually reduced afterward shows that some bacilli species can enhance ammonia from the water column more efficiently under aerobic than under anaerobic conditions.

Various studies have revealed the effectiveness of commercial and indigenous bacilli bacteria as valuable probiotics for enhancing rearing water quality variables in ponds or recirculation biofiltration systems. *Bacillus* species such as *B. subtilis, B. licheniformis, B. cereus*, and *B. coagulans* can act as suitable probiotics for bioremediation of organic wastes (Divya et al., 2015), even though they are not naturally present in sufficient load densities in water column or sediment. Among bacilli bacteria species, *B. subtilis* and *B. licheniformis* have been shown as more suitable candidates for bioremediation of rearing water quality (Singh, 2002). Despite adequate data on the positive effects of bacilli probiotics for bioremediation of water quality in aquaculture, there are minimum reports describing their nil or negative effects. For instance, Gupta et al. (2016) found that an eight-week application of *Paenibacillus polymyxa* (10^3–10^5 cfu/mL) as a probiotic in common carp rearing water had no significant effect on water quality parameters including pH, total ammonia, nitrite, and dissolved oxygen (Table 8.1).

8.3 MECHANISMS USED BY PROBIOTIC *BACILLUS* IN IMPROVING WATER QUALITY

Accumulation of nitrogenous and organic wastes such as ammonia and nitrite increases loads of organic matter as a result of aquaculture activity, and such waste products are toxic to reared fish and shrimp and can cause stress, morbidity, and mortality in the farmed animal (Loh, 2017). Total ammonia nitrogen (TAN), nitrate, and nitrite are different forms of nitrogen and are used by some microorganisms including probiotics for their metabolism, contributing to nitrogen removal from

TABLE 8.1
Some Details of Bioremedatory Effects of *Bacillus* Probiotics Used in Aquaculture

Type of *Bacillus*	Animal Species	Source/Dose	Water temp (°C)	Efficacy	Reference
B. subtilis + *B. megaterium*	Red-parrot fish (*C. citrinellum* × *C. synspilum*)	Commercial (1 × 10^{10} CFU/trail) twice a week	28	Total ammonia nitrogen (D), chemical oxygen demand (D), water transparency (I)	Chen and Chen (2001)
Bacillus sp.	*P. monodon*	Commercial (7 × 10^{11} cells/0.8 ha pond)	28	Water transparency (I), total organic carbon (D), *Vibrio* density (D)	Dalmin et al. (2001)
Bacillus sp. + *Saccharomyces* sp. *Bacillus* sp. + *Nitrosomonas* sp. + *Nitrobacter* sp.	*P. monodon*	Commercial (1 ×10^8 + 5.6 10^5 CFU mL^{-1}) (1 × 10^8 CFU mL^{-1})	30	Total ammonia nitrogen (I), nitrate (I), ammonia (I), transparency (I), hardness (I), chemical oxygen demand (D), pH (NS), hydrogen sulphide (NS)	Matias et al. (2002)
Bacillus sp. + *S. cerevisiae*, + *Nitrosomonas* sp. + photosynthetic bacteria + *Nitrobacter* sp.	*L. vannamei*	Commercial (1 × 10^4–10^9 CFU mL^{-1})	22.34	pH (NS), phosphate (D), water transparency (I), dissolved oxygen (I), total inorganic nitrogen (D), chemical oxygen demand (D)	Wang et al. (2005)
B. subtilis + *B. cereus* + *B. licheniformis*	*C. carpio*	Indigenous (each at 1 × 105 CFU L^{-1})	20 ± 1	Nitrite (D), nitrate (D), ammonia (D), and *A. hydrophila* count (D)	Lalloo et al. (2007)
B. subtilis	*P. reticulata*, *P. sphenops*, *X. helleri*, and *X. maculatus*	Indigenous (5 × 9 10^5–5 × 10^8 cells mL^{-1})	30–32	Dissolved organic matter (D), total ammonia nitrogen (D), total motile *Aeromonas* (D), and total coliforms (D)	Ghosh et al. (2008)
B. pumilus	*P. monodon*	Indigenous (1 × 10^6 CFU mL^{-1} every 3 days)	28	Total ammonia nitrogen (D), nitrate (D), dissolved oxygen (NS), salinity (NS), and pH (NS)	Banerjee et al. (2010)
Bacillus sp.	*M. Rosenberger*	1 × 10^6 cells mL^{-1}	29.5–29.8	Ammonia (D), nitrite (D), pH (D), and dissolved oxygen (NS)	Rahiman et al. (2010)
B. subtilis	*Tilapia nilotica*	0.1 g L^{-1} at unknown density for 67 days	25–28	Total ammonia nitrogen (D), dissolved oxygen (I), pH (I)	Mohamed and Refat. (2011)
Bacillus sp. +*Lactobacillus*	*L. vannamei*	Commercial (1.65 × 10^{10} CFU L^{-1} for 16 weeks	26.7–28.7	Pyrrophyta density (I), quality of water sediment (I), water quality (I)	de Paiva-Maia et al. (2013)

(Continued)

TABLE 8.1 (Continued)
Some Details of Bioremedatory Effects of *Bacillus* Probiotics Used in Aquaculture

Type of *Bacillus*	Animal Species	Source/Dose	Water temp (°C)	Efficacy	Reference
Mixed *Bacillus (B. thuringiensis, B. megaterium, B. polymyxa, B. licheniformis, B. subtilis,* and *B. circulans)*	*L. vannamei*	Commercial (each probiotic at 1×10^9 CFU mL^{-1})	31–32	pH (D), ammonia (D), nitrite (D), nitrate (NS), salinity (NS), dissolved oxygen (NS), and phosphate (NS)	Nimrat et al. (2012)
B. subtilis	*C. idellus*	Indigenous (1×10^9 CFU m^{-3})	Unknown	Ammonia (D), nitrite (D), nitrate (NS), and total nitrogen (D)	Zhang et al. (2013)
B. subtilis	*L. vannamei*	10^5 and 10^8 CFU mL^{-1} (8 weeks)	28	Ammonia (D), nitrite (D), and nitrate (D)	Zokaeifar et al. (2014)
B. pumilus + *B. licheniformis* + *B. subtilis*	*P. monodon*	Indigenous marine water + soil (1×10^6 CFU mL^{-1})	28	Nitrate (D), dissolved oxygen (NS), salinity (NS), and pH (NS)	Devaraja et al. (2013)
B. subtilis	*L. vannamei*	Commercial (5×10^4 cells mL^{-1})	24–29	pH (D), nitrite (D), water transparency (I), soluble reactive phosphorus (D), chemical oxygen demand (I), and C*hlorophyll a* density (I)	Wu et al. (2016)
Paenibacillus polymyxa	*C. carpio*	10^3–10^5 CFU mL^{-1}	29.5	pH (NS), nitrite (NS), ammonia (NS, dissolved oxygen (NS)	Gupta et al. (2016)
B. subtilis + *B. licheniformis* + *B. pumilus*	Nile tilapia (*O. niloticus*)	Commercial (totally at 1×10^{10} CFU g^{-1})	30	Ammonia (D), electric conductivity (I), salinity (I), total dissolved solid (I), and pH (I)	Elsabagh et al. (2018)
B. pumilus + *L. delbrueckii*	*C. carpio*	*B. pumilus* (62.5×10^8 cells mL^{-1}) + *L. delbrueckii* (67.5×10^8 cells mL^{-1})	17.7–20.3	Temperature (NS), pH (NS), dissolved oxygen (NS), total ammonia nitrogen (D), total suspended solid (I), and total dissolved solid (I)	Dash et al. (2018)
B. subtilis	Nile tilapia (*O. niloticus*)	Commercial 1–4 g (1.19×10^8 CFU g^{-1})	27.51–27.37	Total ammonia nitrogen (D), dissolved oxygen (NS), temperature (NS), pH (NS), nitrite (D), nitrate (D), and total suspended solids (I)	Mohammadi et al. (2020)
B. subtilis	*C. carpio*	Commercial (4.94×10^5–1.82×10^6 CFU mL^{-1})	25–30	Total ammonia nitrogen (D)	Sungsirin et al. (2024)
B. subtilis + *B. thuringiensis*	*Carassius auratus*	*Malacosoma neustria* and *Ricania simulans* (1×10^7 CFU mL^{-1})	17–20	Ammonia nitrogen (D), nitrite nitrogen (D), nitrate nitrogen (D), heavy metals concentration (D), sulphate (I), total phosphorus (I), and temperature (D)	Kalaycı Kara et al. (2021)

Bacillus NP5	*Clarias gariepinus*	1×10^9–1×10^{10} CFU mL^{-1}	28.5–29.0	Temperature (NS), ammonia (D), dissolved oxygen (NS), and pH (NS)	Putra et al. (2020)
Bacillus cereus	*Penaeus vannamei*	Commercial (10^6 CFU mL^{-1})	29.79–29.97	Nitrate (D), ammonia (D), biochemical oxygen demand (D), temperature (NS), salinity (NS), pH (NS), and transparency (NS)	Khademzade et al. (2020)
B. licheniformis, B. subtilis, and *B. amyloliquefaciens*	*P. vannamei*	Commercial (1.1×10^9 CFU g^{-1})		Ammonia nitrogen (D), phosphate (D), chemical oxygen demand (D), nitrite nitrogen (D), and nitrate nitrogen (D)	Hu et al. (2023)
B. subtilis	*L. vannamei*	Commercial (2×10^{11} CFU g^{-1})	26.6	Dissolved oxygen (NS), temperature (NS), and total ammonia nitrogen (D)	Abdel-Tawwab et al. (2022)
B. flexus and B. licheniformis	*L. vannamei*	Tropical mariculture systems (10^4 CFU mL^{-1})		Chemical oxygen demand (D), ammonia nitrogen (D), and nitrate (D)	Ren et al. (2021)
B. subtilis		Commercial (3×10^4 CFU mL^{-1})	20	Phosphate (D), and dissolved inorganic nitrogen (D)	Patil et al. (2022)
B. subtilis	*L. vannamei*	Indigenous (1×10^3 –1×10^5 CFU mL^{-1})	27.62–27.93	Temperature (NS), pH (NS), alkalinity (NS), dissolved oxygen (NS), ammonia (D), and nitrite (NS)	Kewcharoen and Srisapoome (2019)
B. subtilis	*L. vannamei*	Indigenous (1×10^6 –1×10^8 CFU mL^{-1})	29.45–29.66	Temperature (NS), pH (NS), salinity (NS), dissolved oxygen (I), nitrate (NS), nitrite (NS), turbidity (I), and ammonia (D)	Mirbakhsh et al. (2021)
B. subtilis	*Ctenopharyngodon idellus*	Soil (5.5×10^7 – 1×10^9 CFU mL^{-1})	29–40	Total nitrogen (D), ammonia (D), nitrate (D), nitrite (D), and turbidity (D)	Shao et al. (2021)
B. licheniformis and *B. subtilis*	*Ctenopharyngodon idellus*	Commercial (3×10^9 CFU g^{-1})	25–32	Total nitrogen (D), total phosphorus (D), nitrite (D), nitrate (D), and chemical oxygen demand (D)	Li et al. (2023)
B. subtilis	NM	200 mg	NM	Total phosphorus (NS), ammonia nitrogen (NS), dissolved oxygen (NS), and chemical oxygen demand (D)	Zhai et al. (2020)
B. licheniformis	*L. vannamei*	Commercial (1×10^{10} CFU g^{-1})	28	Ammonia nitrogen (D), dissolved oxygen (D), pH (NS), salinity (NS), nitrite (D), and nitrate (D)	Zhang et al. (2020)

(D= decrease, I=increase, NS= not significant, NM = not mentioned, U= Unknown).
Source: Modified from Soltani et al. (2019).

the water column (Martínez-Córdova et al., 2015). However, production of heavy loads of these wastes requires extra effort to remove the extra loads. The process of nitrogen cycle includes ammonification, nitrification, and denitrification. The initial form of nitrogen that is in the form of organic nitrogen can be converted to ammonium and ammonia by bacteria such as *Bacillus* species in a process called ammonification. In this process, the ammonium is converted to nitrite and nitrate mainly by *Nitrosomonas* and *Nitrobacter* species, respectively, in a process called nitrification. This process then is followed by the conversion of nitrate to nitrogen gas called denitrification, removing bioavailable nitrogen and returning it to the atmosphere (Bernhard, 2010). Unlike *Nitrosomonas* and *Nitrobacter* which are mainly involved in nitrification and sometimes denitrification (Liu et al., 2020), *Bacillus* species play significant roles in the nitrogen cycle through ammonification, nitrification, denitrification, and nitrogen fixation (Hui et al., 2019; Rout et al., 2017; Verbaendert et al., 2011; Yousuf et al., 2017), As an example, strains of *Bacillus amyloliquefaciens* can convert organic nitrogen into ammonium (Hui et al., 2019) and strains of *Bacillus cereus* can also remove nitrogen from wastewater (Barman et al., 2018). Therefore, *Bacillus* bacteria are able to remove the different forms of nitrogen from aquaculture rearing water.

Gram-positive bacteria are able to decrease the build-up of particulate matters and dissolved organic carbon in water, and heterotrophic bacteria can utilize both organic and inorganic sources of carbon for their growth. Thus, they play a significant role in the decomposition of organic matter and the production of food substances from dissolved organics (Balcázar et al., 2006; Padmavathi et al., 2012). *Bacillus* species can convert organic substances into carbon dioxide effectively, e.g., carbon dioxide is used by β- and γ-proteobacteria as carbon source, respectively (Koops & Pommerening-Röser, 2001), while other bacteria covert most of the organic matters into slime or bacterial biomass (Mohapatra et al., 2013; Zorriehzahra et al., 2016). *Bacillus* are capable of removing the load of organic matters in aquaculture environment via the recycling of the nutrients in water column and decreasing sludge accumulation (Soltani et al., 2019). In aquaculture environment, the loads of organic matter are induced by uneaten food, while probiotic *Bacillus* enhance the fish/shrimp appetite via increasing the level of digestive enzymes' activities giving a better result of feed utilization and less waste production (Hura et al., 2018).

During the mineralization, oxygen is consumed by microorganisms and produce carbon dioxide, water, and nutrients (Bokossa et al., 2014). During the photosynthesis process, the carbon dioxide and nutrients are consumed by phytoplankton resulting in releasing oxygen that its concentration is depended on phytoplankton density (Wang et al., 2005). *Bacillus* bacteria in the form of probiotic can modulate dissolved oxygen through a decrease in fish stress, hence, causing a less consumption of oxygen (Zink et al., 2011). In addition, during the photosynthesis process, the free carbon dioxide and bicarbonates are utilized resulting in an enhancement of dissolved oxygen and carbonates leading to the modification of the water quality (Sunitha & Padmavathi, 2013). Additionally, during the nitrification process, hydrogen ions are produced which is important in the water acidification (Camargo & Alonso, 2006; Gomes et al., 2008; Nimrat et al., 2012a). Thus, during the processes of mineralization and nitrification, bacilli probiotics can play a role in modulation of dissolved oxygen and pH of water. Eutrophication and deterioration of water will cause an improvement in the level of phosphorus, but bacilli probiotics can consume phosphates for their activities resulting in a decrease in phosphorus level in aquaculture waters (Rao, 2002; Sunitha & Padmavathi, 2013) The microbial community of the rearing ponds can also be maintained by bacilli bacteria resulting in microbial balance, particularly inhibition of the proliferation of the harmful bacteria (Soltani et al., 2019). Still more investigations are required to demonstrate the actual mechanisms used by bacilli bacteria in bioremediation and maintaining of the aquaculture water quality. Figure 8.1 summarizes the possible mechanisms used by probiotic *Bacillus* for remediation of water quality.

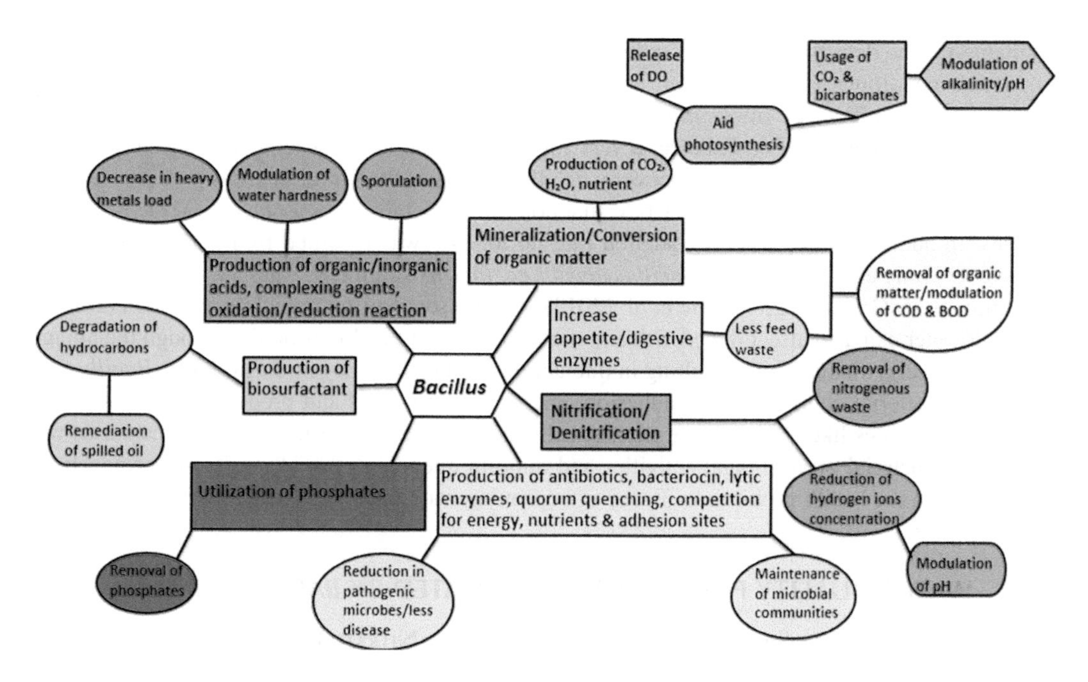

FIGURE 8.1 Mode of actions by probiotic *Bacillus* in modulating water quality. DO = dissolved oxygen; COD = chemical oxygen demand; BOD = biological oxygen demand. Rectangular compartments represent the first stage of the mechanisms; oval compartments represent the second stage of the mechanisms; rounded rectangles represent the third stage of the mechanisms; pointed pentagon represents the fourth stage of the mechanisms; hexagon represents the fifth stage of the mechanisms; tear drop represents intersection between two processes (adapted from Hlordzi et al., 2020).

8.4 FACTORS AFFECTING THE EFFECTIVENESS OF PROBIOTIC *BACILLUS* IN ENHANCING THE QUALITY OF REARING WATER

It has been shown that direct addition of probiotic *Bacillus* in rearing water is the most effective route of administration for improving aquaculture water quality (Jahangiri & Esteban, 2018). However, the efficiency of probiotic bacilli can be affected by several factors such as hardness, pH, temperature, osmotic pressure, dissolved oxygen, and bacilli species and strain. For instance, strains of L10, G1, and E20 of *B. subtilis* exhibited various efficacies in improving water quality (Liu et al., 2009; Zokaeifar et al., 2014) that could be in part due to various environmental factors, e.g., dissolved oxygen, temperature, ammonium, phosphorus, and pH, and nutrient sources, e.g., cellulose, starch, sucrose, and glucose (Barman et al., 2018; Rajakumar et al., 2008). Strains of *B. fusiformis* are able to degrade petroleum hydrocarbon by low levels of Ca, Mg, and Fe ions (Dongfeng et al., 2011) indicating that the efficacy of bacilli bacteria in improving water quality can be affected by some metal ions. *Bacillus* sp. YX-6 was able to enhance the nitrification process via an increase in dissolved oxygen (Song et al., 2011). For instance, growth and denitrification efficacy by bacilli bacteria were reduced in the absence of dissolved oxygen and nitrite oxide (Patureau et al., 2000)). Typically, *Bacillus* W2 reduced 97% of nitrogen in the presence of 2 mg/L dissolved oxygen, but deletion of nitrogen decreased to 85% in the presence of 4–5 mg/L dissolved oxygen (Yu et al., 2005). With an increase in dissolved oxygen, the level of nitrification was enhanced by *Bacillus* cereus PB88 (Barman et al., 2018), suggesting that high levels of dissolved oxygen may not be suitable for the denitrification process by *Bacillus*. It is clear that probiotic *Bacillus* have different optimum conditions for a maximum action associated with the water quality parameters. Thus, to improve the efficiency of *Bacillus* function in enhancing water quality, suitable environmental conditions are required. For instance, a very high fluctuation in water temperature can negatively affect the level of denitrification process by bacilli probiotics, therefore, *Bacillus* probiotics require optimum water temperatures to show best function for the denitrification process (Maag & Vinther, 1996). In addition, the nitrite degradation rate up to 90% was recorded by *Bacillus* sp. YX-6 at 25–40°C (Song et al., 2011); and *Bacillus amyloliquefaciens* strain HN exhibited a higher ammonia deletion at 35°C than at 25°C (Xie et al. (2013). The presence of microbial flora in the column of rearing water can also affect the growth and multiplication of bacilli probiotics through the antagonistic or synergistic functions resulting in fluctuations in the efficiency of the probiotic *Bacillus*. For instance, bacterial competition for adhesion, energy, and nutrients and production of bacteriocins and lytic enzymes like chitinases, proteases, cellulases, lipase, amylase, and β-1,3-glucanases with either antibacterial or antifungal activities and quorum quenching can all positively or negatively interact with the bacilli probiotics efficacy in improving water quality (Kuebutornye et al., 2020).

8.5 WATER QUALITY PARAMETERS MODULATED BY *BACILLUS* SPECIES

8.5.1 Nitrogenous Species (Nitrites, Ammonia, and Nitrates)

Nitrogenous species (i.e., ammonia, nitrites, and nitrates) are important water quality parameters that are toxic to aquatic organisms. In aquaculture, accumulation of nitrogenous wastes as the key end products of protein metabolism results in induction of stress condition, subsequently making the fish/shellfish become more susceptible to pathogenic agents (Faramarzi et al., 2012; Lalloo et al., 2007). In aquaculture, the nitrogenous wastes can be removed via various routes such as rotating biological contactors, fluidized bead filters, sand biofilters, and trickling filters (Crab et al., 2007; Shan et al., 2016). It seems bacilli bacteria are more effective in removing the nitrogenous wastes from aquaculture rearing waters. This is because bacilli bacteria are involved in both nitrification and denitrification processes, thus they can have an economic benefit (Kim et al., 2007; Nimrat et al., 2012). Several studies demonstrated the efficacy of bacilli probiotics in the removal of nitrogenous wastes from the rearing waters. *Bacillus* bacteria are known as heterotrophic nitrifiers (Mével & Prieur, 2000), and such biological function seen in, e.g., *B. subtilis* and *B. cereus* under aerobic and facultative aerobic/anaerobic conditions enable them to change nitrogen metabolism, accelerating both nitrification and denitrification processes. A combination of *B. subtilis* and *B. megaterium* (supplemented twice a week) into recirculating water systems of red-parrot fish (male midas cichlid (*Cichlasoma citrinellum* × *C. synspilum*)) remarkably decreased the TAN parameters compared to control treatment (Chen & Chen, 2001). The mortality of fish reared at high density was also decreased. The level of TAN was reduced after inclusion of *Bacillus* sp. in combination with *Saccharomyces* sp., *Nitrosomonas* sp., and *Nitrobacter* sp. in the rearing water of *P. monodon*, even better than the use of a mixture of *Nitrosomonas* sp. and *Nitrobacter* sp. alone (Matias et al., 2002). Use of a mixture of *Bacillus* sp. and *Saccharomyces cerevisiae, Nitrosomonas* sp., and *Nitrobacter* sp. in the rearing water of *Litopanaeus vannamei* revealed an improvement in the load of beneficial bacteria, but decreased total inorganic nitrogen (Wang et al., 2005). Utilization of an indigenous strain of *B. subtilis* obtained from mrigal intestine in rearing water of guppy (*Poecilia reticulata*), molly (*P. sphenops*), green swordtail (*Xiphophorus helleri*), and platy (*X. maculatus*) at four different concentrations induced a significant reduction in TAN and total load of motile *Aeromonas* and coliforms compared to control groups. This shows the inhibitory activity by *B. subtilis* against Gram-negative bacteria (Ghosh et al., 2008). The concentrations of ammonia and nitrite were decreased in the rearing water of giant freshwater prawn *(Macrobrachium rosenbergii)* after treating with *Bacillus* sp. within 60 days of cultivation (Rahiman et al., 2010). Supplementation of indigenous *B. subtilis* strain isolated from grass carp in the fish rearing water significantly reduced total nitrogen, ammonia, and nitrite compared to the control. The microbial diversity was also improved in the probiotic treated ponds (Zhang et al., 2013). The concentrations of ammonia and nitrite in the rearing water of *L. vannamei* reduced compared to controls, after inclusion of two microencapsulated combinations of *Bacillus* species, i.e., mixture I: including *B. licheniformis, B. subtilis, B. thuringiensis, B. megaterium*, and *B. polymyxa*, and mixture II: *B. subtilis, B. megaterium, B. circulans, B. pumilus*, and *B. polymyxa*, applied in the shrimp rearing water (Nimrat et al. (2012). Inclusion of *B. subtilis* in the rearing water of Nile tilapia previously affected by *Flavobacterium columnare* improved TAN and fish survival (Mohamed & Refat, 2011), showing an antagonistic activity by *B. subtilis* to the cause of columnaris disease. Use of a mixture of *Bacillus* spp. in the rearing water of *L. vannamei* exhibited a remarkable decrease in ammonia, nitrite, and nitrate ions (Zokaeifar et al., 2014). With such enhancement in water quality the culture shrimp demonstrated an increase in the growth, digestive enzyme activity, immune response, and resistance against *V. harveyi* infection (Zokaeifar et al. (2014). A decrease in nitrite was seen in the rearing water of *L. vannamei* after inclusion of an indigenous probiotic *B. subtilis* once a week (Wu et al., 2016). The probiotic could affect the bacterial community of the cultured water and the greater impact in the early culture could probably be attributed to faster initial colonization and multiplication of the probiotic bacteria through peptone addition in the water column. A combination of *B. pumilus* and *L. delbrueckii* used in the rearing water of common carp in a bio-floc culture system with molasses as the carbon source induced a reduction in TAN while no significant change was seen in water temperature, pH, or dissolved oxygen compared to the control (Dash et al., 2018). Use of *Bacillus velezensis* AP193 in catfish rearing water could reduce 75% of nitrate-nitrogen and 43% of total nitrogen levels (Thurlow et al., 2019). A mixture supplementation of *B. subtilis, B. mycoides*, and *B. licheniformis* isolated from mud sediment and common carp intestinal tract could significantly decrease ammonia, nitrite, and nitrate in the rearing water of common carp kept in recirculation tanks at 20°C (Lalloo et al. (2007). The enzymes of nitrate and nitrite reductases are secreted by *B. subtilis* as this bacilli species can use nitrate or nitrite as sole sources of nitrogen and also as electron acceptors for anaerobic respiration (Lalloo et al., 2007). After addition of *B. pumilus* in combination with periphytic algae in the rearing water of *P. monodon* (bio-floc system), total ammonia and nitrite were significantly reduced with no significant change in pH, salinity, dissolved oxygen, and temperature (Banerjee et al., 2010). Regardless of nitrite and nitrate forms, a reduced level has been demonstrated by some researchers. For instance, a reduction in nitrate and nitrite levels (Hura et al., 2018; Lalloo et al., 2007; Nimrat et al., 2012; Zokaeifar et al., 2014) and a decrease in nitrite-N level (Song et al., 2011; Xie et al., 2013) have been shown by various researchers (e.g., Hura et al., 2018; Lalloo et al., 2007; Nimrat et al., 2012; Song et al., 2011; Xie et al., 2013; Zokaeifar et al., 2014). Also, bacilli probiotics can reduce the level of ammonia toxicity. For example, use of *B. megaterium* (Hura et al., 2018), *B. amyloliquefaciens* (Xie et al., 2013), and *B. subtilis* (Cha et al., 2013; Zokaeifar et al., 2014) exhibited a decrease in ammonia level. When strains of indigenous *B. pumilus, B. licheniformis*, and *B. subtilis* isolated from marine water and soil were added in the rearing water of shrimp, a reduction was seen in the level of TAN, and optimum growth was obtained at pH 7.5, 1.5% NaCl, and 30°C (Devaraja et al., 2013). Inclusion of *Bacillus* sp. in the rearing water of common carp significantly decreased TAN levels compared to the control, though nitrate level was enhanced (Naderi Samani et al., 2016). Similar result was seen by Reddy et al. (2018) after inclusion of *Bacillus* sp. into aquaculture ponds. Thus, *Bacillus* probiotics are a suitable biological tool to mineralize nitrogenous wastes through nitrification and denitrification processes for removal of nitrogen species (Nimrat et al., 2012).

8.5.2 Temperature

Water temperature is one of the crucially significant variables of the aquaculture rearing waters because the immune-physiological functions of fish/shellfish are temperature-dependent (Fernandes et al., 2018; Martinez et al., 2018). Many microflora of the water column are mesophilic favouring a water temperature at 25–35°C that is quite suitable for processes of nitrification and denitrification

(Klotz & Stein, 2011; Paudel et al., 2015), thus, use of an optimal water temperature is critically important for fish/shellfish growth performance and microbial proliferation. There are no adequate data regarding the effect of bacilli probiotics on the modulation of temperature in the rearing waters. A study on the use of mix bacilli probiotics including *B. subtilis*, *B. licheniformis*, *B. megaterium*, and *B. laterosporous* in the transport water of yellowfin tuna *Thunnus albacares* yolk sac larva noticed relatively decreased water temperature that can reduce un-ionized ammonia in the water column (Zink et al., 2011). Also, reduced temperature was documented during the bioremediation of aquaculture wastewater using *Bacillus* species (Kalaycı Kara et al., 2021). In contrast, many other researchers exhibited no significant effect of *Bacillus* probiotics on water temperature (Banerjee et al., 2010; Ghosh et al., 2008; Mohammadi et al., 2020; Nimrat et al., 2012), suggesting temperature is not affected by biological processes such as bacilli probiotics. However, the growth, multiplication, and extracellular secretions of bacilli bacteria are temperature-dependent, thus, the functions of these probiotic bacteria are maximum at the optimum temperature required for their growth.

8.5.3 Biological Oxygen Demand (BOD) and Chemical Oxygen Demand (COD)

BOD as the amount of dissolved oxygen needed by microorganisms to degrade organic matter exists in water column, while COD is an indicator for the oxygen required to oxidize soluble organic matter and particles in water. Thus, a higher BOD and COD are indicative of poor water quality. The breakdown of organic matter by microorganisms require oxygen demand resulting in reduction of dissolved oxygen that can cause hypoxia and anoxia in the reared aquatic animals. The efficacy of bacilli probiotics as the reducer of BOD and COD has been demonstrated in several studies resulting in a better feed consumption by fish, thus less organic matter is decomposed by DO. Inclusion of *Bacillus* sp. probiotic in the aerated rearing water of *P. monodon* reduced level of COD and increased the final production (Porubcan, 1991). Utilization of a mixture of commercial *Bacillus* sp. and *Saccharomyces* sp., *Nitrosomonas* sp., and *Nitrobacter* sp. in rearing water of *P. monodon* exhibited a lower level of COD and BOD in the shrimp ponds treated with the *Bacillus* sp. combined with *Saccharomyces* sp. than in ponds treated with mixture of *Nitrosomonas* sp. and *Nitrobacter* sp. alone (Matias et al., 2002). Again, a combination of *Bacillus* sp., *Saccharomyces cerevisiae, Nitrosomonas* sp., and *Nitrobacter* sp. in *Litopanaeus vannamei* ponds exhibited a reduction in COD level (Wang et al., 2005). A mixture of *B. subtilis* and *B. megaterium* in recirculating water systems of red-parrot fish significantly reduced the concentration of COD (Chen & Chen, 2001). Use of *B. megaterium* was able to reduce more BOD of rearing water of major carps compared to control (Hura et al., 2018). Utilization of mix *Bacillus* including *B. subtilis, B. mojavensis*, and *B. cereus* decreased above 90% of BOD in the water (Reddy et al., 2018); and a mixture of *B. cereus* and *Aeromonas vernoii* resulted in reduced BOD in the water (Divya et al., 2015). Additionally, the COD of rearing waters was decreased after inclusion of *B. subtilis* (Wen-jun, 2011), *B. megaterium* (Hura et al., 2018), and *Bacillus* sp. YB1701 (Zhou et al., 2018) in the rearing waters. Conversely, an increase in COD was measured in the rearing water of *L. vannamei* after adding probiotic *B. subtilis* once a week that probably was due in part to an increase in concentration of chlorophyll in the water (Wu et al., 2016).

8.5.4 Dissolved Oxygen

In aquaculture activity, dissolved oxygen is one of the most important water quality factors; thus, under a low oxygen level, fish/shellfish have low performance and become more susceptible to pathogenic agents (Dabrowski et al., 2018; Manahan, 2017). Also, in aquaculture practice, a reduction in dissolved oxygen results in the increasing levels of organic matter, ammonia, phosphorous, copper, and other nutrients (dos Santos Simões et al., 2008) that can suppress the animal immunity. There is little data regarding the modulation and effect of bacilli probiotics on dissolved oxygen. Use of *Bacillus* sp. together with *Saccharomyces cerevisiae, Nitrosomonas* sp., and *Nitrobacter* sp.

in the rearing water of *Litopanaeus vannamei* enhanced dissolved oxygen level (Wang et al., 2005). Inclusion of *B. megaterium* in rearing water of cultured major carps exhibited a higher dissolved oxygen level compared to the control (Hura et al., 2018). An improvement in dissolved oxygen level was recorded by the inclusion of a mixture of *Bacillus* species in the rearing water of tilapia larvae (Hainfellner et al., 2018). Inclusion of *Bacillus* species including *B. subtilis, B. licheniformis, B. megaterium*, and *B. laterosporous* in the transport water of yellowfin tuna yolk sac larvae and *Carnegiella strigata* revealed a higher dissolved oxygen level than control ones (Gomes et al., 2008; Zink et al., 2011). Use of *B. subtilis* in the rearing water of Nile tilapia increased dissolved oxygen of the water (Mohamed & Refat, 2011). Contrastingly, Mohammadi et al., 2020), did not observe any difference in the dissolved oxygen levels after *B. subtilis* treatment, while Kalaycı Kara et al. (2021) observed a decrease in dissolved oxygen when *Bacillus thuringiensis* and *Bacillus subtilis* were used for the bioremediation of aquaculture wastewater. However, studies evaluating the effect of bacilli probiotics on the value of dissolved oxygen are few compared to its effects in balancing nitrogenous species in aquaculture waters and thus, more research is needed to assess the efficacy of bacilli probiotics on the modulation of dissolved oxygen in aquaculture rearing water.

8.5.5 Total Dissolved Solids (TDS)

TDS (i.e., accumulation of organic molecules and minerals, organic pollutants, and toxic metals) is another parameter of aquaculture rearing water quality that assesses the organic matter, inorganic, salts and other dissolved substances in water. Although TDS is not considered as a primary pollutant parameter, it is used as an indicator for a wide range of chemical contaminants and thus, can cause toxicity through differences in the ionic composition of the water, increases in salinity, and toxicity of the individual ions (Barman et al., 2018). The sources of dissolved minerals as art of TDS include agriculture/aquaculture, coal mines, residual runoffs, and discharge from industrial or sewage treatment plants ((Daniels et al., 2016; Shi et al., 2014; Wu & Maskaly, 2018). TDS concentration can affect the spatial distribution of freshwater invertebrates via inducing an imbalance between water dissolved oxygen and the water ions (Cormier et al., 2013; Mueller et al., 2017; Olson & Hawkins, 2017; Pendashteh et al., 2012). Some *Bacillus* species demonstrated an ability in removing TDS in aquaculture waters. For instance, use of a mixture of *B. subtilis, B. licheniformis*, and *B. pumilus* exhibited a maintenance in TDS concentration at the acceptable range in the rearing water of tilapia (Elsabagh et al., 2018). In addition, an enhancement in TDS and total suspended solids (TSS) of rearing water of common carp culture was seen under a bio-floc system (Dash et al., 2018). Use of *B. megaterium* reduced TDS level of the rearing water that was attributed to an improvement in feed utilization and digestion by fish (Hura et al., 2018). Application of *B. cereus* PB88 or *B. subtilis* HS1 in culturing of shrimp and *Dicentrarchus labrax* larvae reduced TDS level (Barman et al., 2018; Md et al., 2015). More researches are required to show bacilli probiotics' efficacy on the modulation of TDS in aquaculture waters.

8.5.6 pH and Alkalinity

Alkalinity (capacity factor) as an indicator of water ability to neutralize strong acids and pH (intensity factor) as the indicator of hydrogen ion concentrations in water are important water quality factors (Boyd et al., 2011) that are involved in all biological/chemical processes (Summerfelt et al., 2015). *Bacillus* probiotics affect the variation of alkalinity and pH by promoting the process of mineralization of organic matter that enhances photosynthetic processes. The pH in rearing water of giant freshwater prawn (*Macrobrachium rosenbergii*) exhibited a reduction after inclusion of *Bacillus* sp. for two months (Rahiman et al., 2010). The pH of water of Nile tilapia culture was enhanced after inclusion of *B. subtilis* in the water (Mohamed & Refat, 2011). The water pH of *L. vannamei* ponds was lower than control groups after addition of *B. licheniformis, B. subtilis, B. thuringiensis, B. megaterium, B. polymyxa, B. circulans*, and *B. pumilus* in forms of mixtures in the shrimp rearing water (Nimrat et al. (2012). Use of *B. subtilis* in the form of probiotic in water of *L. vannamei* reduced pH level (Wu et al., 2016). Addition of *B. megaterium* exhibited a better maintenance of alkalinity (basic condition) than acidic condition in rearing water of carp (Hura et al. (2018), and an enhancement in the pH of tilapia rearing water was seen after inclusion of *Bacillus* species (Elsabagh et al., 2018) suggesting the beneficial effect of some bacilli probiotics for improving the pH of acidic waters that is more suitable for growing of aquatic organisms. The results of Mohammadi et al. (2020) showed no significant difference in the pH values of tilapia culture water after *B. subtilis* treatment. However, some bacilli may reduce the basic pH to neutral condition (Gomes et al., 2008; Nimrat et al., 2012; Wu et al., 2016).

8.5.6.1 Hardness

Water hardness as the amount of calcium and magnesium ions (also sometimes including iron, manganese, zinc, aluminium, strontium, and hydrogen ions) dissolved in water is one of the important water quality parameters (George et al., 2017; Swann, 1997). The level of water hardness can be influenced by bacilli probiotics in softening hard aquaculture waters (George et al., 2016). For instance, use of a commercial bacilli probiotic (Ecotrax® contains a mix of 7 *Bacillus* species) significantly decreased the total hardness of rearing water of white leg shrimp culture compared to the control (George et al., 2016). Utilization of *B. subtilis, B. mojavensis*, and *B. cereus* reduced the total hardness of rearing water (Reddy et al., 2018). In contrast *B. megaterium* induced a higher total hardness (Hura et al., 2018). More studies are required to elucidate the efficacy of bacilli probiotics on the level of water hardness in the rearing waters.

8.5.6.2 Organic Matter

An increase in the level of organic matter through the uneaten feed is one of the serious water quality issues of aquaculture. Under an optimum condition, a less organic matter load is released in aquaculture rearing waters and *Bacillus* probiotics can enhance water quality via a decomposition of organic matter into smaller molecules (Loh, 2017) resulting in a decrease in organic matter loads (Dalmin et al., 2001). Use of *B. subtilis* in rearing water of guppy (*Poecilia reticulata*), molly (*P. sphenops*), green swordtail (*Xiphophorus helleri*), and platy (*X. maculatus*) at four different concentrations significantly decreased the level of dissolved organic substances (Ghosh et al., 2008).

8.5.6.3 Phosphates

Phosphate ions are necessary for the immunophysiological functions of aquatic organisms in the water column; however, the condition of eutrophication is related to the surplus of phosphates (Luo et al., 2016; Reddy et al., 2018). The accumulation of phosphate in aquaculture waters increases the growth of algae resulting in algal bloom that is fatal to fish/shellfish (Lalloo et al., 2007). The aquatic animals including fish feed and fertilizers are the main sources of phosphorus (it exists in the form of phosphate ions in water) (Querijero & Mercurio, 2016; Tovar et al., 2000). The level of phosphate accumulation needs to be controlled and modulated in the rearing water; otherwise its high concentration can negatively affect both fish health status and the environment resulting in induction of stress condition which increases the susceptibility of the animals to disease outbreaks (Jana & Jana, 2003; Lalloo et al., 2007). In aquaculture systems, early studies have shown that bacilli bacteria are a strong phosphate reducer in water column (Porubcan, 1991). Inclusion of a mixture of *Bacillus* sp. and *Saccharomyces cerevisiae, Nitrosomonas sp.*, and *Nitrobacter* sp. in *L. vannamei* water decreased the amount of phosphate concentration in the rearing water (Wang et al., 2005). The phosphate of rearing water of *L. vannamei* was decreased after inclusion of probiotic *B. subtilis* once a week (Wu et al., 2016). Strains of *B. subtilis, B. mojavensis*, and *B. cereus* were able to reduce 81% of phosphate ions (Reddy et al., 2018). Use of a combination of *B. subtilis, B. mycoides*, and *B. licheniformis* significantly decreased phosphate ions in the rearing water of common carp kept in recirculation tanks at 20°C (Lalloo et al. (2007). Strains of *B. subtilis, B. cereus*, and *B. licheniformis* reduced phosphate ions in the rearing water of shrimp culture (Lalloo et al., 2007), and strains of *B. velezensis* could decrease total phosphorus in rearing water of catfish ponds (Thurlow et al.,

2019). Contrastingly, a 2% increase in total phosphorus was recorded when *B. thuringiensis* and *B. subtilis* was used for bioremediation (Kalaycı Kara et al., 2021).

8.5.6.4 Transparency

Transparency, as an indicator of phytoplankton population and suspended inorganic and organic matters (Mahmud et al., 2016), is a primary indicator of quality water for aquaculture. A low transparency means the existence of high levels of suspended particles and nutrient loads that may be harmful for the rearing of aquatic organisms. Few studies have focused on the efficacy of *Bacillus* bacteria on the transparency of rearing waters. Treatment of rearing water of major carp culture with *B. megaterium* increased water transparency (Hura et al. (2018). Use of bacilli probiotic in a recirculation system of red-parrot culture could maintain the transparency of rearing water at 30–50 cm (Chen & Chen, 2001). The transparency of shrimp rearing water was increased after treatment with bacilli bacteria (Matias et al., 2002). In addition, the transparency of rearing water of *L. vannamei* was increased after inclusion of *Bacillus* sp., *Saccharomyces cerevisiae, Nitrosomonas* sp., and *Nitrobacter* sp. in a mixture form (Wang et al., 2005). Further, water transparency of rearing water of *L. vannamei* was increased following inclusion of *B. subtilis* in the rearing water (Wu et al., 2016). *Bacillus* probiotics can maintain the transparency of the rearing waters through the modulation of organic and inorganic matters resulting in a decrease in nutrient loads.

8.5.6.5 Water Conductivity

Data on the modulation of conductivity as a water quality parameter by *Bacillus* species is scarce. No significant modulation on water conductivity was seen by inclusion of *B. cereus* in the rearing waters (Bhatnagar & Lamba, 2015; Bhatnagar & Lamba, 2017). Similarly, no difference was seen in water conductivity after inclusion of Efinol®L product (mixture of *B. subtilis, B. licheniformis, Lactobacillus acidophilus*, and *Saccharomyces cerevisiae*) into the transport water of cardinal tetra, *Paracheirodon axelrodi* (Gomes et al., 2008). In contrast, an enhancement in water conductivity was seen after inclusion of a mixture of *Bacillus* spp. into water (Hainfellner et al., 2018). Also, a higher conductivity was measured in the water after adding the mixture of *B. subtilis, B. licheniformis*, and *B. pumilus* (Elsabagh et al., 2018). Such controversial data require more attention and more investigations.

8.5.6.6 Heavy Metals

Heavy metals including silver, copper, chromium, mercury, lead, cadmium, cobalt, zinc, and iron are one of the serious pollution sources of the aquatic ecosystems if the acceptable levels are exceeded (Stefanescu, 2015). Accumulation of heavy metals in fish/shellfish tissues causes toxicity in the target animals and also in human when they are fish/shellfish consumers; thus, their removal or their decrease to acceptable level in the aquatic ecosystems is highly necessary (Chatterjee et al., 2010; Issazadeh et al., 2011; Stefanescu, 2015). Use of microbiota is one of the most cost-effective routes for the removal of heavy metals (Chatterjee et al., 2010). Microorganisms like bacteria can use their metabolic processes to mobilize heavy metals through the production of inorganic and organic acids, complexing agent excretion, and decreasing or oxidation reactions (Stefanescu, 2015). *Bacillus* bacteria can reduce the level of heavy metals through sporulation process (heavy metals are used by bacilli bacteria during sporulation phase) (Kolodziej & Slepecky, 1964; Stefanescu, 2015), thus, bacilli bacteria are considered a suitable effective remediation of heavy metals in the aquaculture waters. For instance, prevention of an experimental lead poisoning in *Carassius gibelio* culture was effective by inclusion of *B. subtilis* (10^9 CFU/g) (Yin et al., 2018). An effective bioaccumulation of cadmium, zinc, copper, and lead ions was achieved by use of *B. subtilis* and *B. cereus* (Costa & Duta, 2001; Issazadeh et al., 2011). Use of *B. megaterium* could bio-accumulate copper, iron, zinc, and manganese (Stefanescu, 2015). Similar observation was made by Kalaycı Kara et al. (2021) when *B. thuringiensis* and *B. subtilis* were used in the bioremediation of aquaculture water singularly or in combination. Thus, it seems that application of bacilli probiotics can combat heavy metals in aquaculture, although more studies are needed to elucidate the actual functions.

8.5.6.7 Oil Spillage

The leakage of oils can sometimes happen in aquaculture sector due to various activities, e.g., via manufactures and oil transport and refinery products, and this is one of the aquaculture challenges (Banerjee & Ghoshal, 2016) that can interfere with the aquaculture production (Edet et al., 2018; Joo & Kim, 2013). Use of effective and cheap methods is therefore important to remove the oil spill from the rearing waters, and the use of microbes can be considered as one of the alternatives. Microorganisms can use oils as a source of food and energy and break down oils into smaller and harmless substances (Joo & Kim, 2013). Many microorganisms can utilize hydrocarbons as a carbon source resulting in the production of carbon dioxide, water, and biomass (Cunha & Leite, 2000). Many studies have shown the effectiveness of *Bacillus* bacteria as the bioremediation of spilled oils. For instance, *B. polymyxa, B. subtilis*, and *B. megaterium* are used to remove the diesel oil by producing bio-surfactants through emulsification process (Tariq et al., 2016). *B. thuringiensis* and *B. cereus* are able to effectively degrade hydrocarbons and, thus, are prefect candidates for the remediation of oil-polluted aquatic environments (Kebria et al., 2009; Maddela et al., 2015). Some strains of bacilli bacteria such as *B. subtilis* JK-1 are good degraders of crude oil (Joo & Kim, 2013). Although more works are required to assess the potential use of *Bacillus* probiotics in reduction of oil spill from the aquaculture rearing waters, these evidences clearly show that *Bacillus* species can be applied in aquaculture for the bioremediation of oil pollution.

8.6 CONCLUSION

A great attention has been paid to the application of *Bacillus* for the bioremediation of aquaculture waters, and the obtained data are promising. *Bacillus* probiotics can modulate various factors of water quality including TDS, dissolved oxygen, BOD, COD, pH, carbon dioxide, alkalinity, hardness, transparency, heavy metals, phosphates, nitrogenous species, and oil spillage with a subsequent reduction in stress and disease outbreaks. The efficacy of *Bacillus* probiotics in the bioremediation process can be affected by some parameters such as species/strain of *Bacillus*, route of application, sources of nutrients, dissolved oxygen, metal ions, pH, and temperature. Therefore, in aquaculture practice, to achieve a maximum beneficiary of bacilli probiotics as a bioremediatory tool, attention needs to be paid for the management of these factors at an optimum level. Still, more studies are required to elucidate the actual mode of action used by *Bacillus* probiotics in the bioremediation process.

BIBLIOGRAPHY

Abdel-Tawwab, M., Khalil, R.H., Nour, A.M., Elkhayat, B.K., Khalifa, E., Abdel-Latif, H.M.R., 2022. Effects of Bacillus subtilis-fermented rice bran on water quality, performance, antioxidants/oxidants, and immunity biomarkers of White leg shrimp (Litopenaeus vannamei) reared at different salinities with zero water exchange. *J. Appl.* Aquac. 34, pp.332–357.

Balcázar, J.L., Blas, I. de, Ruiz-Zarzuela, I., Cunningham, D., Vendrell, D., Múzquiz, J.L., 2006. The role of probiotics in aquaculture. *Vet. Microbiol.* 114, pp.173–186. https://doi.org/10.1016/j.vetmic.2006.01.009

Baldwin, B.R., Peacock, A.D., Park, M., Ogles, D.M., Istok, J.D., McKinley, J.P., Resch, C.T., White, D.C., 2008. Multilevel samplers as microcosms to assess microbial response to biostimulation. *Groundwater* 46, pp.295–304.

Banerjee, A., Ghoshal, A.K., 2016. Biodegradation of real petroleum wastewater by immobilized hyper phenol-tolerant strains of Bacillus cereus in a fluidized bed bioreactor. *3 Biotech* 6, p.137.

Banerjee, S., Khatoon, H., Shariff, M., Yusoff, F.M., 2010. Enhancement of Penaeus monodon shrimp postlarvae growth and survival without water exchange using marine Bacillus pumilus and periphytic microalgae. *Fish. Sci.* 76, pp.481–487.

Barman, P., Bandyopadhyay, P., Kati, A., Paul, T., Mandal, A.K., Mondal, K.C., Mohapatra, P.K.D., 2018. Characterization and Strain Improvement of Aerobic Denitrifying EPS Producing Bacterium Bacillus cereus PB88 for Shrimp Water Quality Management. *Waste Biomass Valori.* 9, pp.1319–1330.

Bernhard, A., 2010. The nitrogen cycle: Processes, players, and human impact. *Nat. Educ. Knowl.* 2, p.12.

Bhatnagar, A., Lamba, R., 2017. Molecular characterization and dosage application of autochthonous potential probiotic bacteria in Cirrhinus mrigala. *J. Fish. Com.* 11, p.46.

Bhatnagar, A., Lamba, R., 2015. Antimicrobial ability and growth promoting effects of feed supplemented with probiotic bacterium isolated from gut microflora of Cirrhinus mrigala. *J. Integr. Agric.* 14, pp.583–592.

Bokossa, H.K.J., Saïdou, A., Sossoukpe, E., Fiogbé, D.E., Kossou, D., 2014. Decomposition and mineralization effect of various sources of Pig manure on water quality and nutrients availability for agro-Fish System in Benin. *Agric. Sci.* 5, p.1194.

Boyd, C.E., Tucker, C.S., Viriyatum, R., 2011. Interpretation of pH, acidity, and alkalinity in aquaculture and fisheries. *N. Am. J. Aquac.* 73, pp.403–408.

Brim, H., Venkateswaran, A., Kostandarithes, H.M., Fredrickson, J.K., Daly, M.J., 2003. Engineering Deinococcus geothermalis for bioremediation of high-temperature radioactive waste environments. *Appl. Environ. Microbiol.* 69, pp.4575–4582.

Camargo, J.A., Alonso, Á., 2006. Ecological and toxicological effects of inorganic nitrogen pollution in aquatic ecosystems: A global assessment. *Environ. Int.* 32, pp.831–849. https://doi.org/10.1016/j.envint.2006.05.002

Cha, J.H., Rahimnejad, S., Yang, S.Y., Kim, K.W., Lee, K.J., 2013. Evaluations of Bacillus spp. As dietary additives on growth performance, innate immunity and disease resistance of olive flounder (Paralichthys olivaceus) against streptococcus iniae and as water additives. *Aquaculture* 402–403, pp.50–5. https://doi.org/10.1016/j.aquaculture.2013.03.030

Chatterjee, S.K., Bhattacharjee, I., Chandra, G., 2010. Biosorption of heavy metals from industrial waste water by Geobacillus thermodenitrificans. *J. Hazard. Mater.* 175, pp.117–125. https://doi.org/10.1016/j.jhazmat.2009.09.136

Chávez-Crooker, P., Obreque-Contreras, J., 2010. Bioremediation of aquaculture wastes. *Curr. Opin. Biotechnol.* 21, pp.313–317. https://doi.org/10.1016/j.copbio.2010.04.001

Chen, C.-C., Chen, S.-N., 2001. Water quality management with Bacillus spp. in the high-density culture of red-parrot fish *Cichlasoma citrinellum*× *C. synspilum*. *N. Am. J. Aquac.* 63, pp.66–73.

Cormier, S.M., Suter, G.W., Zheng, L., 2013. Derivation of a benchmark for freshwater ionic strength. *Environ. Toxicol. Chem.* 32, pp.263–271.

Costa, A.C.A. da, Duta, F.P., 2001. Bioaccumulation of copper, zinc, cadmium and lead by Bacillus sp., *Bacillus cereus*, *Bacillus sphaericus* and *Bacillus subtilis*. *Brazilian J. Microbiol.* 32, pp.1–5.

Crab, R., Avnimelech, Y., Defoirdt, T., Bossier, P., Verstraete, W., 2007. Nitrogen removal techniques in aquaculture for a sustainable production. *Aquaculture* 270, pp.1–14.

Cunha, C.D. da, Leite, S.G.F., 2000. Gasoline biodegradation in different soil microcosms. *Brazilian J. Microbiol.* 31, pp.45–49.

Dabrowski, J.J., Rahman, A., George, A., Arnold, S., McCulloch, J., 2018. State space models for forecasting water quality variables: An application in aquaculture prawn farming, in: Proceedings of the *24th ACM SIGKDD International Conference on Knowledge Discovery & Data Mining. ACM*, pp.177–185.

Dalmin, G., Kathiresan, K., Purushothaman, A., 2001. Effect of probiotics on bacterial population and health status of shrimp in culture pond ecosystem. *Indian J. Exp. Biol.* 39, pp.939–942.

Daniels, W.L., Zipper, C.E., Orndorff, Z.W., Skousen, J., Barton, C.D., McDonald, L.M., Beck, M.A., 2016. Predicting total dissolved solids release from central Appalachian coal mine spoils. *Environ. Pollut.* 216, pp.371–379. https://doi.org/10.1016/j.envpol.2016.05.044

Dash, P., Tandel, R.S., Bhat, R.A.H., Mallik, S., Pandey, N.N., Singh, A.K., Sarma, D., 2018. The addition of probiotic bacteria to microbial floc: Water quality, growth, non-specific immune response and disease resistance of Cyprinus carpio in mid-Himalayan altitude. *Aquaculture* 495, pp.961–969.

de Paiva-Maia, E., Alves-Modesto, G., Otavio-Brito, L., Vasconcelos-Gesteira, T.C., Olivera, A., 2013. Effect of a commercial probiotic on bacterial and phytoplankton concentration in intensive shrimp farming (*Litopenaeus vannamei*) recirculation systems. *Lat. Am. J. Aquat. Res.* 41, pp.126–137.

Deschamps, A.M., Mahoudeau, G., Lebeault, J.M., 1980. Fast degradation of kraft lignin by bacteria. *Eur. J. Appl. Microbiol. Biotechnol.* 9, pp.45–51.

Devaraja, T., Banerjee, S., Yusoff, F., Shariff, M., Khatoon, H., 2013. A holistic approach for selection of *Bacillus* spp. as a bioremediator for shrimp postlarvae culture. *Turkish J. Biol.* 37, pp.92–100. https://doi.org/10.3906/biy-1203-19

Divya, M., Aanand, S., Srinivasan, A., Ahilan, B., 2015. Bioremediation–an eco-friendly tool for effluent treatment: A review. *Int. J. Appl. Res.* 1, pp.530–537.

Dongfeng, Z., Weilin, W., Yunbo, Z., Qiyou, L., Haibin, Y., Chaocheng, Z., 2011. Study on isolation, identification of a petroleum hydrocarbon degrading bacterium Bacillus fusiformis sp. and influence of environmental factors on degradation efficiency. *China Pet. Process. Petrochemical Technol.* 13, pp.74–82.

dos Santos Simões, F., Moreira, A.B., Bisinoti, M.C., Gimenez, S.M.N., Yabe, M.J.S., 2008. Water quality index as a simple indicator of aquaculture effects on aquatic bodies. *Ecol. Indic.* 8, pp.476–484.

Edet, U.O., Antai, S.P., Brooks, A.A., Asitok, A.D., 2018. Microbiological examination and physicochemical analysis of estuary water used as a point of source drinking water. *Int. J. Pathog. Res.* pp.1–13.

Elsabagh, M., Mohamed, R., Moustafa, E.M., Hamza, A., Farrag, F., Decamp, O., Dawood, M.A.O., Eltholth, M., 2018. Assessing the impact of *Bacillus* strains mixture probiotic on water quality, growth performance, blood profile and intestinal morphology of Nile tilapia, *Oreochromis niloticus*. *Aquac. Nutr.* 24, pp.1613– 1622. https://doi.org/10.1111/anu.12797

Faramarzi, M., Jafaryan, H., Roozbehfar, R., Jafari, M., Biria, M., 2012. Influences of probiotic bacilli on ammonia and urea excretion in two conditions of starvation and satiation in Persian sturgeon (*Acipenser persicus*) larvae. *Glob. Vet.* 8, pp.185–189.

Fernandes, E.M., de Almeida, L.C.F., Hashimoto, D.T., Lattanzi, G.R., Gervaz, W.R., Leonardo, A.F., Neto, R.V.R., 2018. Survival of purebred and hybrid Serrasalmidae under low water temperature conditions. *Aquaculture* 497, pp.97–102.

Gadd, G.M., 2000. Bioremedial potential of microbial mechanisms of metal mobilization and immobilization. *Curr. Opin. Biotechnol.* 11, pp.271–279.

George, E.G.J., Jeyaraj, G.P., Balaraman, D., 2017. *Bacillus* Probiotic Strains of Ecotoxnil® as Eco-friendly and Efficient Bio-decomposing Agent in Curbing Sludge and Toxic Gases from *Litopenaeus vannamei* (Boone, 1931) Shrimp Culture Ponds. *Int. J. Fish. Aquat. Stud.* 5, pp.283–291.

George, E.G.J., Jayaraj, G.P., Balaraman, D., Soundarapandy, A., 2016. Augmenting efficacy of the commercial probiotic consortium, Ecotrax® on soil, water quality, survival, growth and feed transformation on the semi-intensive pond culture system of the white leg shrimp, *Litopenaeus vannamei* (Boone, 1931). *Pelagia Res. Libr. Adv. Appl. Sci. Res.* 7, pp.32–42.

Gheewala, S.H., Annachhatre, A.P., 1997. Biodegradation of aniline. *Water Sci. Technol.* 36, pp.53–63.

Ghosh, S., Sinha, A., Sahu, C., 2008. Bioaugmentation in the growth and water quality of livebearing ornamental fishes. *Aquac. Int.* 16, pp.393–403.

Gomes, L.C., Brinn, R.P., Marcon, J.L., Dantas, L.A., Brandão, F.R., de Abreu, J., McComb, D.M., Baldisserotto, B., 2008. Using Efinol®L during transportation of marbled hatchetfish, *Carnegiella strigata* (Günther). *Aquac. Res.* 39, pp.1292–1298. https://doi.org/10.1111/j.1365-2109.2008.01993.x

Gupta, A., Gupta, P., Dhawan, A., 2016. *Paenibacillus polymyxa* as a water additive improved immune response of *Cyprinus carpio* and disease resistance against *Aeromonas hydrophila*. *Aquac. Reports* 4, pp.86–92.

Gupta, V.K., Shrivastava, A.K., Jain, N., 2001. Biosorption of chromium (VI) from aqueous solutions by green algae Spirogyra species. *Water Res.* 35, pp.4079–4085.

Hainfellner, P., Cardozo, M.V., Borzi, M.M., Almeida, C.C., José, L., Pizauro, L., Schocken-Iturrino, R.P., Costa, G.N., de Ávila, F.A., 2018. Commercial probiotic increases survival rate and water quality in aquariums with high density of nile tilapia larvae (*Oreochromis niloticus*). *Int. J. Probiotics Prebiotics* 13, pp.139–142.

Heitzer, A., Sayler, G.S., 1993. Monitoring the efficacy of bioremediation. *Trends Biotechnol.* 11, pp.334–343.

Hlordzi, V., Kuebutornye, F.K.A., Afriyie, G., Abarike, E.D., Lu, Y., Chi, S., Anokyewaa, M.A., 2020. The use of Bacillus species in maintenance of water quality in aquaculture: A review. *Aquac. Reports* 18, p.100503.

Hu, X., Xu, Y., Su, H., Xu, W., Wen, G., Xu, C., Yang, K., Zhang, S., Cao, Y., 2023. Effect of a bacillus probiotic compound on Penaeus vannameisurvival, water quality, and microbial communities. *Fishes* 8, p.362.

Hui, C., Wei, R., Jiang, H., Zhao, Y., Xu, L., 2019. Characterization of the ammonification, the relevant protease production and activity in a high-efficiency ammonifier Bacillus amyloliquefaciens DT. *Int. Biodeterior. Biodegradation* 142, pp.11–17.

Hura, M.U.D., Zafar, T., Borana, K., Prasad, J.R., Iqbal, J., 2018. Effect of commercial probiotic bacillus megaterium on water quality in composite culture of major carps. *Int. J. Curr. Agric. Sci.* 8, pp.268–273.

Issazadeh, K., Pahlaviani, M., Massiha, A., 2011. Bioremediation of toxic heavy metals pollutants by *Bacillus* spp. isolated from Guilan bay sediments, north of Iran, in: *International Conference on Biotechnology and Environment Management ICBEE*. pp. 67–71.

Jahangiri, L., Esteban, M.Á., 2018. Administration of probiotics in the water in finfish aquaculture systems: A Review. *Fishes* 3, pp.30–33. https://doi.org/10.3390/fishes3030033

Jana, B.B., Jana, S., 2003. The potential and sustainability of aquaculture in India. *J. Appl. Aquac.* 13, pp.283–316.

Joo, M.H., Kim, J.Y., 2013. Characteristics of crude oil biodegradation by biosurfactant-producing bacterium Bacillus subtilis JK-1. *J. Korean Soc. Appl. Biol. Chem.* 56, pp.193–200.

Kalaycı Kara, A., Fakıoğlu, Ö., Kotan, R., Atamanalp, M., Alak, G., 2021. The investigation of bioremediation potential of *Bacillus subtilis* and *B. thuringiensis* isolates under controlled conditions in freshwater. *Arch. Microbiol.* 203, pp.2075–2085.

Kebria, D.Y., Khodadadi, A., Ganjidoust, H., Badkoubi, A., Amoozegar, M.A., 2009. Isolation and characterization of a novel native Bacillus strain capable of degrading diesel fuel. *Int. J. Environ. Sci. Technol.* 6, pp.435–442.

Kewcharoen, W., Srisapoome, P., 2019. Probiotic effects of *Bacillus* spp. from Pacific white shrimp (*Litopenaeus vannamei*) on water quality and shrimp growth, immune responses, and resistance to *Vibrio parahaemolyticus* (AHPND strains). *Fish Shellfish Immunol.* 94, pp.175–189. https://doi.org/10.1016/j.fsi.2019.09.013

Khademzade, O., Zakeri, M., Haghi, M., Mousavi, S.M., 2020. The effects of water additive *Bacillus cereus* and *Pediococcus acidilactici* on water quality, growth performances, economic benefits, immunohematology and bacterial flora of whiteleg shrimp (*Penaeus vannamei* Boone, 1931) reared in earthen ponds. *Aquac. Res.* 51, pp.1759–1770.

Kim, S.U., Cheong, Y.H., Seo, D.C., Hur, J.S., Heo, J.S., Cho, J.S., 2007. Characterisation of heavy metal tolerance and biosorption capacity of bacterium strain CPB4 (*Bacillus* spp.). *Water Sci. Technol.* 55, pp.105–111.

Klotz, M.G., Stein, L.Y., 2011. *Research on nitrification and related processes*. Academic Press.

Kolodziej, B.J., Slepecky, R.A., 1964. Trace metal requirements for sporulation of *Bacillus megaterium*. *J. Bacteriol.* 88, pp.821–830.

Koops, H.-P., Pommerening-Röser, A., 2001. Distribution and ecophysiology of the nitrifying bacteria emphasizing cultured species. *FEMS Microbiol. Ecol.* 37, pp.1–9.

Kuebutornye, F.K.A., Abarike, E.D., Lu, Y., Hlordzi, V., Sakyi, M.E., Afriyie, G., Wang, Z., Li, Y., Xie, C.X., 2020. Mechanisms and the role of probiotic Bacillus in mitigating fish pathogens in aquaculture. *Fish Physiol. Biochem.* pp.1–23.

Kumar, B.R., Prasad, M.L., Srinivasrao, D., Rao, K.R.S. 2013. Bioremediation of sewage using specific consortium of microorganisms. *IJRANSS* 1(6), pp.15–26.

Kumar, V., Roy, S., Meena, D.K., Sarkar, U.K., 2016. Application of probiotics in shrimp aquaculture: Importance, mechanisms of action, and methods of administration. *Rev. Fish. Sci. Aquac.* 24, pp.342–368.

Labbé, N., Laurin, V., Juteau, P., Parent, S., Villemur, R., 2007. Microbiological community structure of the biofilm of a methanol-fed, marine denitrification system, and identification of the methanol-utilizing microorganisms. *Microb. Ecol.* 53, pp.621–630.

Lalloo, R., Ramchuran, S., Ramduth, D., Görgens, J., Gardiner, N., 2007. Isolation and selection of Bacillus spp. as potential biological agents for enhancement of water quality in culture of ornamental fish. *J. Appl. Microbiol.* 103, pp.1471–1479. https://doi.org/10.1111/j.1365-2672.2007.03360.x

Li, Z., Gao, Y., Lu, Z., Xie, J., Liu, Y., Wang, G., Cheng, X., 2023. Strengthening and microbial regulation mechanism of *Bacillus* on purification device for grass carp culture wastewater. *Front. Environ. Sci.* https://doi.org/10.3389/fenvs.2023.1128329

Liu, C.H., Chiu, C.S., Ho, P.L., Wang, S.W., 2009. Improvement in the growth performance of white shrimp, *Litopenaeus vannamei*, by a protease-producing probiotic, *Bacillus subtilis* E20, from natto. *J. Appl. Microbiol.* 107, pp.1031–1041. https://doi.org/10.1111/j.1365-2672.2009.04284.x

Liu, T., He, X., Jia, G., Xu, J., Quan, X., You, S., 2020. Simultaneous nitrification and denitrification process using novel surface-modified suspended carriers for the treatment of real domestic waste water. *Chemosphere* 125831.

Loh, J.-Y., 2017. The role of probiotics and their mechanisms of action: An aquaculture perspective. *World Aquac.* pp.19–23.

Luo, W., Hai, F.I., Price, W.E., Guo, W., Ngo, H.H., Yamamoto, K., Nghiem, L.D., 2016. Phosphorus and water recovery by a novel osmotic membrane bioreactor–reverse osmosis system. *Bioresour. Technol.* 200, pp.297–304. https://doi.org/10.1016/j.biortech.2015.10.029

Ma, X., Novak, P.J., Ferguson, J., Sadowsky, M., LaPara, T.M., Semmens, M.J., Hozalski, R.M., 2007. The impact of H2 addition on dechlorinating microbial communities. *Bioremediat. J.* 11, pp.45–55.

Maag, M., Vinther, F.P., 1996. Nitrous oxide emission by nitrification and denitrification in different soil types and at different soil moisture contents and temperatures. *Appl. Soil Ecol.* 4, pp.5–14.

Maddela, N.R., Masabanda, M., Leiva-Mora, M., 2015. Novel diesel-oil-degrading bacteria and fungi from the Ecuadorian Amazon rainforest. *Water Sci. Technol.* 71, pp.1554–1561.

Mahmud, S., Ali, M.L., Alam, M.A., Rahman, M.M., Jørgensen, N.O.G., 2016. Effect of probiotic and sand filtration treatments on water quality and growth of tilapia (*Oreochromis niloticus*) and pangas (*Pangasianodon hypophthalmus*) in earthen ponds of southern Bangladesh. *J. Appl. Aquac.* 28, pp.199–212.

Manahan, S., 2017. *Environmental chemistry*. CRC Press.

Martínez-Córdova, L.R., Emerenciano, M., Miranda-Baeza, A., Martínez-Porchas, M., 2015. Microbial-based systems for aquaculture of fish and shrimp: An updated review. *Rev. Aquac.* 7, pp.131–148.

Martinez, M., Mangano, M.C., Maricchiolo, G., Genovese, L., Mazzola, A., Sarà, G., 2018. Measuring the effects of temperature rise on Mediterranean shellfish aquaculture. *Ecol. Indic.* 88, pp.71–78.

Matias, H.B., Yusoff, F.M., Shariff, M., Azhar, O., 2002. Effects of commercial microbial products on water quality in tropical shrimp culture ponds. *Asian Fish. Sci.* 15, pp.239–248.

Md, S.A., Nour, A.M., Srour, T.M., Assem, S.S., Ibrahim, H.A., El-Sayed, H.S., 2015. Greenwater, Marine *Bacillus subtilis* HS1 Probiotic and synbiotic enriched artemia and rotifers improved European seabass *Dicentrarchus labrax* larvae early weaning length growth, survival, water and bacteriology quality. *Am. J. Life Sci.* 3, pp.45–52.

Mével, G., Prieur, D., 2000. Heterotrophic nitrification by a thermophilic *Bacillus* species as influenced by different culture conditions. *Can. J. Microbiol.* 46, pp.465–473.

Mirbakhsh, M., Mahjoub, M., Afsharnasab, M., Kakoolaki, S., Sayyadi, M., Hosseinzadeh, S., 2021. Effects of *Bacillus subtilis* on the water quality, stress tolerance, digestive enzymes, growth performance, immune gene expression, and disease resistance of white shrimp (*Litopenaeus vannamei*) during the early hatchery period. *Aquac. Int.* 29, pp.2489–2506.

Mohamed, M.H., Refat, N., 2011. Pathological evaluation of probiotic, *Bacillus subtilis*, against *Flavobacterium columnare* in tilapia nilotica (*Oreochromis niloticus*) fish in Sharkia Governorate, Egypt. *J. Am. Sci.* 7, pp.244–256.

Mohammadi, G., Adorian, T.J., Rafiee, G., 2020. Beneficial effects of *Bacillus subtilis* on water quality, growth, immune responses, endotoxemia and protection against lipopolysaccharide-induced damages in *Oreochromis niloticus* under biofloc technology system. *Aquac. Nutr.* 26, pp.1476–1492.

Mohapatra, S., Chakraborty, T., Kumar, V., DeBoeck, G., Mohanta, K.N., 2013. Aquaculture and stress management: A review of probiotic intervention. *J. Anim. Physiol. Anim. Nutr. (Berl).* 97, pp.405–430.

Moriarty, D.J.W., 1998. Control of luminous *Vibrio* species in penaeid aquaculture ponds. *Aquaculture* 164, pp.351–358. https://doi.org/10.1016/S0044-8486(98)00199-9

Mueller, J.S., Grabowski, T.B., Brewer, S.K., Worthington, T.A., 2017. Effects of temperature, total dissolved solids, and total suspended solids on survival and development rate of larval Arkansas River Shiner. *J. Fish Wildl. Manag.* 8, pp.79–88.

Naderi Samani, M., Jafaryan, H., Gholipour, H., Harsij, M., Farhangi, M., 2016. Effect of different concentration of profitable *Bacillus* on bioremediation of common carp (*Cyprinus carpio*) pond discharge. *Iran. J. Aquat. Anim. Heal.* 2, pp.44–54.

Narmadha, D., Kavitha M. S. V. J., 2012. Treatment of domestic wastewater using natural flocculants. *Int. J. Life Sci. Biotechnol. Pharma. Res.* 1(3), pp.206–213.

Nimrat, S., Suksawat, S., Boonthai, T., Vuthiphandchai, V., 2012. Potential *Bacillus* probiotics enhance bacterial numbers, water quality and growth during early development of white shrimp (Litopenaeus vannamei). *Vet. Microbiol.* 159, pp.443–450. https://doi.org/10.1016/j.vetmic.2012.04.029

Olson, J.R., Hawkins, C.P., 2017. Effects of total dissolved solids on growth and mortality predict distributions of stream macroinvertebrates. *Freshw. Biol.* 62, pp.779–791. https://doi.org/10.1111/fwb.12901

Padmavathi, P., Sunitha, K., Veeraiah, K., 2012. Efficacy of probiotics in improving water quality and bacterial flora in fish ponds. *African J. Microbiol. Res.* 6, pp.7471–7478.

Panigrahi, A., Mohaatra, A., Panigrahi, A. 2005. Bioremediation: An environmental friendly approach for sustainable aquaculture. In: Thomas, G. M., Ward, C. H., Raymond, R. L., Wilson, J. T., Loehr, R. C. (1992). Bioremediation, pp. 369–385. In: *Encyclopedia of Microbiology* (Leperberg, J. Ed.), London: Academic Press.

Patil, M.P., Jeong, I., Woo, H.-E., Oh, S.-J., Kim, H.C., Kim, K., Nakashita, S., Kim, K., 2022. Effect of *Bacillus subtilis* zeolite used for sediment remediation on sulfide, phosphate, and nitrogen control in a microcosm. *Int. J. Environ. Res. Public Health* 19, p.4163.

Patureau, D., Bernet, N., Delgenes, J.P., Moletta, R., 2000. Effect of dissolved oxygen and carbon–nitrogen loads on denitrification by an aerobic consortium. *Appl. Microbiol. Biotechnol.* 54, pp.535–542.

Paudel, S.R., Choi, O., Khanal, S.K., Chandran, K., Kim, S., Lee, J.W., 2015. Effects of temperature on nitrous oxide (N2O) emission from intensive aquaculture system. *Sci. Total Environ.* 518, pp.16–23.

Pendashteh, A.R., Abdullah, L.C., Fakhru'l-Razi, A., Madaeni, S.S., Abidin, Z.Z., Biak, D.R.A., 2012. Evaluation of membrane bioreactor for hypersaline oily wastewater treatment. *Process Saf. Environ. Prot.* 90, pp.45–55.

Porubcan, R.S., 1991. Reduction in chemical oxygen demand and improvement in Penaeus monodon yield in ponds inoculated with aerobic Bacillus bacteria, in: Program and Abstract of the 22nd Annual Conference and Exposition of the World Aquaculture Socity, 1991. World Aquaculture Society, pp. 16–20.

Putra, A.N., Bayu Syamsunarno, M., Ningrum, W., Jumyanah, J., Mustahal, M., 2020. Effect of the administration of probiotic Bacillus NP5 in the rearing media on water quality, growth, and disease resistance of African catfish (*Clarias gariepinus*). *Biodiversitas J. Biol. Divers.* 21.

Querijero, B.L., Mercurio, A.L., 2016. Water quality in aquaculture and non-aquaculture sites in Taal Lake, Batangas, Philippines. *J. Exp. Biol. Agric. Sci.* https://doi.org/10.18006/2016.4109

Rahiman, K.M., Jesmi, Y., Thomas, A.P., Mohamed Hatha, A.A., 2010. Probiotic effect of *Bacillus* NL110 and *Vibrio* NE17 on the survival, growth performance and immune response of *Macrobrachium rosenbergii* (de Man). *Aquac. Res.* 41, pp.120–134. https://doi.org/10.1111/j.1365-2109.2009.02473.x

Rajakumar, S., Ayyasamy, P.M., Shanthi, K., Thavamani, P., Velmurugan, P., Song, Y.C., Lakshmanaperumalsamy, P., 2008. Nitrate removal efficiency of bacterial consortium (*Pseudomonas* sp. KW1 and Bacillus sp. YW4) in synthetic nitrate-rich water. *J. Hazard. Mater.* 157, pp.553–563. https://doi.org/10.1016/j.jhazmat.2008.01.020

Rao, V.A., 2002. Bioremediation technology to maintain healthy ecology in aquaculture ponds. *Fish. Chimes. Sept.* 22, pp.39–42.

Reddy, K.V., Reddy, A.V.K., Babu, B.S., Lakshmi, T.V., 2018. Applications of Bacillus sp in Aquaculture Waste Water Treatment. *Int. J. S. Res. Sci. Tech.* 4, pp.1806–1812.

Ren, W., Wu, H., Guo, C., Xue, B., Long, H., Zhang, X., Cai, X., Huang, A., Xie, Z., 2021. Multi-strain tropical *Bacillus* spp. as a potential probiotic biocontrol agent for large-scale enhancement of mariculture water quality. *Front. Microbiol.* 12.

Rout, P.R., Bhunia, P., Dash, R.R., 2017. Simultaneous removal of nitrogen and phosphorous from domestic wastewater using *Bacillus cereus* GS-5 strain exhibiting heterotrophic nitrification, aerobic denitrification and denitrifying phosphorous removal. *Bioresour. Technol.* 244, pp.484–495. https://doi.org/10.1016/j.biortech.2017.07.186

Shan, H.W., Bao, W.Y., Ma, S., Wei, D.P., Gao, L., 2016. Ammonia and nitrite nitrogen removal in shrimp culture by *Vibrio alginolyticus* VZ5 immobilized in SA beads. *Aquac. Int.* 24, pp.357–372.

Shao, Y., Zhong, H., Wang, L., Elbashier, M.M.A., 2021. Use of *Bacillus subtilis* D9 to purify coastal aquaculture wastewater and improve grass carp resistance to *Vibrio* infection. *Aquac. Environ. Interact.* 13, pp.249–258.

Shapleigh, J.P., 2011. Oxygen control of nitrogen oxide respiration, focusing on α-proteobacteria. *Biochem. Soc. Trans.* 39, pp.179–183.

Sharma, R., 1999. Probiotics: A new horizon in aquaculture. *Fish. World. Febr.* 1999, pp.1–8.

Shi, X., Lefebvre, O., Ng, K.K., Ng, H.Y., 2014. Sequential anaerobic–aerobic treatment of pharmaceutical wastewater with high salinity. *Bioresour. Technol.* 153, pp.79–86.

Singh, B.J.S., 2002. Bioremediation in prawn grow out systems. *Tech. Pap. Cent. Mar. Fish. Res. Inst.*

Soltani, M., Ghosh, K., Hoseinifar, S.H., Kumar, V., Lymbery, A.J., Roy, S., Ringø, E., 2019. Genus bacillus, promising probiotics in aquaculture: Aquatic animal origin, bio-active components, bioremediation and efficacy in fish and shellfish. *Rev. Fish. Sci. Aquac.* pp.1–49.

Song, Z.-F., An, J., Fu, G.-H., Yang, X.-L., 2011. Isolation and characterization of an aerobic denitrifying *Bacillus* sp. YX-6 from shrimp culture ponds. *Aquaculture* 319, pp.188–193. https://doi.org/10.1016/j.aquaculture.2011.06.018

Stefanescu, I.A., 2015. Bioaccumulation of heavy metals by *Bacillus megaterium* from phosphogypsum waste. *Sci. Study Res. Chem. Chem. Eng. Biotechnol. Food Ind.* 16, p.93.

Summerfelt, S.T., Zühlke, A., Kolarevic, J., Reiten, B.K.M., Selset, R., Gutierrez, X., Terjesen, B.F., 2015. Effects of alkalinity on ammonia removal, carbon dioxide stripping, and system pH in semi-commercial scale water recirculating aquaculture systems operated with moving bed bioreactors. *Aquac. Eng.* 65, pp.46–54. https://doi.org/10.1016/j.aquaeng.2014.11.002

Sungsirin, N., Songsuk, A., Jaisupa, N., Kulabtong, S., 2024. Efficacy of a probiotic *Bacillus subtilis* strain in fish culture water for ammonia removal and enhancing survival of juvenile common carps (*Cyprinus carpio*). *Int. J. Agric. Technol.* 20, pp.355–364.

Sunitha, K., Padmavathi, P., 2013. Influence of probiotics on water quality and fish yield in fish ponds. *Int. J. Pure Appl. Sci. Technol* 19, pp.48–60.

Swann, L., 1997. A fish farmer's guide to understanding water quality. *Aquac. Ext.* pp.1–8.

Tang, C.Y., Fu, Q.S., Criddle, C.S., Leckie, J.O., 2007. Effect of flux (transmembrane pressure) and membrane properties on fouling and rejection of reverse osmosis and nanofiltration membranes treating perfluorooctane sulfonate containing wastewater. *Environ. Sci. Technol.* 41, pp.2008–2014.

Tariq, A.L., Sudha, S., Reyaz, A.L., 2016. Isolation and screening of *Bacillus* species from sediments and application in bioremediation. *Int. J. Curr. Microbiol. App. Sci* 5, pp.916–924.

Thomas, J.M., Ward, C.H., 1992. Subsurface microbial ecology and bioremediation. *J. Hazard. Mater.* 32, pp.179–194. https://doi.org/10.1016/0304-3894(92)85091-E

Thurlow, C.M., Williams, M.A., Carrias, A., Ran, C., Newman, M., Tweedie, J., Allison, E., Jescovitch, L.N., Wilson, A.E., Terhune, J.S., Liles, M.R., 2019. *Bacillus velezensis* AP193 exerts probiotic effects in channel catfish (*Ictalurus punctatus*) and reduces aquaculture pond eutrophication. *Aquaculture* 503, pp.347–356. https://doi.org/10.1016/j.aquaculture.2018.11.051

Tovar, A., Moreno, C., Mánuel-Vez, M.P., García-Vargas, M., 2000. Environmental impacts of intensive aquaculture in marine waters. *Water Res.* 34, pp.334–342. https://doi.org/10.1016/S0043-1354(99)00102-5

Verbaendert, I., Boon, N., De Vos, P., Heylen, K., 2011. Denitrification is a common feature among members of the genus Bacillus. *Syst. Appl. Microbiol.* 34, pp.385–391.

Verstraete, W., Focht, D.D., 1977. Biochemical ecology of nitrification and denitrification, in: *Adv. Microbial Ecology*, pp.135–214.

Wang, Y.-B., Xu, Z.-R., Xia, M.-S., 2005. The effectiveness of commercial probiotics in northern white shrimp *Penaeus vannamei* ponds. *Fish. Sci.* 71, pp.1036–1041. https://doi.org/10.1111/j.1444-2906.2005.01061.x

Wen-Jun, W., 2011. Purification Ability of Bacillus subtilis on Eutrophic Water [J]. *Hubei Agric. Sci.* p.10.

Wilks, J.C., Kitko, R.D., Cleeton, S.H., Lee, G.E., Ugwu, C.S., Jones, B.D., BonDurant, S.S., Slonczewski, J.L., 2009. Acid and base stress and transcriptomic responses in *Bacillus subtilis. Appl. Environ. Microbiol.* 75, pp.981–990.

Wu, D.X., Zhao, S.M., Peng, N., Xu, C.P., Wang, J., Liang, Y.X., 2016. Effects of a probiotic (Bacillus subtilis FY99-01) on the bacterial community structure and composition of shrimp (*Litopenaeus vannamei*, Boone) culture water assessed by denaturing gradient gel electrophoresis and high-throughput sequencing. *Aquac. Res.* 47, pp.857–869. https://doi.org/10.1111/are.12545

Wu, S., Maskaly, J., 2018. Study on the effect of total dissolved solids (TDS) on the performance of an SBR for COD and nutrients removal. *J. Environ. Sci. Heal. Part A* 53, pp.146–153.

Xie, F., Zhu, T., Zhang, F., Zhou, K., Zhao, Y., Li, Z., 2013. Using Bacillus amyloliquefaciens for remediation of aquaculture water. *Springerplus* 2, p.119. https://doi.org/10.1186/2193-1801-2-119

Yin, Y., Zhang, P., Yue, X., Du, X., Li, W., Yin, Y., Yi, C., Li, Y., 2018. Effect of sub-chronic exposure to lead (Pb) and *Bacillus subtilis* on *Carassius auratus* gibelio: Bioaccumulation, antioxidant responses and immune responses. *Ecotoxicol. Environ. Saf.* 161, pp.755–762. https://doi.org/10.1016/j.ecoenv.2018.06.056

Yousuf, J., Thajudeen, J., Rahiman, M., Krishnankutty, S., Alikunj, A.P., Abdulla, M.H., 2017. Nitrogen fixing potential of various heterotrophic *Bacillus* strains from a tropical estuary and adjacent coastal regions. *J. Basic Microbiol.* 57, pp.922–932.

Yu, A., Li, Y., Yu, J., 2005. Denitrification of a newly isolated *Bacillus* strain W2 and its application in aquaculture. *Eur. PMC* 25, pp.77–81.

Yu, C.-H., Wang, Y., Guo, T., Shen, W.-X., Gu, M.-X., 2012. Isolation and identification of ammonia nitrogen degradation strains from industrial wastewater.

Zhai, J., Jiumu, L., Cong, L., Wu, Y., Dai, L., Zhang, Z., Zhang, M., 2020. Reed decomposition under Bacillus subtilis addition conditions and the influence on water quality. *Ecohydrol. Hydrobiol.* 20, pp.504–512.

Zhang, X., Fu, L., Deng, B., Liang, Q., Zheng, J., Sun, J., Zhu, H., Peng, L., Wang, Y., Wenying, S., 2013. *Bacillus subtilis* SC02 supplementation causes alterations of the microbial diversity in grass carp water. *World J. Microbiol. Biotechnol.* 29, pp.1645–1653.

Zhang, Z., Yang, Z., Zheng, G., Lin, Q., Zhuo, X., Zhang, G., 2020. Effects of addition of sucrose and probiotics on whiteleg shrimp *Litopenaeus vannamei* postlarvae performance, water quality, and microbial community. *N. Am. J. Aquac.* 82, pp.43–53.

Zhou, S., Xia, Y., Zhu, C., Chu, W., 2018. Isolation of marine *Bacillus* sp. with antagonistic and organic-substances-degrading activities and its potential application as a fish probiotic. *Mar. Drugs* 16, p.196.

Zink, I.C., Benetti, D.D., Douillet, P.A., Margulies, D., Scholey, V.P., 2011. Improvement of water chemistry with *Bacillus* probiotics inclusion during simulated transport of yellowfin tuna yolk sac larvae. *N. Am. J. Aquac.* 73, pp.42–48. https://doi.org/10.1080/15222055.2011.544622

Zokaeifar, H., Babaei, N., Saad, C.R., Kamarudin, M.S., Sijam, K., Balcazar, J.L., 2014. Administration of *Bacillus subtilis* strains in the rearing water enhances the water quality, growth performance, immune response, and resistance against *Vibrio harveyi* infection in juvenile white shrimp, *Litopenaeus vannamei. Fish Shellfish Immunol.* 36, pp.68–74. https://doi.org/10.1016/j.fsi.2013.10.007

Zorriehzahra, M.J., Delshad, S.T., Adel, M., Tiwari, R., Karthik, K., Dhama, K., Lazado, C.C., 2016. Probiotics as beneficial microbes in aquaculture: An update on their multiple modes of action: A review. *Vet. Q.* 36, pp.228–241. https://doi.org/10.1080/01652176.2016.1172132

9 *Bacillus* Effects on the Immune System

Ajay Valiyaveettil Salimkumar
Research Centre for Experimental Marine Biology and Biotechnology (PIE-UPV/EHU), Areatza Pasealekua, Plentzia- Biskaia, Basque Country, Spain

Athira Ambili Sasikumar
CSIR NIIST, Thiruvananthapuram, Kerala, India

Mohammad Shafiqur Rahman
Institute of Urban Environment, Chinese Academy of Sciences, Xiamen, China

Preetham Elumalai
Cochin University of Science and Technology (CUSAT), Cochin, India

9.1 INTRODUCTION

Among the diverse array of microorganisms inhabiting the human body, *Bacillus* species have emerged as attractive candidates due to their distinctive traits and potential immune-modulatory abilities. The Gram-positive, spore-forming *Bacillus* species are common in nature and present in a range of environments, including soil, water, and both human and animal gastrointestinal tracts. Studies over the years have revealed the complicated connections between *Bacillus* species and the immune system, demonstrating both advantageous and complex interactions. *Bacillus* strains can enhance host defense systems against infections and perhaps reduce the risk of immunological-mediated illnesses by favorably modulating immune responses (Liu et al. 2020; Han et al. 2023). Studies proved that *Bacillus* strains can be used as a therapeutic agent for preventing intestinal inflammations in mice (Pesarico et al. 2022; Américo et al. 2023). These advantageous outcomes are frequently linked to the synthesis of immunomodulatory molecules that have been found to have immunomodulatory properties, meaning they can influence the immune response positively. For example, some *Bacillus* strains have been shown to stimulate the production of cytokines, which are proteins that help regulate the immune response (Alcaraz et al. 2010). They are well known to express a variety of genes involved in different cellular processes. For example, *Bacillus subtilis* is a model organism widely used in genetic and molecular studies, and its genome has been extensively characterized (Rahman et al. 2020).

In addition, various studies have shown that *Bacillus* species can enhance the action of immune cells, such as natural killer cells and macrophages, which are important components of the innate immune system. These cells play a critical role in identifying and destroying pathogens in the body. The effects of *Bacillus* on the immune system can vary depending on the strain and the individual. While some strains may have immunomodulatory properties, others may cause infections and negatively impact immune function (Taverniti and Guglielmetti 2011; Alton et al. 2015). The host's immunological state,

DOI: 10.1201/9781003503811-9

the dosage and mode of administration of *Bacillus*, and the presence of other resident bacteria within the host's microbiota can all have a big impact on how immune interactions turn out.

Bacillus species have demonstrated potential for modulating the activity of antioxidant enzymes. These enzymes are essential for maintaining cellular redox harmony and protecting it from oxidative harm. Through a variety of processes, including enzyme stimulation, gene expression upregulation, and the creation of bioactive compounds, *Bacillus* strains have been discovered to affect the activity and expression of antioxidant enzymes, including superoxide dismutase (SOD), catalase (CAT), glutathione peroxidase (GPx), and glutathione reductase (GR) (Kuebutornye et al. 2019; Pei et al. 2023). Understanding how the *Bacillus* species affect antioxidant enzymes may help to promote cellular antioxidant defense and avoid diseases linked to oxidative stress. Individuals, animals, and plants have all noticed how effective they are in reducing stress. These effects include increased stress tolerance, a decrease in stress-related behaviors, and an improvement in general well-being. The production of stress-responsive chemicals, alteration of the neurotransmitter system, control of the hypothalamic–pituitary–adrenal axis, and interactions with gut microbiota are only a few of the mechanisms used by these species to modulate stress (Eissa et al. 2018).

By studying the gene expression patterns of *Bacillus* species under different conditions, researchers can gain insights into the biochemical pathways and regulatory networks that allow these bacteria to thrive in diverse environments. This information can be used for a range of applications, including the development of new antibiotics, bioremediation strategies, and biotechnological processes. *Bacillus* species are known for their ability to adapt to different environmental conditions and produce a wide range of biologically active compounds. To understand the mechanisms underlying this adaptability and diversity, researchers have studied the gene expression patterns of these bacteria (Shen 2020). Thus, it is crucial to comprehend how different *Bacillus* species interact with the immune system, as this will help us better understand host–microbe dynamics and have implications for maintaining health and preventing diseases.

9.2 PRODUCTION AND MODULATION OF ANTIOXIDANTS BY *BACILLUS* SPECIES

The important oxidants that are present in almost all the species are reactive oxygen species (ROS), lipid peroxidation (LPO), and reactive nitrogen species (RNS). If these oxidants' production exceeds the amount of antioxidants in any organism, it results in oxidative stress in the cell (Hasan et al. 2017; Gowtham et al. 2022). These antioxidants protect the organism from oxidative stresses due to ROS by producing enzymes that counteract the damage by ROS (Bermejo-Nogales et al. 2016). Glutathione, along with SOD and CAT, acts as an antioxidant enzyme to reduce ROS (Mouthuy et al. 2016). While SOD helps in the decomposition of O_2^- to H_2O_2, and catalase helps in the conversion of H_2O_2 to H_2O and O_2. *Bacillus* species, as a probiotic, have the ability to produce and modulate the antioxidants such as SOD, CAT, and GPx to remove the free radicals formed (Kuebutornye et al. 2019; Pei et al. 2023).

Bacillus species help in the increased response of antioxidants in the different groups of fishes. For example, the antioxidant response in an Asian catfish, *Pangasius hypothalamus*, was increased by *B. licheniformis Dahb1*. Carp's immunity and antioxidant activities were studied when treated with a different combination of *Bacillus* species (Wang et al. 2017). The combination of *Bacillus* species, such as *B. subtilis* and *B. licheniformis*, controls the production and clearance of free radicals in the liver of common carp; thus, an oxidant–antioxidant balance is maintained. *Bacillus* can produce fermentative hydrolysates, resulting in a new peptide profile (Hindu et al. 2018). These fermented protein hydrolysates from different fish meat have antioxidant properties and 2,2-diphenyl-1-picrylhydrazyl (DPPH) scavenging activity (Li et al. 2012). Also, in tilapia, *O. mossambicus*, the growth, immune parameters, and the activity of the antioxidant enzymes in the serum increased when *B. licheniformis* Dahb was included in the fish diet. This results in the resistance of the fish against *Aeromonas hydrophila*. So, *Bacillus* can protect fish from other pathogens through its antioxidant properties (Gobi et al. 2018).

Bacteria produce some secondary metabolites called biosurfactants. The various physiological functions of these biosurfactants are bacterial pathogenesis, heavy metal binding, enhancing the solubility of hydrophobic or water-insoluble compounds, cell adhesion quorum sensing, and biofilm formation (Gudiña et al. 2010). The ability of a biological system to detoxify the reactive intermediates due to the production of ROS by the producer microorganisms is known to be oxidative stress (Zhang et al. 2013). The biological system will follow different methods to withstand these conditions by producing antioxidants against them. One of the promising antioxidants is polysaccharides. They are getting further developed to be more promising and potent antioxidants. *Bacillus subtilis*, *Lactobacillus plantarum* C88, etc. are good at producing antioxidants, which can also show high antibacterial activity (mainly *Bacillus*). Also, there is a need for new antimicrobial substances because of the pharmacological limitations of antibiotics and the increased prevalence of antibiotic-resistant pathogens (Krishna et al. 2011).

As mentioned above, *Bacillus subtilis* shows antioxidant activity. When *Bacillus subtilis* was subjected to mutation using UV radiation, the antibacterial activity of the strain was increased, and its activity varied under different time intervals. This is due to the production of certain antibiotics, which may be because of the activation of specific genes (Krishna et al. 2011). The lactic acid bacteria usually produce exopolysaccharides with unique rheological and physical properties. *Lactobacillus plantarum* C88 is a lactic acid-producing bacteria that produces a neutral exopolysaccharide (EPS) and is designated as LPC1. Studies showed that LPC1 shows good scavenging ability on hydroxyl radicals and DPPH free radicals. This made LPC1 show a protective effect on H2O2-induced LPO in Caco-2 cells. It can also raise the activities of SOD and total antioxidant capacities (T-AOC) and inhibits malondialdehyde (MDA) formation in a dose-dependent manner. These findings show that the EPS from *L. plantarum* C88 exhibits antioxidant effects, including reducing LPO, up-regulating enzymatic and non-enzymatic antioxidant activities, and scavenging ROS (Zhang et al. 2013).

As discussed above, *Bacillus* species can produce exopolysaccharides, showing antioxidant activity. This EPS has emerged as a promising antioxidant in pharmaceutics and biomedicine. One of the *Bacillus* species, which can produce antioxidants that have the quality of producing potential economic EPS, is the *B. velezensis* VTX20. The antioxidants from this species can even be used as an alternative to commercial antioxidants (Vu et al. 2021). *Lactobacillus helveticus* MB2-1 is another *Bacillus* species that produces EPS, which has antioxidant properties. EPS1, EPS2, and EPS3 are three different purified forms of crude EPS from *L. helveticus* MB2-1. Both crude and purified EPS showed similar antioxidant activity (correct order: EPS>EPS3>EPS2>EPS1). These EPS can also be a potent natural antioxidant in therapeutics and the food industry (Li et al. 2014).

The biosurfactant produced by the *B. subtilis* RW-1 also shows antioxidant activity. It helps them to protect against oxidative stress by scavenging free radicals. The studies on DPPH scavenging, reducing power, and chelating agents confirm the antioxidant activities. It also relates the reciprocal correlation of antioxidant activities of natural components with their reducing powers (Benitez et al. 2012). Also, the FTIR and HPLC studies show that the biosurfactant has a lipopeptide structure (Yalcin and Cavusoglu 2010). The food industry and academia are interested in replacing synthetic antioxidants with these natural compounds, which depict their antioxidant activity and possible usage as a natural additive (Yalcin and Cavusoglu 2010).

Bacillus mojavensis is another *Bacillus* species that produces antioxidants as well as lipopeptides which can heal wounds. The four assays used to study *in vitro* antioxidant activities are reducing power, LPO, DPPH scavenging, and β-carotene bleaching by linoleic acid assay. Studies showed that A21 lipopeptide has a good scavenging effect on DPPH and reducing power and significantly inhibited LPO. The wound-healing activity of A21 lipopeptide gel was studied on rats, and it seems to have an accelerated healing activity compared to control rats (Ayed et al. 2015). Also, those animals treated with A21 lipopeptide gel were free from inflammatory reactions such as redness and swelling. This wound-healing activity results from the potential antioxidant ability of A21 lipopeptide gel. The free radical scavenging property of the gel helps reduce inflammation and improve tissue formation (Thiem and Goślińska 2004). *Bacillus cirulans*, a marine microbe, can produce potent antimicrobe against Gram-negative and Gram-positive pathogenic strains (Das et al. 2008).

Burma (a significant proportion of traditional Burmese cuisine consists of fish and fish-derived products) is famous for having fish products as a significant source of animal protein. Fermented/salted fish or shrimp paste is their dietary supplement and flavor enhancer. And also, these products have some functional roles in health. These fermented fish also have sufficient polyunsaturated fatty acids, including docosahexaenoic acid (DHA) (Montaño et al. 2001). These metabolites show antioxidant activity. Studies on the bacterial flora in the Ngapi, a fermented shrimp paste, showed that the antibacterial compounds produced by the flora prevent the spoilage of food from pathogens and reduce microbial contamination during fermentation. Through screening, two strains, NP1-1 and NP3-2, showed high free radical scavenging and antibacterial activities and belonged to *Bacillus subtilis* and *Bacillus* sp., respectively. The antioxidant activity of these two strains makes the food safer to consume (Aung et al. 2004).

Two *Bacillus* species were used for the biodegradation of Nile perch wastewater, utilizing their antioxidant activity (Mhina et al. 2020). High levels of protein breakdown were achieved through biodegradation employing two different *Bacillus* species, and the tiny (2 kDa) peptides and amino acids generated from biodegraded Nile perch effluent displayed strong antibacterial and antioxidant properties. The use of Nile perch wastewater as a natural antibacterial and antioxidant component for nutraceuticals is being reported for the first time in this chapter (Mhina et al. 2020).

Bacillus subtilis A26 is a proteolytic bacterium that produces protein hydrolysates during the fermentation of fish proteins. The antioxidant and antibacterial properties of the protein hydrolysates are evaluated and studied through β-carotene bleaching, reducing power assay, DNA nicking assay, and DPPH radical method. This antioxidant property shows that there are some electron donors in the form of peptides in hydrolysates that stabilize the product by reacting with free radicals (Jemil et al. 2014).

Bacillus methylotrophicus DCS1 is another strain of *Bacillus*, which produces a surfactant DCS1. This lipopeptide DCS1 shows high antioxidant activities, and they are assayed through different tests. These tests include scavenging effect on DPPH radicals, reducing power, β-carotene bleaching by linoleic acid assay, etc. Its antioxidation mechanism includes hydrogen or electron donation, radical scavenging during peroxidation, and metal ion chelating. Moreover, DCS1 lipopeptide shows high antifungal and antibacterial activities. In addition to this, they also showed anti-adhesive activities against biofilm formation and also the potential to destroy the pre-formed biofilm (Jemil et al. 2017).

B38 is a strain of *Bacillus subtilis*, which produces C14, C15, and C16 bacillomycin D-like lipopeptides. Studies on reducing power, inhibition of LPO, and superoxide and hydroxyl anion radical scavenging activities showed that the three bacillomycin D-like lipopeptides show high scavenging and antioxidant activities. These antioxidant and scavenging abilities show their protective function on DNA from damage due to hydroxyl radical oxidation. This antioxidant ability of the strain also has applications in therapeutics, pharmaceuticals, and cosmetics to slow down/prevent oxidation (Tabbene et al. 2012).

Organisms can cause many problems due to oxidative stress. Mycotoxin like deoxynivalenol (DON) can result in inducing this oxidative stress. An acidophilic and thermostable SOD is developed, which is a potent antioxidant enzyme (Dong et al. 2021). *Bacillus subtilis* expressed this antioxidant enzyme as an extracellular protein (Jemil et al. 2014), which can overcome oxidative damage due to mycotoxins like DON.

The *Bacillus* species can produce antioxidants during stress conditions. Also, this is one of the cheapest methods for producing antioxidants. One of the stress conditions is salinity stress. ST1 and ST2 are the two strains of *Bacillus* that produce antioxidants due to salinity stress (Hassan et al. 2020a). As a result of the induction of antioxidants in ST1 and ST2, there is an increase in phenols, flavonoids, peroxidase, glutaredoxin, and/or SOD. Pharmaceuticals and the food industry have excellent applications for these antioxidants such as postbiotics (Aguilar-Toalá et al. 2018). Moreover, four strains (HT1 to HT4) of the *Bacillus* genus can also produce antioxidants due to heat stress. This is also a cost-effective method to produce antioxidants. These antioxidants have a promising application in the pharmaceuticals and food industry as postbiotics (Hassan et al. 2020b).

One of the main problems the food industry faces is the fast oxidation of food products, which results in lousy flavor, odor, and even the development of poisonous materials (Scott 1997; Tseng et al. 2009). So, these antioxidants can delay or prevent the oxidation of food for some more time. Introducing bacteria into this food with a stress condition to the bacteria makes them produce antioxidants to withstand these conditions. Stress conditions can be unfavorable pH, temperature, salinity conditions, etc. (Aung et al. 2004).

Bacillus subtilis produces a bioactive lipopeptide called plipastatin, which the food industry uses because of its high antioxidant capacity. A combination of plipastatin and surfactants is also used for the same activity (Tseng et al. 2009). In Zebrafish larvae (using *in vitro* studies), they also measure GPx, SOD, etc. levels. The antioxidant activity of surfactants produced by the *Bacillus* species is already shown (Knapen et al. 2013). Phenolic, benzoic acid (Safronova et al. 2021), or even exopolysaccharides like Levan (Pei et al. 2020) are some metabolites produced by the *Bacillus* genes, which also have antioxidant activity.

Also, *Bacillus* belongs to plant growth-promoting rhizobacteria (PGPR), which helps plants like lettuce to increase the yield of plants under stress as well as during normal conditions. Rhizobium, Azotobacter, and Serratia also belong to the PGPR family, which helps in the same function. They help in producing vitamins, antioxidants, and many phytohormones during stress conditions (T et al. 2020; Dussert et al. 2022). The study by AlKahtani et al. (2021) demonstrates that applying *Bacillus thuringiensis* as a seed treatment and silicon as a foliar spray can effectively alleviate the negative impacts of salinity on lettuce plants and enhance their yield production (AlKahtani et al. 2021). Increased salinity stress resists the plant's water uptake along with the other nutrients essential for the plant's growth, such as the growth of root and shoot (Zhao et al. 2016; AlKahtani et al. 2021), which ultimately results in the yield. So, this *B. thuringiensis* helps mitigate the adverse effect of salinity by modulating antioxidants (AlKahtani et al. 2020) and regulating the concentration of plant hormones such as ethylene and indole acetic acid.

Chemicals such as nitric oxide, nitrous oxide, and peroxynitrite can damage cells by free radical chemical reactions. It can be avoided by producing antioxidants such as thiols and ascorbic acid

TABLE 9.1
Antioxidant Assays for Different *Bacillus* Species

Species	Test	Reference
Bacillus subtilis RW-I	Reducing power assay, DPPH radical scavenging assay, and ferrous ion-chelating assay	Yalcin and Cavusoglu (2010)
Bacillus mojavensis A21	DPPH radical scavenging assay, reducing power assay, β-carotene bleaching by linoleic acid assay, and lipid peroxidation inhibition assay	Das et al. (2008)
Bacillus velezensis VTX20	Hydroxyl radical scavenging assay and DPPH radical scavenging assay	Vu et al. (2021)
Bacillus subtilis A26	DPPH radical scavenging assay, β-carotene bleaching by linoleic acid assay, reducing power assay, and DNA nicking assay	Jemil et al. (2014)
Bacillus anthracis	DPPH radical scavenging assay, 2,2-azino-bis3-ethylbenzothiazoline-6 sulfonic acid radical cation decolorization assay, and ferrous ion-chelating assay	Mhina et al. (2020)

(*Continued*)

TABLE 9.1 (Continued)
Antioxidant Assays for Different *Bacillus* Species

Species	Test	Reference
Bacillus fusiformis	DPPH radical scavenging assay, 2,2-azino-bis3-ethylbenzothiazoline-6 sulfonic acid radical cation decolorization assay, and ferrous ion-chelating assay	Mhina et al. (2020)
*Bacillus methylotrophicus*DCS1	DPPH radical scavenging assay, ferric-reducing assay, ferrous ion-chelating assay, β-carotene bleaching assay, and linoleic acid peroxidation assay (by inhibition)	Jemil et al. (2017)
Lactobacillus helveticus MB2-1	DPPH radical scavenging assay, ferrous ion-chelating assay, superoxide anion scavenging assay, and hydroxyl radical scavenging assay	Li et al. (2014)
B. subtilis strain B38	DPPH radical scavenging assay, ferric-reducing assay, ferrous iron-chelating assay, lipid peroxidation assay, superoxide radical scavenging assay, and hydroxyl radical scavenging assay	Tabbene et al. (2012)
Lactobacillus plantarum C88	Hydroxyl radical scavenging assay, DPPH free radical scavenging assay, and H2O2-induced oxidative stress in Caco-2 cells	Zhang et al. (2013)

through microorganisms such as *B. subtilis, B. pumilus*, and *B. licheniformis* (Hafez et al. 2020). Various antioxidant assays used to detect *Bacillus* species are given in Table 9.1.

The antioxidant property of *Bacillus* is used in pharmaceuticals, cosmetics, food industries, and so on. Here we see the production and modulation of antioxidants and also how it is beneficial. Besides these, *Bacillus* has many other functions also, such as the production of cheese (Meng et al. 2018), substitutes for antibiotics against *Clostridium perfringes* which is a threat to poultry farming (Cheng et al. 2018), and some vitamins produced by them are used as medicine for osteoporosis (Knapen et al. 2013).

9.3 STRESS MITIGATION BY *BACILLUS* SPECIES

One of the major advantages of bacilli bacteria is the capability to produce endospores so they can be resistant to very adverse conditions. The *Bacillus* genus is divergent and consists of both pathogenic and non-pathogenic bacteria. One of the major diseases, anthrax, was caused by the species *B. anthracis*. *Bacillus* can survive in various adverse conditions and this ability is used in different sectors such as pharmaceuticals and agriculture. The important function of *Bacillus* is to enhance the yield and production of plant growth through the mechanisms of nitrogen fixation, nutrient solubilization, phytohormone production, and siderophore production (Dame et al. 2021). The most important advantage of *Bacillus* spp. is stress toleration in biotic and abiotic conditions such as phyto-pathogenicity, salinity, heavy metal toxicity, UV radiations, heat, and desiccation (Figure 9.1). Microbial inoculants of *Bacillus* are generally used for controlling plant pathogen disease and improving nutrient availability and increasing the plant hormone-like indole acid acetic, gibberellins, and cytokinins production, improving plant defense mechanism (Ruiz et al. 2021; María et al. 2022). The strains are also used in wheat and rice fields to improve the yield and growth under saline conditions (Ibarra-Villarreal et al. 2021; Ali et al. 2022; Etesami et al. 2023).

9.3.1 *Bacillus* Species against Biotic Stress

Different strains of *Bacillus* are used as a biocontrol against phytopathogen. One of the major causes of yield and growth reduction is caused by disease-causing microbes. For controlling these microbes, *Bacillus* can produce endospores, antibiotic compounds, and biofilm formation. Phytopathogen

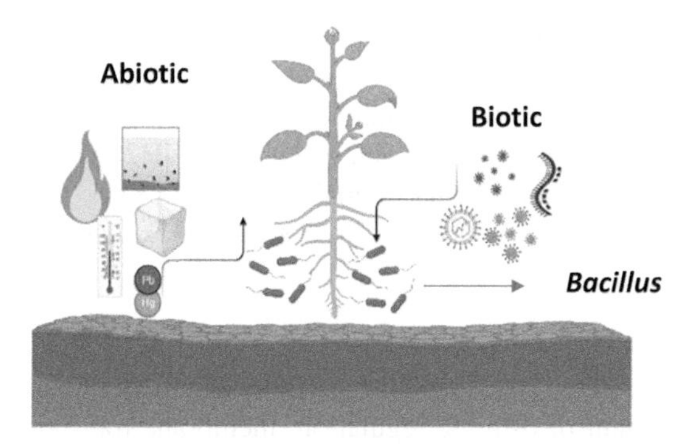

FIGURE 9.1 Representation of *Bacillus* spp. in the mitigation of biotic and abiotic stresses.

produces ROS and redox imbalance in the cell, but in this system, the *Bacillus* can produce antioxidants like peroxidase and SOD, which can scavenge the ROS (Yasmin et al. 2016). These species produce secondary metabolites and chitinase, protease, and glucanase-like enzymes can have the ability to suppress the growth of pathogens (Tyagi et al. 2018). 2,3-Butanediol and acetoin are important volatile chemical compounds that protect maize against *Pseudomonas syringae* (Kierul et al. 2015). Cyclic lipopeptides are also secreted by the *Bacillus* and they have antimicrobial properties. Surfactin is an example of a cyclic lipopeptide applied on the root of, tomato, peanut, and tobacco against *B. cinerea* and *Podosphaera fusca* (Ongena et al. 2007; García-Gutiérrez et al. 2013; Cawoy et al. 2014).

9.3.2 *Bacillus* Species against Abiotic Stress

9.3.2.1 Salinity Tolerance

The salinity of soil is an important environmental stress, and it negatively affects the growth and yield of wheat- and rice-like crops. Mainly the salinity issue affects the crop's spike development, seed germination, reproductive growth, etc. (Mukherjee et al. 2019; Ke et al. 2021). The accumulation of Cl^- and Na^+ negatively affects the quality of soil, porosity, aeration, and water conductivity, which leads to osmotic stress and cell death in plants. This stress will affect the reduction of photosynthesis rate, necrosis of root, reduction of leaf area, thickness, etc. (Hassan et al. 2021; Stassen et al. 2021). Pesticides and fertilizers are widely used in modern agricultural systems to resolve problems. This will alter the fertility, nutrient supply, and diversity of soil. In our balanced ecosystem, the microbial community plays an important role in nutrient cycling, aeration of the soil, transformation of soil aggregates, xenobiotic bioremediation, and tolerance of abiotic stresses. *Bacillus* spp. can reduce the salinity issues and they have to tolerate high salinity and changes in osmolarity. Different mechanisms were used to reduce stress management including promoting antioxidant enzyme production, production of phytohormone, improving K, Ca, N, and Zn-like nutrient availability, production of ACC deaminase (1-aminocyclopropane-1-carboxylic acid), accumulating osmolytes in crops, etc.

The stress condition reduces the production of phytohormone-like cytokinins, abscisic acid, auxins, salicylic acids, and gibberellic acid. These hormones help to adapt plants to stress by regulating plant growth and increasing the expression of stress-related genes. Strains of *B. aryabhattai, B. amyloliquefaciens, B. velezensis, B. megaterium, B. subtilis*, and *B. pumilus* are able to produce abscisic acid, jasmonic acid auxins, cytokinins, ethylene, and gibberellins that can help to increase growth hormones (El-Tayeb 2005; Abeer et al. 2015; Bjelić et al. 2020).

9.3.2.2 Organic and Inorganic Pollutant Toxicity Tolerance

Microorganisms can promote the growth of plants that grow in heavy metal contaminated areas using the mechanisms like production of ACC deaminase, the synthesis of chelating agents for heavy metals, promoting the accessibility of minerals (such as N, P, K, Ca, Fe, Mn, Cu, and Zn), inducing genes that confer resistance to environmental pollutants, modification of root biomass and morphology, etc. (Etesami 2018). Treesubsuntorn et al. (2017) demonstrated the effect of *B. subtilis* and *B. cereus* in paddy fields (Treesubsuntorn et al. 2017). It reduces the accumulation of Cd in different parts of rice root, shoot, and grain. *Bacillus* sp. JH2-2 helps to reduce the chromium contamination stress in the mustard field (Radhakrishnan and Lee 2016). *B. altitudinis* produces indole-3-acetic acid (IAA) and it reduces the iron stress in *Triticum aestivum L.* seedlings (Sun et al. 2017).

9.3.2.3 Cold Tolerance

Bacillus species can tolerate cold stress with the use of mechanisms like the production of cold shock proteins, the commencement of osmotic regulation, membrane transport, an antioxidant enzyme, and initiation of signal transduction pathway (Zubair et al. 2019). *B. velezensis FZB42* and *Bacillus* strains RJGP41 and CJCL2 have the genes that can combat cold stress. These strains contain genes including KatA, trxA, per, and SodA, which help to scavenge free radicals. Some of the bacteria can produce biofilm under cold conditions. The most important factors of cold stress-tolerant bacteria are the features of regulating the expression of proteins and their metabolic pathway (Abd El-Daim et al. 2019).

9.3.2.4 Nutritional Imbalance Tolerance

Bacillus improves nutrient availability to plants by increasing the production of metabolites. When these bacterial cultures are added to fertilizers, they help reduce the harmful effects of synthetic fertilizers and fix nitrogen (*B. circulans, B. cereus B. aquimaris, B. licheniformis, B. pumilus, B. vietnamensis, B. aerophilus*, and *B. altitudin*) and solubilize K, Zn, and P (Etesami and Maheshwari 2018). They have the capacity to produce organic acids especially lactic acid, formic acid, and glycolic acid and improve the uptake of phosphates (Wilson et al. 2016). Some of the species help to generate a multitude of metabolites while enhancing the accessibility of essential nutrients such as Zn, K, N, P, and Fe.

9.3.2.5 Crop Heat Tolerance

Plant growth-promoting bacteria can improve the heat tolerance in plants like tomatoes, potatoes, wheat, sorghum, and chickpeas (Ali et al. 2009; Mukhtar et al. 2023) by regulating abscisic acid levels, boosting heat-shock proteins (HSPs) levels, and enhancing biofilm formation. *B. safensis* (SCAL1) have the potential to produce kinetin IAA and gibberellic acid exopolysaccharide, which helps to reduce the stress in tomatoes (Mukhtar et al. 2023). *B. amyloliquefaciens* UCMB5113 is inoculated in wheat seedlings and observed to decrease the level of HSP, which is essential for stabilizing heat stress problems (Abd El-Daim et al. 2019).

9.4 GENE EXPRESSION BY *BACILLUS* SPECIES

Gene expression discusses the process by which the genetic information deposited in DNA is transformed into functional proteins that carry out several cellular functions. Gene expression is also used to produce other products, such as a protein or non-coding RNA within a cell. Fundamentally, the whole process of gene expression is highly regulated and contains multiple steps, together with transcription, RNA processing, and translation (Keen et al. 2017; Shen 2020). However, gene expression signs based on statistical techniques and responses in public databases can be used to categorize compounds into different toxicity classes and prioritize compounds for further testing, improving the efficiency of the assessment pattern (Shen 2020).

Bacillus species are commonly used as probiotics, which are live microorganisms that provide health benefits when consumed. One way of how *Bacillus* species exert their beneficial effects as probiotics is through the regulation of gene expression. The ability of *Bacillus* species to modulate gene expression in fish and shellfish is an important mechanism by which they exert their beneficial effects as probiotics in aquaculture. By regulating the expression of key genes involved in immune function, gut health, nutrient metabolism, and stress response, *Bacillus* species can improve the health and productivity of aquatic animals (Hamdy et al. 2018, 2020; Rahman et al. 2020).

The genome of *Bacillus* species that are used as probiotics contains several important features. These probiotic strains have been extensively studied and are known to provide a wide range of benefits to human health. One of the primary features of *Bacillus* probiotics is their ability to produce various enzymes and antimicrobial compounds, which can help to promote digestive health and support the immune system. Additionally, these *Bacillus* probiotics can help to maintain healthy gut flora by inhibiting the growth of harmful bacteria. Another important feature of *Bacillus* probiotics is their ability to survive in harsh environments. These probiotics can withstand extremes of pH, temperature, and pressure, which allows them to survive passage through the digestive system and colonize the gut effectively (Alcaraz et al. 2010; Hamdy et al. 2020).

The *Bacillus* genus of bacteria is known to express a variety of genes involved in different cellular processes. For example, *Bacillus subtilis* is a model organism widely used in genetic and molecular studies, and its genome has been extensively characterized. This bacterium is capable of adapting to various environmental conditions, such as nutrient depletion, by altering its gene expression patterns. It is also capable of producing and secreting enzymes and other proteins, which can be used in various applications such as food processing, agriculture, and environmental remediation. Similarly, *Bacillus thuringiensis* produces insecticidal proteins that are widely used as a biological pesticide in agriculture. Insecticidal genes are expressed in certain conditions, such as during the sporulation stage, and can be transferred to other organisms through genetic engineering (Rapoport and Klier 1990; Shen 2020; Hu et al. 2004). Overall, *Bacillus* species are known for their versatility in gene expression and their capability to manufacture a variety of useful chemicals.

9.4.1 Importance of Gene Expression

Gene expression can be critical for the proper function and growth of all living organisms. Any abnormalities or mutations in the genetic sequence can affect reformed or even dysfunctional gene expression, leading to genetic diseases and disorders. The whole method has several processes that can be modulated at any phase, allowing cells to regulate the appearance of functional proteins as needed for normal functioning or persistence.

Quantitative analysis of gene expression is significant for empathetic the relationship between gene expression profiles and cellular or organism phenotypes. During the translation, several post-translation steps in gene expression can modify the activity of a protein, including covalent modifications, protein splicing, and specific covalent attachment of small molecules.

9.4.2 Mechanism of Gene Expression

Gene regulation is an essential and complex process that plays a vital role in how organisms function. It is essential for many biological processes, including morphogenesis, cellular differentiation, development, and the overall versatility and adaptability of every living organism. The modulation of gene expressions in all stages is crucial for this regulation. Gene expression is primarily controlled at two levels. At first, restricting the quantity of mRNA (messenger ribonucleic acid) generated by a particular gene regulates transcription. The translation of mRNA into proteins is then controlled by post-transcriptional processes.

Transcription marks the initial stage of gene expression and involves the synthesis of mRNA through the enzymatic activity of RNA polymerase, which copies the DNA sequence of a gene. Following transcription, the mRNA molecule is processed to mature mRNA and then transported out of the nucleus. The subsequent level of gene expression is translation, in which the ribosome decodes the mRNA sequence, resulting in the production of a distinct protein molecule. The precise arrangement of nucleotides in the mRNA determines the arrangement of amino acids in the resultant

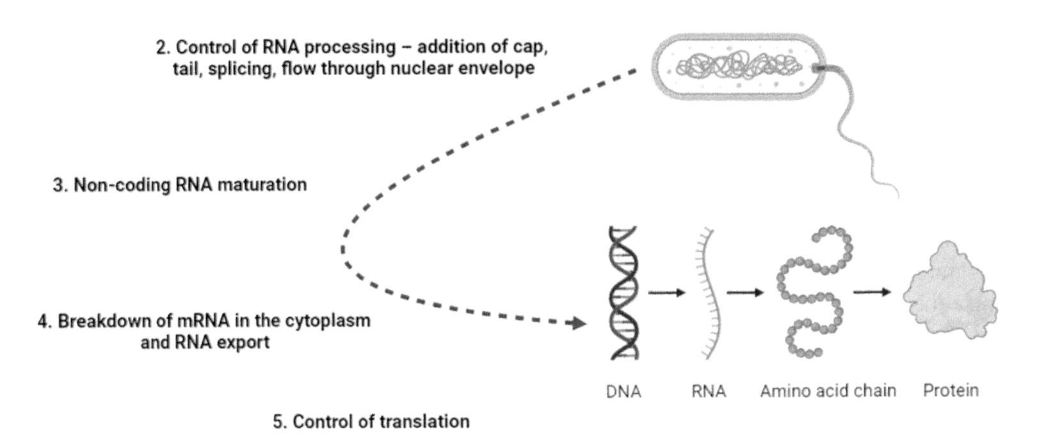

FIGURE 9.2 Six steps in gene expression.

protein molecule (Rapoport and Klier 1990; Shen 2020). Gene expression takes place primarily through six steps (Figure 9.2).

9.4.3 Gene Expression in Different *Bacillus* Species

Ferdinand Cohn (1872) first described and classified the *Bacillus* genus with *Bacillus subtilis* assigned as the type species. While various *Bacilli* may be responsible for opportunistic infections, the pathogenicity primarily resides in bacteria belonging to the *Bacillus cereus* group, which includes *B. cereus, B. anthracis*, and *B. thuringiensis*. The Gram-positive, spore-forming *Bacillus* species are common in nature and present in a range of environments, including soil (predominant), water, and both human and animal gastrointestinal tracts (Mandic-Mulec et al. 2015).

Bacillus species are renowned for their capacity to express a diverse set of genes, such as in metabolism, stress response, and antibiotic synthesis. Some of the key factors that influence gene expression in *Bacillus* species include environmental conditions, nutrient availability, and the presence of specific regulatory proteins. Understanding the genetic regulation of their gene expression is a crucial step toward harnessing their potential for biotechnological applications. The investigation of gene expression has been made easier by the accessibility of their full genome sequences and the advancements in high-throughput technologies for transcriptomics, proteomics, and metabolomics (Schallmey et al. 2004; Hu et al. 2023). However, they are used as model organisms to study gene expression, sporulation, and other biological processes (Schallmey et al. 2004; Nikolaidis et al. 2022).

Meanwhile, *Bacillus subtilis*, commonly known as the hay or grass *Bacillus*, is a Gram-positive bacterium with catalase-positive properties. It is often found in the soil and gastrointestinal tracts of various organisms. It can serve as a host for the synthesis and secretion of several heterologous proteins, including growth factors and toxins. Protease signal sequences can aid in the secretion of these heterologous proteins. The regulation of promoter activity can lead to high-level expression of these proteins in *Bacillus subtilis* (Tarquini et al. 2021; Nikolaidis et al. 2022).

Bacillus subtilis is a host organism for the production of heterologous proteins. Several studies have been conducted to express and secrete various proteins, such as ribonuclease, pertussis toxin, and pneumolysin, using *B. subtilis*. The use of protease signal sequences and the levansucrase system has also been employed to aid in the secretion of these proteins. Additionally, the Study mentions the production of diphtheria toxin and α galactosidase from *Cyamopsis tetragonoloba* using *B. subtilis* as the host organism. A particular strain of bacterium, *B. subtilis* strain WS1A, was isolated and identified from a marine sponge found in Saint Martin's Island region of the Bay of Bengal (Overbeeke et al. 1990; Rahman et al. 2020).

Several studies have investigated the molecular mechanisms underlying gene expression in *Bacillus* species. These studies have revealed the roles of transcription factors, RNA polymerases, sigma factors, and other regulatory proteins in controlling gene expression. They have also provided insights into the impact of environmental factors, such as temperature, nutrient availability, and stress conditions, on the expression of *Bacillus* genes. Overall, the understanding of gene expression in *Bacillus* species is still evolving and is likely to continue to advance with the development of new techniques and a deeper understanding of bacterial physiology (Eichenberger et al. 2004).

9.4.4 Role of Gene Expression in *Bacillus* Species

One of the most important regulatory proteins in *Bacillus* species is the transcription factor Spo0A, which is crucial for controlling gene expression in processes such as sporulation, biofilm formation, and stress response. Other regulatory proteins that play important roles in gene expression in *Bacillus* species include the two-component systems, sigma factors, and RNA polymerase. In addition to these regulatory proteins, *Bacillus* species also have a range of metabolic pathways that allow them to utilize a variety of carbon and nitrogen sources. An intricate gene regulatory network controls these processes, including those involved in the uptake and metabolism of sugars, amino acids, and other nutrients (Liu et al. 2003; Schilling et al. 2007).

Moreover, numerous physiological processes in fish and shellfish, including immune function, gut health, stress response, and nutrient metabolism, are regulated by genes that are expressed by certain *Bacillus* species. For example, *Bacillus subtilis* has demonstrated upregulating the gene expression of immune response in rainbow trout, such as interleukin-1β and tumor necrosis factor-α (Bermejo-Nogales et al. 2016; Mouthuy et al. 2016; Gobi et al. 2018). *Bacillus coagulans* have demonstrated upregulating the gene expression of immune response in mice, such as interferon-γ and interleukin-2 (Hiramoto et al. 2023).

9.4.5 Significance of Gene Expression by *Bacillus* Species

Bacillus species are renowned for producing different kinds of industrially important enzymes, antibiotics, and other metabolites. The ability of *Bacillus* species to express a wide range of genes is an important aspect of their capacity to survive and thrive in a wide range of conditions. By adapting their gene expression patterns to changing conditions, these bacteria can maintain their metabolic activity and continue to grow and reproduce even in challenging environments. In addition, some strains have shown the ability to promote the proliferation of beneficial bacteria in the gut, which can benefit the immune system. These bacteria help to maintain a healthy gut microbiome, which is important for overall immune function (Turnbull et al. 1996).

Several *Bacillus* species are important as probiotics. Some of the most commonly used ones include *Bacillus coagulans*, *Bacillus clausii*, and *Bacillus subtilis*. *Bacillus coagulans* is commonly used as a probiotic due to its ability to survive harsh digestive conditions and colonize the gut. It has also been shown to improve gastrointestinal health and reduce inflammation in the gut. *Bacillus subtilis* is used to produce the enzyme amylase, which is used in the production of starch-based products. *Bacillus thuringiensis* produces a toxin that is toxic to many insect pests. *Bacillus clausii* is another common probiotic that has been shown to have positive effects on gut health, particularly in cases of antibiotic-associated diarrhea. It is also known to help improve immune function and support the growth of beneficial gut bacteria. *Bacillus subtilis* is a spore-forming probiotic that has been studied for its ability to improve immune function and reduce inflammation in the gut. It is also commonly used in animal feed as a probiotic to improve digestive health. It has been shown to have a positive impact on aquaculture, particularly in terms of improving growth, survival, and disease resistance in fish and shellfish. One of the ways in which *Bacillus* species exert their beneficial effects is through the regulation of gene expression (Palma et al. 2014; Elshaghabee et al. 2017; Acosta-Rodríguez-Bueno et al. 2022).

Bacillus species have demonstrated regulating gene expression of gut health and metabolism. *Bacillus subtilis*, for example, was recently revealed to increase the expression of genes involved in the creation of short-chain fatty acids in the gut, which are essential for gut health and immunological function. However, some *Bacillus* species have the potential to cause infections in humans, especially among individuals with compromised immune systems. *Bacillus* infections can range from mild to severe and may require antibiotic treatment (Xie et al. 2022).

Bacillus sp. strains have been used to degrade petroleum hydrocarbons and other contaminants in soil and water (Wang et al. 2019). They were found to regulate the expression of stress-related genes in fish and shellfish. *Bacillus licheniformis* has been shown to upregulate the expression of HSP genes in Pacific white shrimp, which helps the shrimp cope with environmental stressors such as high temperature and low oxygen (García-Medel et al. 2020). Furthermore, they can also influence the gene expression engaged in nutrient metabolism, which can improve growth and feed utilization in fish and shellfish. *Bacillus amyloliquefaciens* has the ability to enhance the gene expression associated with lipid metabolism in Nile tilapia, which improves growth and feed conversion efficiency (Xu et al. 2022). Similarly in obese mice, *Bacillus clausii* improves insulin sensitivity and reduces inflammation using the same mechanism (Ghelardi et al. 2022).

Overall, the capability of *Bacillus* species to modulate gene expression in fish and shellfish is an important mechanism by which they exert their beneficial effects in aquaculture. By regulating the expression of key genes involved in immune function, stress response, and nutrient metabolism, *Bacillus* species can improve the health and productivity of aquatic animals.

9.5 ENHANCEMENT OF MUCOSAL IMMUNITY BY *BACILLUS* SPECIES

Even though we are unclear on the exact mechanism by which probionts confer their beneficial effect, studies and pieces of evidence show a myriad of mechanisms. One of the proposed mechanisms by which probionts perform this function is by colonizing the gut microbiota, bringing in homeostasis, and preventing pathogenic infection via the secretion of antimicrobials and bacteriocins and competing for adhesion and nutrition. They also enhance mucosal barrier properties along with immunomodulatory functions and gut–brain axis modulation (Acosta-Rodríguez-Bueno et al. 2022). *Bacillus* species are efficient probionts and are used as a feed additive since most of them are non-pathogenic to the host, and their administration to farm animals showed an enhancement of growth, immunity, and disease resistance (Abarike et al. 2018). Additionally, the ability of *Bacillus* to secrete antimicrobials that are effective against many pathogenic microbes, however not being toxic to the animal to which they were administered, and the ability to form endospores rapidly make them very versatile microbes that make it applicable in different fields unlike other probiotics like *Lactobacillus* sp. (Kuebutornye et al. 2019). These spores are very hardy and metabolically inactive. It has also been found that these spores can tolerate extremities such as bile salts and the acidic conditions of the gastrointestinal tract since *Bacillus* specimens have been isolated from feces showing passage through the tract (*Bacillus clausii* and *Bacillus licheniformis* were present in adult human feces) (Ghelardi et al. 2022). The reason why *Bacillus* shows such an efficient function as a probiotic can be attributed to several factors. Apart from the sporulation capacity, studies show that they inhibit pathogenic microbial growth by vying for attachment sites within the host's body.

Bacillus also inhibits the production of bacteriocins by these pathogens. They can also effectively lyse out pathogenic bacterium cell walls by the secretion of lytic enzymes including proteases, chitinases, cellulases, and β-1,3-glucanases. It was also discovered that *Bacillus* has the capability to inhibit the expression of virulence genes through a process known as quorum quenching (Kuebutornye et al. 2019). In another instance, the administration of mixed *Bacillus* species (*Bacillus subtilis, Bacillus pumlis*, *Bacillus amyloliqueficiens*, and *Bacillus licheniformis*) induced the hepatic gene expression of several early innate immune modulators while reducing the plasma and hepatic Hsp70 responses when the host was exposed to external stressors (Eissa et al. 2018). Moreover, *Bacillus* bacteria also promote the growth of the administered host by supplying the animal with nutrients and enzymatic digestion (Kuebutornye et al. 2019). Administration of *Bacillus* seems also to increase and modulate the levels of serum immune parameters and many of them have potent bacteriocidal properties against pathogenic microbes that infect the host (Kuebutornye et al. 2019).

Bacillus subtilis, B.licheniformis, and *B.cereus* are the most often used *Bacillus* species as effective probionts. However, whether given separately or in combination, each of these species appears to regulate immunity differently. *B. subtilis* supplementation increased serum immunoglobulin levels and non-specific immune parameters, whereas *B.licheniformis* was involved in immune system modulation by assisting in the maturation and development of the spleen and thymus, and *B. cereus* was involved in overall growth and microbiota balance in the gastrointestinal tract (Wang et al. 2017).

A potential probiotic, *B. subtilis* spores, is composed of a peptidoglycan cortex, a genome in its inner core, and a unique muramic delta-lactam modification. The spores also seem to express pattern recognition receptor (PRR) ligand(s) that can trigger and differentially affect Th1/Th2 balance and promote the polyfunctional immune response seen in the animals to which they are administered. Additionally, these spores can enhance the priming and boosting of antibodies produced by the pathogenic/co-inoculated antigens (Barnes et al. 2007). The ability of the *Bacillus* species to effectively colonize the gastrointestinal tract is also an important factor in eliciting its inhibitory effect on the pathogenic microflora (Abarike et al. 2018). White shrimps supplemented with *Bacillus subtilis* E20 demonstrated concentration-dependent immune response enhancement, along with selective inhibition of *Vibrio alginolyticus* infection and increases in phenoloxidase activity, phagocytic activity, and clearance efficiency (Tseng et al. 2009).

B. amyloliquefaciens supplementation, in addition to increasing the total protein content in both the mucous and serum of the host and showing immunocompetence, was able to efficiently eliminate infection caused by diverse pathogens through its activity to produce a wide variety of secondary metabolites. They also enhanced the lysozyme levels of the serum and mucus.

Bacillus species can be supplemented individually or in combination with other microbes (that may or may not be of the same genus *Bacillus*) or probiotic materials. When supplemented, a combination of *Lactobacillus acidophilus* and *Bacillus subtilis* was shown to complement the probiotic activity of each bacterial strain, enhance gut immunity, and prevent infection effectively. Since both strains occupy and colonize different niches of the gut microbiota, they could modulate and enhance gut immunity with great efficacy (Xie et al. 2022). Supplementation of *B. subtilis* and *B. licheniformis* was able to successfully modulate the specific and non-specific components of the immune system, such as an increase in leukocyte phagocytosis activity, lysozyme activity, IgM, SIgA, and SI levels. Moreover, the fish showed an increase in the intestinal mucosal fold height (Wang et al. 2017).

These probiotics containing *Bacillus* are usually delivered to animals via their feeds. Recently published studies have demonstrated that dead probiotics named para-probiotics modulate gut microbiota by triggering the activation of the mucosal immune system of the host instead of secreting inhibitory substances or competing for space and/or nutritional substances (Yang et al. 2014). Details of the efficacy of bacilli bacteria in the form of probiotics or para-probiotics on the immune systems of aquatic animals are given in Eissa et al. (2018), Kuebutornye et al. (2019), and Shen (2020).

9.6 CONCLUSION

The most important advantage of *Bacillus* is its divergent nature. By producing fermentative hydrolysates, biosurfactants, antibiotics, exopolysaccharides, lipopeptides, and other substances used in the food, pharmaceutical, and cosmetics industries, *Bacillus* plays a crucial role in stress management, including managing drought, salt cold, pollution, phytopathogens, and heat. According to the studies, it is also used to promote growth and increase the production of crops like wheat and rice. By modulating the gene expression of *Bacillus*, it can have potential applications in aquaculture and production of industrially essential enzymes and antibiotics. As a feed additive, it helps to increase the growth, immunity, and disease resistance of farmed animals.

ACKNOWLEDGEMENT

The first author acknowledges the Erasmus Mundus Joint Master's Scholarship for support.

CONFLICT OF INTEREST

The authors declare no conflict of interest.

ABBREVIATIONS

ACC 1-aminocyclopropane-1-carboxylic acid
CAT Catalase
DHA Docosahexaenoic acid
DNA Deoxyribonucleic acid
DON Deoxynivalenol
DPPH 2,2-Diphenyl-1-picrylhydrazyl
EPS Exopolysaccharide
FTIR Fourier-transform infrared spectroscopy
GPx Glutathione peroxidase
GR Glutathione reductase
HPLC High-performance liquid chromatography
HSPs Heat-shock proteins
IAA Indole-3-acetic acid
LPO Lipid peroxidation
mRNA Messenger ribonucleic acid
PGPR Plant growth-promoting rhizobacteria
PRR Pattern recognition receptor
RNA Ribonucleic acid
RNS Reactive nitrogen species
ROS Reactive oxygen species
SOD Superoxide dismutase
T-AOC Total antioxidant capacities
UV Ultraviolet

BIBLIOGRAPHY

Abarike, E. D., Cai, J., Lu, Y., Yu, H., Chen, L., Jian, J., Tang, J., Jun, L., & Kuebutornye, F. K. A. (2018). Effects of a commercial probiotic BS containing Bacillus subtilis and Bacillus licheniformis on growth, immune response and disease resistance in Nile tilapia, Oreochromis niloticus. *Fish & Shellfish Immunology*, *82*, pp.229–238.

Abd El-Daim, I. A., Bejai, S., & Meijer, J. (2019). *Bacillus velezensis* 5113 Induced Metabolic and Molecular Reprogramming during Abiotic Stress Tolerance in Wheat. *Scientific Reports*, *9*(1), p.16282.

Abeer, H., Abdallah, E. F., Alqarawi, A. A., Al-Huqail, A. A., Alshalawi, S. R. M., Wirth, S., & Dilfuza, E. (2015). Impact of plant growth promoting *Bacillus subtilis* on growth and physiological parameters of *Bassia indica* (*Indian bassia*) grown udder salt stress. *Pakistan Journal of Botany*, *47*(5), pp.1735–1741.

Acosta-Rodríguez-Bueno, C. P., Abreu Y Abreu, A. T., Guarner, F., Guno, M. J. V., Pehlivanoğlu, E., & Perez, M., 3rd (2022). Bacillus clausii for gastrointestinal disorders: A narrative literature review. *Advances in Therapy*, *39*(11), pp.4854–4874.

Aguilar-Toalá, J., Garcia-Varela, R., Garcia, H., Mata-Haro, V., González-Córdova, A., Vallejo-Cordoba, B., & Hernández-Mendoza, A. (2018). Postbiotics: An evolving term within the functional foods field. *Trends in Food Science & Technology*, *75*, pp.105–114.

Alcaraz, L. D., Moreno-Hagelsieb, G., Eguiarte, L. E., Souza, V., Herrera-Estrella, L., & Olmedo, G. (2010). Understanding the evolutionary relationships and major traits of Bacillus through comparative genomics. *BMC Genomics*, 11, p.332.

Ali, Q., Ayaz, M., Mu, G., Hussain, A., Yuanyuan, Q., Yu, C., Xu, Y., Manghwar, H., Gu, Q., Wu, H., & Gao, X. (2022). Revealing plant growth-promoting mechanisms of *Bacillus* strains in elevating rice growth and its interaction with salt stress. *Frontiers in Plant Science*, *13*, p.994902.

Ali, S. Z., Sandhya, V., Grover, M., Kishore, N., Rao, L. V., & Venkateswarlu, B. (2009). Pseudomonas sp. strain AKM-P6 enhances tolerance of sorghum seedlings to elevated temperatures. *Biology and Fertility of Soils*, *46*, pp.45–55.

AlKahtani, M. D., Attia, K. A., Hafez, Y. M., Khan, N., Eid, A. M., Ali, M. A., & Abdelaal, K. A. (2020). Chlorophyll fluorescence parameters and antioxidant defense system can display salt tolerance of salt acclimated sweet pepper plants treated with chitosan and plant growth promoting Rhizobacteria. *Agronomy*, *10*(8), p.1180.

AlKahtani, M., Hafez, Y., Attia, K., Al-Ateeq, T., Ali, M. A. M., Hasanuzzaman, M., & Abdelaal, K. (2021). *Bacillus thuringiensis* and silicon modulate antioxidant metabolism and improve the physiological traits to confer salt tolerance in lettuce. *Plants (Basel, Switzerland)*, *10*(5), p.1025.

Alton, E. W., Armstrong, D., Ashby, D., Bayfield, K. J., Bilton, D., Bloomfield, E. V., Boyd, A. C., Brand, J., Buchan, R., Calcedo, R., Carvelli, P., Chan, M., Cheng, S. H., Collie, D., Cunningham, S., Davidson, H., Davies, G., Davies, J., Davies, L. A., & Wolstenholme-Hogg, P. (2015). Repeated nebulisation of non-viral CFTR gene therapy in patients with cystic fibrosis: A randomised, double-blind, placebo-controlled, phase 2b trial. *The Lancet Respiratory Medicine*, *3*(9), pp.684–691.

Américo, M. F., Freitas, A. D., da Silva, T. F., de Jesus, L. C., Barroso, F. A., Campos, G. M., Santos, R. C., Gomes, G. C., Assis, R., Ferreira, Ê., Laguna, J. G., Chatel, J., Carvalho, R. D., & Azevedo, V. (2023). Growth differentiation factor 11 delivered by dairy Lactococcus lactis strains modulates inflammation and prevents mucosal damage in a mice model of intestinal mucositis. *Frontiers in Microbiology*, *14*, p.1157544.

Aung, W., Naylin, N., Zheng, Z., Watanabe, Y., & Hashinaga, F. (2004). Antioxidant and antibacterial activities of bacteria from Ngapi, a Burmese salted and fermented shrimp paste. *Biocontrol Science*, *9*, pp.117–122.

Ayed, H. B., Bardaa, S., Moalla, D., Jridi, M., Maâlej, H., Sahnoun, Z., Rebai, T., Jacques, P., Nasri, M., & Hmidet, N. (2015). Wound healing and in vitro antioxidant activities of lipopeptides mixture produced by Bacillus mojavensis A21. *Process Biochemistry*, *50*(6), pp.1023–1030.

Barnes, A. G., Cerovic, V., Hobson, P. S., & Klavinskis, L. S. (2007). Bacillus subtilis spores: A novel microparticle adjuvant which can instruct a balanced Th1 and Th2 immune response to specific antigen. *European Journal of Immunology*, *37*(6), pp.1538–1547.

Benitez, L. B., Velho, R. V., de Souza da Motta, A., Segalin, J., & Brandelli, A. (2012). Antimicrobial factor from Bacillus amyloliquefaciens inhibits Paenibacillus larvae, the causative agent of American foulbrood. *Archives of Microbiology*, *194*(3), pp.177–185.

Bermejo-Nogales, A., Fernández, M., Fernández-Cruz, M. L., & Navas, J. M. (2016). Effects of a silver nanomaterial on cellular organelles and time course of oxidative stress in a fish cell line (PLHC-1). *Comparative Biochemistry and Physiology. Toxicology & Pharmacology: CBP*, *190*, pp.54–65.

Bjelić, D., Marinković, J., & Balešević-Tubić, S. (2020). The Significance of *Bacillus* spp. in disease suppression and growth promotion of field and vegetable crops. *Microorganisms*, *8*(7), p.1037.

Cawoy, H., Mariutto, M., Henry, G., Fisher, C., Vasilyeva, N., Thonart, P., Dommes, J., & Ongena, M. (2014). Plant defense stimulation by natural isolates of bacillus depends on efficient surfactin production. *Molecular Plant-microbe Interactions: MPMI*, *27*(2), pp.87–100.

Cheng, Y. H., Zhang, N., Han, J. C., Chang, C. W., Hsiao, F. S., & Yu, Y. H. (2018). Optimization of surfactin production from Bacillus subtilis in fermentation and its effects on Clostridium perfringens-induced necrotic enteritis and growth performance in broilers. *Journal of Animal Physiology and Animal Nutrition*, *102*(5), pp.1232–1244.

Cohn, F. (1872) Untersuchungen über Bacterien. *Beitrage zur Biologie der Pflanzen 1*, pp.127–244.

Dame, Z. T., Rahman, M., & Islam, T. (2021). Bacilli as sources of agrobiotechnology: Recent advances and future directions. *Green Chemistry Letters and Reviews*, *14*(2), pp.246–271.

Das, P., Mukherjee, S., & Sen, R. (2008). Antimicrobial potential of a lipopeptide biosurfactant derived from a marine Bacillus circulans. *Journal of Applied Microbiology*, *104*(6), pp.1675–1684.

Dong, X., Wang, W., Jiang, T., Zhang, Y., Han, H., Zhang, Y., & Yang, C. (2021). Construction and potential application of bacterial superoxide dismutase expressed in Bacillus subtilis against mycotoxins. *PloS one*, *16*(11), p.e0260047.

Dussert, E., Tourret, M., Dupuis, C., Noblecourt, A., Behra-Miellet, J., Flahaut, C., Ravallec, R., & Coutte, F. (2022). Evaluation of antiradical and antioxidant activities of lipopeptides produced by *Bacillus subtilis* Strains. *Frontiers in Microbiology*, *13*, p.914713.

Eichenberger, P., Fujita, M., Jensen, S. T., Conlon, E. M., Rudner, D. Z., Wang, S. T., Ferguson, C., Haga, K., Sato, T., Liu, J. S., & Losick, R. (2004). The program of gene transcription for a single differentiating cell type during sporulation in Bacillus subtilis. *PLoS Biology*, *2*(10), p.e328.

Eissa, N., Wang, P., Yao, H., & Abou-ElGheit, E. (2018). Mixed Bacillus species enhance the innate immune response and stress tolerance in yellow perch subjected to hypoxia and air-exposure stress. *Scientific Reports*, *8*, p.6891.

Elshaghabee, F. M. F., Rokana, N., Gulhane, R. D., Sharma, C., & Panwar, H. (2017). *Bacillus* as potential probiotics: Status, concerns, and future perspectives. *Frontiers in Microbiology*, *8*, p.1490.

El-Tayeb, M. A. (2005). Response of barley grains to the interactive effect of salinity and salicylic acid. *Plant Growth Regulation*, *45*, pp.215–224.

Etesami H. (2018). Bacterial mediated alleviation of heavy metal stress and decreased accumulation of metals in plant tissues: Mechanisms and future prospects. *Ecotoxicology and Environmental Safety*, *147*, pp.175–191.

Etesami, H., Jeong, B. R., & Glick, B. R. (2023). Potential use of *Bacillus* spp. as an effective biostimulant against abiotic stresses in crops—A review. *Current Research in Biotechnology*, *5*, p.100128.

Etesami, H., & Maheshwari, D. K. (2018). Use of plant growth promoting rhizobacteria (PGPRs) with multiple plant growth promoting traits in stress agriculture: Action mechanisms and future prospects. *Ecotoxicology and Environmental Safety*, *156*, pp.225–246.

García-Gutiérrez, L., Zeriouh, H., Romero, D., Cubero, J., de Vicente, A., & Pérez-García, A. (2013). The antagonistic strain Bacillus subtilis UMAF6639 also confers protection to melon plants against cucurbit powdery mildew by activation of jasmonate- and salicylic acid-dependent defence responses. *Microbial Biotechnology*, *6*(3), pp.264–274.

García-Medel, D. I., Angulo, C., Escamilla-Montes, R., Fierro-Coronado, J. A., Diarte-Plata, G., Gámez-Jiménez, C., & Luna-González, A. (2020). Bacillus licheniformis BCR 4-3 increases immune response and survival of Litopenaeus vannamei challenged with Vibrio parahaemolyticus IPNGS16. *Aquaculture International*, *28*, pp.2303–2318.

Ghelardi, E., Abreu Y Abreu, A. T., Marzet, C. B., Álvarez Calatayud, G., Perez, M., 3rd, & Moschione Castro, A. P. (2022). Current progress and future perspectives on the use of Bacillus clausii. *Microorganisms*, *10*(6), p.1246.

Gobi, N., Vaseeharan, B., Chen, J. C., Rekha, R., Vijayakumar, S., Anjugam, M., & Iswarya, A. (2018). Dietary supplementation of probiotic Bacillus licheniformis Dahb1 improves growth performance, mucus and serum immune parameters, antioxidant enzyme activity as well as resistance against Aeromonas hydrophila in tilapia Oreochromis mossambicus. *Fish & Shellfish Immunology*, *74*, pp.501–508.

Gowtham, H. G., Singh, S. B., Shilpa, N., Aiyaz, M., Nataraj, K., Udayashankar, A. C., Amruthesh, K. N., Murali, M., Poczai, P., Gafur, A., Almalki, W. H., & Sayyed, R. Z. (2022). Insight into recent progress and perspectives in improvement of antioxidant machinery upon PGPR augmentation in plants under drought stress: A review. *Antioxidants (Basel, Switzerland)*, *11*(9), p.1763.

Gudiña, E. J., Rocha, V., Teixeira, J. A., & Rodrigues, L. R. (2010). Antimicrobial and antiadhesive properties of a biosurfactant isolated from Lactobacillus paracasei ssp. paracasei A20. *Letters in Applied Microbiology*, *50*(4), pp.419–424.

Hafez, Y. M., Attia, K. A., Kamel, S., Alamery, S. F., El-Gendy, S., Al-Doss, A. A., Mehiar, F., Ghazy, A. I., Ibrahim, E. I., & Abdelaal, K. A. (2020). *Bacillus subtilis* as a bio-agent combined with nano molecules can control powdery mildew disease through histochemical and physiobiochemical changes in cucumber plants. *Physiological and Molecular Plant Pathology*, *111*, p.101489.

Hamdy, A. A., Elattal, N. A., Amin, M. A., Ali, A. E., Mansour, N. M., Awad, G. E., Farrag, A. R. H., & Esawy, M. A. (2018). In vivo assessment of possible probiotic properties of Bacillus subtilis and prebiotic properties of levan. *Biocatalysis and Agricultural Biotechnology*, *13*, pp.190–197.

Hamdy, A. A., Esawy, M. A., Elattal, N. A., Amin, M. A., Ali, A. E., Awad, G. E. A., Connerton, I., & Mansour, N. M. (2020). Complete genome sequence and comparative analysis of two potential probiotics Bacillus subtilis isolated from honey and honeybee microbiomes. *Journal, Genetic Engineering & Biotechnology*, *18*(1), p.34.

Han, Y., Xu, X., Wang, J., Cai, H., Li, D., Zhang, H., Yang, P., & Meng, K. (2023). Dietary Bacillus licheniformis shapes the foregut microbiota, improving nutrient digestibility and intestinal health in broiler chickens. *Frontiers in Microbiology*, *14*, p.1113072.

Hasan, K., El Sabagh, A., Sikdar, S. I., Alam, J., Ratnasekera, D., Barutcular, C., & Islam, M. S. (2017). Comparative adaptable agronomic traits of blackgram and mungbean for saline lands. *Plant Archives*, *17*(1), pp.589–593.

Hassan, A., Fasiha Amjad, S., Hamzah Saleem, M., Yasmin, H., Imran, M., Riaz, M., Ali, Q., Ahmad Joyia, F., Mobeen, Ahmed, S., Ali, S., Abdullah Alsahli, A., & Nasser Alyemeni, M. (2021). Foliar application of ascorbic acid enhances salinity stress tolerance in barley (*Hordeum vulgare* L.) through modulation of morpho-physio-biochemical attributes, ions uptake, osmo-protectants and stress response genes expression. *Saudi Journal of Biological Sciences*, *28*(8), pp.4276–4290.

Hassan, A. H., Alkhalifah, D. H., Al Yousef, S. A., Beemster, G. T., Mousa, A. S., Hozzein, W. N., & AbdElgawad, H. (2020a). Salinity stress enhances the antioxidant capacity of bacillus and Planococcus species isolated from Saline lake environment. *Frontiers in Microbiology*, *11*, p.561816.

Hassan, A. H. A., Hozzein, W. N., Mousa, A. S. M., Rabie, W., Alkhalifah, D. H. M., Selim, S., & AbdElgawad, H. (2020b). Heat stress as an innovative approach to enhance the antioxidant production in Pseudooceanicola and Bacillus isolates. *Scientific Reports*, *10*(1), p.15076.

Hindu, S. V., Thanigaivel, S., Vijayakumar, S., Chandrasekaran, N., Mukherjee, A., & Thomas, J. (2018). Effect of microencapsulated probiotic Bacillus vireti 01-polysaccharide extract of Gracilaria folifera with alginate-chitosan on immunity, antioxidant activity and disease resistance of Macrobrachium rosenbergii against Aeromonas hydrophila infection. *Fish & Shellfish Immunology*, *73*, pp.112–120.

Hiramoto, K., Kubo, S., Tsuji, K., Sugiyama, D., Iizuka, Y., & Hamano, H. (2023). Bacillus coagulans (species of lactic acid-forming Bacillus bacteria) ameliorates azoxymethane and dextran sodium sulfate-induced colon cancer in mice. *Journal of Functional Foods*, *100*, p.105406.

Hu, Y., Xu, R., Feng, J., Zhang, Q., Zhang, L., Li, Y., Sun, X., Gao, J., Chen, X., Du, M., Chen, Z., Liu, X., Fan, Y., & Zhang, Y. (2024). Identification of potential pathogenic hepatic super-enhancers regulatory network in high-fat diet induced hyperlipidemia. *The Journal of Nutritional Biochemistry*, *126*, p.109584.

Hu, Y., Xu, R., Jing, F., Zhang, Q., Zhang, L., Chen, X., Du, M., Chen, Z., Liu, X., & Zhang, Y. (2023). Identification of super-enhancers regulatory network mediates the pathogenesis of hyperlipidemia. *Research Square (Research Square)*. https://doi.org/10.21203/rs.3.rs-2979998/v1

Ibarra-Villarreal, A. L., Gándara-Ledezma, A., Godoy-Flores, A. D., Herrera-Sepúlveda, A., Díaz-Rodríguez, A. M., Parra-Cota, F. I., & de los Santos-Villalobos, S. (2021). Salt-tolerant Bacillus species as a promising strategy to mitigate the salinity stress in wheat (Triticum turgidum subsp. durum). *Journal of Arid Environments*, *186*, p.104399.

Jemil, I., Jridi, M., Nasri, R., Ktari, N., Ben Slama, R., Mehiri, M., Hajji, M., & Nasri, M. (2014). Functional, antioxidant and antibacterial properties of protein hydrolysates prepared from fish meat fermented by Bacillus subtilis A26. *Process Biochemistry*, *49*(6), pp.963–972.

Jemil, N., Ben Ayed, H., Manresa, A., Nasri, M., & Hmidet, N. (2017). Antioxidant properties, antimicrobial and anti-adhesive activities of DCS1 lipopeptides from Bacillus methylotrophicus DCS1. *BMC Microbiology*, *17*(1), p.144.

Ke, J., Wang, B., & Yoshikuni, Y. (2021). Microbiome engineering: Synthetic biology of plant-associated microbiomes in sustainable agriculture. *Trends in Biotechnology*, *39*(3), pp.244–261.

Keen, E. C., Bliskovsky, V. V., Adhya, S. L., & Dantas, G. (2017). Draft genome sequence of the naturally competent *Bacillus simplex* Strain WY10. *Genome Announcements*, *5*(46), pp.e01295–17.

Kierul, K., Voigt, B., Albrecht, D., Chen, X. H., Carvalhais, L. C., & Borriss, R. (2015). Influence of root exudates on the extracellular proteome of the plant growth-promoting bacterium Bacillus amyloliquefaciens FZB42. *Microbiology (Reading, England)*, *161*(Pt 1), pp.131–147.

Knapen, M. H. J., Drummen, N. E., Smit, E. S., Vermeer, C., & Theuwissen, E. (2013). Three-year low-dose menaquinone-7 supplementation helps decrease bone loss in healthy postmenopausal women. *Osteoporosis International*, *24*(9), pp.2499–2507.

Krishna, E.R., Kumar, P.S., & Kumar, B.V. (2011). Study on Antioxidant activity and strain development of Bacillus subtilis (MTCC No.10619). *International Journal of Agricultural Technology*, *7*, pp.1693–1703.

Kuebutornye, F. K. A., Abarike, E. D., & Lu, Y. (2019). A review on the application of Bacillus as probiotics in aquaculture. *Fish & Shellfish Immunology*, 87, pp.820–828.

Li, W. F., Deng, B., Cui, Z. W., Fu, L. Q., Chen, N. N., Zhou, X. X., ... & Yu, D. (2012). Several indicators of immunity and antioxidant activities improved in grass carp given a diet containing Bacillus additive. *Asian Journal of Animal and Veterinary Advances*, *11*(14), pp.2392–7.

Li, W., Ji, J., Chen, X., Jiang, M., Rui, X., & Dong, M. (2014). Structural elucidation and antioxidant activities of exopolysaccharides from Lactobacillus helveticus MB2-1. *Carbohydrate Polymers*, *102*, pp.351–359.

Liu, B., Zhou, W., Wang, H., Li, C., Wang, L., Li, Y., & Wang, J. (2020). Bacillus baekryungensis MS1 regulates the growth, non-specific immune parameters and gut microbiota of the sea cucumber Apostichopus japonicus. *Fish & Shellfish Immunology*, 102, pp.133–139.

Liu, J., Tan, K., & Stormo, G. D. (2003). Computational identification of the Spo0A-phosphate regulon that is essential for the cellular differentiation and development in Gram-positive spore-forming bacteria. *Nucleic Acids Research*, *31*(23), pp.6891–6903.

Mandic-Mulec, I., Štefanič, P., & Van Elsas, J. D. (2015). Ecology of bacillaceae. *Microbiology Spectrum*, *3*(2), pp.59–85.

María, F. C., Fannie, P., Santoyo, G., Del Carmen, O. M., & De Los Santos-Villalobos, S. (2022). Draft genome sequence of *Bacillus* sp. strain FSQ1, a biological control agent against white mold in common bean (*Phaseolus vulgaris* L.). *Current Research in Microbial Sciences*, *3*, p.100138.

Meng, F., Chen, R., Zhu, X., Lu, Y., Nie, T., Lu, F., & Lu, Z. (2018). Newly effective milk-clotting enzyme from Bacillus subtilis and its application in cheese making. *Journal of Agricultural and Food Chemistry*, *66*(24), pp.6162–6169.

Mhina, C. F., Jung, H. Y., & Kim, J. K. (2020). Recovery of antioxidant and antimicrobial peptides through the reutilization of Nile perch wastewater by biodegradation using two Bacillus species. *Chemosphere*, *253*, p.126728.

Montaño, N., Gavino, G., & Gavino, V. C. (2001). Polyunsaturated fatty acid contents of some traditional fish and shrimp paste condiments of the Philippines. *Food Chemistry*, *75*(2), pp.155–158.

Mouthuy, P. A., Snelling, S. J. B., Dakin, S. G., Milković, L., Gašparović, A. Č., Carr, A. J., & Žarković, N. (2016). Biocompatibility of implantable materials: An oxidative stress viewpoint. *Biomaterials*, *109*, pp.55–68.

Mukherjee, A., Gaurav, A. K., Singh, S., Chouhan, G. K., Kumar, A., & Das, S. (2019). Role of potassium (K) solubilising microbes (KSM) in growth and induction of resistance against biotic and abiotic stress in plant: A book review. *Climate Change and Environmental Sustainability*, *7*(2), pp.212–214.

Mukhtar, T., Ali, F., Rafique, M., Ali, J., Afridi, M. S., Smith, D., & Chaudhary, H. J. (2023). Biochemical Characterization and Potential of Bacillus safensis Strain SCAL1 to Mitigate Heat Stress in Solanum lycopersicum L. *Journal of Plant Growth Regulation*, *42*(1), pp.523–538.

Nikolaidis, M., Hesketh, A., Mossialos, D., Iliopoulos, I., Oliver, S. G., & Amoutzias, G. D. (2022). A comparative analysis of the core proteomes within and among the *Bacillus subtilis* and *Bacillus cereus* evolutionary groups reveals the patterns of lineage- and species-specific adaptations. *Microorganisms*, *10*(9), p.1720.

Ongena, M., Jourdan, E., Adam, A., Paquot, M., Brans, A., Joris, B., Arpigny, J. L., & Thonart, P. (2007). Surfactin and fengycin lipopeptides of Bacillus subtilis as elicitors of induced systemic resistance in plants. *Environmental Microbiology*, *9*(4), pp.1084–1090.

Overbeeke, N., Termorshuizen, G. H., Giuseppin, M. L., Underwood, D. R., & Verrips, C. T. (1990). Secretion of the alpha-galactosidase from Cyamopsis tetragonoloba (guar) by Bacillus subtilis. *Applied and Environmental Microbiology*, *56*(5), pp.1429–1434.

Ozabor, P .T., Olaitan, J. O., Olaosun, O. S., & Fadahunsi, I.F.. (2020). Antibacterial and Antioxidant Activity of Bacillus Species Isolated from Fermented Parkia biglobosa (IRU) and Ricinus communis (OGIRI)-African Traditionally Fermented Food Condiments. *The Asia Journal of Applied Microbiology*, 7(1), pp.19–29.

Palma, L., Muñoz, D., Berry, C., Murillo, J., & Caballero, P. (2014). Bacillus thuringiensis toxins: An overview of their biocidal activity. *Toxins*, *6*(12), pp.3296–3325.

Pei, F., Ma, Y., Chen, X., & Liu, H. (2020). Purification and structural characterization and antioxidant activity of levan from *Bacillus megaterium* PFY-147. *International Journal of Biological Macromolecules*, *161*, pp.1181–1188.

Pei, J., Pan, X., Wei, G., & Hua, Y. (2023). Research progress of glutathione peroxidase family (GPX) in redoxidation. *Frontiers in Pharmacology*, 14, p.1147414.

Pesarico, A. P., Jesus, G. F., Córneo, E., Borges, H. D., Calixto, K. D., Garcez, M. L., Voytena, A. P., Rossetto, M., Ramlov, F., & Michels, M. (2022). Bacillus strains prevent lipopolysaccharide-induced inflammation in gut and blood of male mice. *Journal of Applied Microbiology*, *134*(1).

Radhakrishnan, R., & Lee, I. J. (2016). Gibberellins producing *Bacillus methylotrophicus* KE2 supports plant growth and enhances nutritional metabolites and food values of lettuce. *Plant Physiology and Biochemistry: PPB*, *109*, pp.181–189.

Rahman, M. M., Paul, S. I., Akter, T., Tay, A. C. Y., Foysal, M. J., & Islam, M. T. (2020). Whole-Genome Sequence of Bacillus subtilis WS1A, a Promising Fish Probiotic Strain Isolated from Marine Sponge of the Bay of Bengal. *Microbiology Resource Announcements*, *9*(39), pp.e00641–20.

Rapoport, G., & Klier, A. (1990). Gene expression using Bacillus. *Current Opinion in Biotechnology*, *1*(1), pp.21–27.

Ruiz, V. V., Luzania, R. A. C., Parra-Cota, F. I., Santoyo, G., & De Los Santos Villalobos, S. (2021). Extracellular polymeric substances from agriculturally important microorganisms. in *Springer eBooks* (pp. 217–234).

Safronova, L. S., Skorochod, I. A., & Ilyash, V. M. (2021). Antioxidant and Antiradical Properties of Probiotic Strains *Bacillus amyloliquefaciens* ssp. *plantarum*. *Probiotics and Antimicrobial Proteins*, *13*(6), pp.1585–1597.

Schallmey, M., Singh, A., & Ward, O. P. (2004). Developments in the use of Bacillus species for industrial production. *Canadian Journal of Microbiology*, *50*(1), pp.1–17.

Schilling, O., Frick, O., Herzberg, C., Ehrenreich, A., Heinzle, E., Wittmann, C., & Stülke, J. (2007). Transcriptional and metabolic responses of Bacillus subtilis to the availability of organic acids: Transcription regulation is important but not sufficient to account for metabolic adaptation. *Applied and Environmental Microbiology*, *73*(2), pp.499–507.

Scott, G. (1997). *Antioxidants in science, technology, medicine and nutrition*. Elsevier.

Shen G. (2020). Campbell biology (edited by Lisa Urry, Michael Cain, Steven Wasserman, Peter Minorsky and Jane Reece). *Journal of Biological Research (Thessaloniki, Greece)*, *27*(1), p.19.

Stassen, M. J. J., Hsu, S. H., Pieterse, C. M. J., & Stringlis, I. A. (2021). Coumarin communication along the microbiome-root-shoot axis. *Trends in Plant Science*, *26*(2), pp.169–183.

Sun, Z., Liu, K., Zhang, J., Zhang, Y., Xu, K., Yu, D., & Li, C. (2017). IAA producing Bacillus altitudinis alleviates iron stress in Triticum aestivum L. seedling by both bioleaching of iron and up-regulation of genes encoding ferritins. *Plant and Soil*, *419*, pp.1–11.

Tabbene, O., Gharbi, D., Slimene, I. B., Elkahoui, S., Alfeddy, M. N., Cosette, P., Mangoni, M. L., Jouenne, T., & Limam, F. (2012). Antioxidative and DNA protective effects of bacillomycin D-like lipopeptides produced by b38 strain. *Applied Biochemistry and Biotechnology*, *168*(8), pp.2245–2256.

Tarquini, G., Ermacora, P., & Firrao, G. (2021). Polymorphisms at the 3'end of the movement protein (MP) gene of grapevine Pinot gris virus (GPGV) affect virus titre and small interfering RNA accumulation in GLMD disease. *Virus Research*, *302*, p.198482.

Taverniti, V., & Guglielmetti, S. (2011). The immunomodulatory properties of probiotic microorganisms beyond their viability (ghost probiotics: Proposal of paraprobiotic concept). *Genes & Nutrition*, *6*(3), pp.261–274.

Thiem, B., & Goślińska, O. (2004). Antimicrobial activity of Rubus chamaemorus leaves. *Fitoterapia*, *75*(1), pp.93–95.

Treesubsuntorn, C., Dhurakit, P., Khaksar, G., & Thiravetyan, P. (2017). Effect of microorganisms on reducing cadmium uptake and toxicity in rice (*Oryza sativa* L.). *Environmental Science and Pollution Research International*, *25*(26), pp.25690–25701.

Tseng, D. Y., Ho, P. L., Huang, S. Y., Cheng, S. C., Shiu, Y. L., Chiu, C. S., & Liu, C. H. (2009). Enhancement of immunity and disease resistance in the white shrimp, Litopenaeus vannamei, by the probiotic, Bacillus subtilis E20. *Fish & Shellfish Immunology*, *26*(2), pp.339–344.

Turnbull, P. C., Kramer, J. M., & Melling, J. (1996). Bacillus. *Medical Microbiology*, *4*, p.233.

Tyagi, S., Mulla, S. I., Lee, K. J., Chae, J. C., & Shukla, P. (2018). VOCs-mediated hormonal signaling and cross-talk with plant growth promoting microbes. *Critical Reviews in Biotechnology*, *38*(8), pp.1277–1296.

Vu, T. H., Quach, N. T., Nguyen, N. A., Nguyen, H. T., Ngo, C. C., Nguyen, T. D., Ho, P., Hoang, H., Chu, H. H., & Phi, Q. (2021). Genome Mining Associated with Analysis of Structure, Antioxidant Activity Reveals the Potential Production of Levan-Rich Exopolysaccharides by Food-Derived Bacillus velezensis VTX20. *Applied Sciences*, *11*(15), p.7055.

Wang, D., Lin, J., Lin, J., Wang, W., & Li, S. (2019). Biodegradation of Petroleum Hydrocarbons by *Bacillus subtilis* BL-27, a Strain with Weak Hydrophobicity. *Molecules (Basel, Switzerland)*, *24*(17), p.3021.

Wang, L., Ge, C., Wang, J., Dai, J., Zhang, P., & Li, Y. (2017). Effects of different combinations of Bacillus on immunity and antioxidant activities in common carp. *Aquaculture International*, *25*(6), pp.2091–2099.

Wilson, B. R., Bogdan, A. R., Miyazawa, M., Hashimoto, K., & Tsuji, Y. (2016). Siderophores in Iron Metabolism: From Mechanism to Therapy Potential. *Trends in Molecular Medicine*, *22*(12), pp.1077–1090.

Xie, Z., Li, M., Qian, M., Yang, Z., & Han, X. (2022). Co-cultures of *Lactobacillus acidophilus* and *Bacillus subtilis* enhance mucosal barrier by modulating gut microbiota-derived short-chain fatty acids. *Nutrients*, *14*(21), p.4475.

Xu, R., Ding, F., Zhou, N., Wang, T., Wu, H., Qiao, F., Chen, L., Du, Z., & Zhang, M. (2022). Bacillus amyloliquefaciens protects Nile tilapia against Aeromonas hydrophila infection and alleviates liver inflammation induced by high-carbohydrate diet. *Fish & Shellfish Immunology*, *127*, pp.836–842.

Yalcin, E., & Cavusoglu, K. (2010). Structural analysis and antioxidant activity of a biosurfactant obtained from Bacillus subtilis RW-I. *Turkish Journal of Biochemistry-Turk Biyokimya Dergisi*, *35*(3), pp.243–247.

Yang, H. L., Xia, H. Q., Ye, Y. D., Zou, W. C., & Sun, Y. Z. (2014). Probiotic Bacillus pumilus SE5 shapes the intestinal microbiota and mucosal immunity in grouper Epinephelus coioides. *Diseases of aquatic organisms*, *111*(2), pp.119–127.

Yasmin, S., Zaka, A., Imran, A., Zahid, M. A., Yousaf, S., Rasul, G., Arif, M., & Mirza, M. S. (2016). Plant growth promotion and suppression of bacterial leaf blight in rice by inoculated bacteria. *PloS one*, *11*(8), p.e0160688.

Zhang, L., Liu, C., Li, D., Zhao, Y., Zhang, X., Zeng, X., Yang, Z., & Li, S. (2013). Antioxidant activity of an exopolysaccharide isolated from Lactobacillus plantarum C88. *International Journal of Biological Macromolecules*, *54*, pp.270–275.

Zhao, S., Zhou, N., Zhao, Z. Y., Zhang, K., Wu, G. H., & Tian, C. Y. (2016). Isolation of endophytic plant growth-promoting bacteria associated with the Halophyte Salicornia europaea and evaluation of their promoting activity under salt stress. *Current Microbiology*, *73*(4), pp.574–581.

Zubair, M., Hanif, A., Farzand, A., Sheikh, T. M. M., Khan, A. R., Suleman, M., Ayaz, M., & Gao, X. (2019). Genetic Screening and Expression Analysis of Psychrophilic *Bacillus* spp. Reveal Their Potential to Alleviate Cold Stress and Modulate Phytohormones in Wheat. *Microorganisms*, *7*(9), p.337.

10 *Bacillus* Probiotics in Fish Culture

Mehdi Soltani
Murdoch University, Perth, Australia
University of Tehran, Tehran, Iran

Nermeen Abu-Elala
King Salman International University (KSIU), South Saini, Egypt
Cairo University, Egypt

Einar Ringø
UiT the Arctic University of Norway, Tromsø, Norway

10.1 INTRODUCTION

To improve the growth performance, decrease mortality, and enhance the reproductive performance of animals including commercial aquatic organisms, chemotherapeutics including antibiotics have been used as part of a supplementary diet since many years ago. The banning of antibiotics as a feed additive in 2003 made, however, an urgent demand by the scientific communities particularly in the European Union to search for alternatives to decrease the abuse of antibiotics, and probiotic therapy has been raised as a promising feed additive. In present aquaculture activity, the diet price alone is about 50% of the production cost (Amiri et al., 2017), thus, there is much interest in different approaches to reduce the costs of production, especially using growth enhancers such as prebiotics (Ng & Koh, 2016; Hoseinifar et al., 2017). The efficacy of probiotics depends on various parameters including type of probiotic, dosage, vector of administration, and duration of application. So far known functions of probiotics in aquaculture are extracellular secretions that act as an antagonist for quorum sensing function, adhesion and colonization to the intestinal mucosa, competitive exclusion of harmful bacteria, enhancement of gastrointestinal tract functions, modulation of the microbiota of gastrointestinal tract, competition for some major and minor nutrients such as iron, vitamins, and sources of nutrients and enzymes for digestion, and improvement of animal immune functions. In fish culture, different species/strains of bacteria as potential probiotics have been assessed in three past decays (e.g., Ringø et al., 2020, Ringø et al., 2018; Soltani et al., 2018; Nayak, 2021), and most studies of *Bacillus* probiotics have exhibited promising effects on the performance and disease resistance in different fish species (Dawood et al., 2017; Fečkaninová et al., 2017; Dawood & Koshio, 2016; Hoseinifar et al., 2016, Sumon et al., 2022), with the focus being on *B. subtilis* (Zhang et al., 2014). The present chapter addressed the efficacy and potency of different species of *Bacillus* on growth performance, immune parameters, and resistance against diseases in fish culture.

10.2 EFFICACY OF *BACILLUS* PROBIOTICS ON GROWTH PERFORMANCE OF FISH

Growth promotion by supplementation of probiotics in the diet has been studied in various species of fish (Hai, 2015), with the proposed mode of action as an enhancer in intestinal physiology intermediated by the secretion of exogenous enzymes (Hoseinifar et al., 2017). Concerning the efficacy of probiotics of different *Bacillus* species, there are many publications with various results, even

DOI: 10.1201/9781003503811-10

using the same probiotic and fish species (Table 10.1); this can probably be due to differences in intestinal microbiota, life stages, and culture conditions of the target aquatic animals. Also, among *Bacillus* species, more attention has been paid to *B. subtilis* in assessing its efficacy on fish growth performance than other bacilli species.

In the first study by Kumar et al. (2006), the efficacy of *B. subtilis* as a growth enhancer was assessed in rohu (*L. rohita*) for 15 days and exhibited a significant improvement in weight gain. Later, in a study by Bagheri et al. (2008), supplementation of a commercial *B. subtilis* at different concentrations was carried out in the diet of rainbow trout of fry size (Table 10.1). A short period (13 days) of oral use of the probiotic exhibited an improvement in weight gain and specific growth rate (SGR), and a decline in the feed conversion ratio (FCR), and the best growth rate was seen in the fish fed with 3.8×10^9 cfu/g probiotic. Liu et al. (2012) studied the oral efficacy of *B. subtilis* E20 on the growth of orange-spotted grouper (*Epinephelus coioides*) and exhibited a significant improvement in weight gain and feeding efficiency. It was suggested that such an enhancement in fish growth could be in part due to the nutrients and exogenous enzymes including protease and lipase produced by *B. subtilis*. Publications are reporting that *Bacillus* spp. can supply some essential nutrients including amino acids, and vitamins K and B_{12} that have beneficial effects on fish growth performance (Sun et al., 2012; Liu et al., 2012). Feeding grass carp with various concentrations (1.0×10^9, 3.0×10^9, and 5.0×10^9 cfu/kg) of *B. subtilis* strain Ch9 exhibited a significant enhancement in weight gain and SGR and a decrease in FCR, and the best results were obtained in fish fed 3.0×10^9 CFU kg^{-1} of *B. subtilis* (Wu et al., 2012). A significant improvement was also seen in the activity of digestive enzymes and intestinal morphology in the treated fish. Similarly, the inclusion of *B. subtilis* (1×10^{10} cfu/g) in the diet of Olive flounder significantly improved the final weight and protein efficiency ratio and reduced the FCR compared to control fish (Cha et al., 2013). Some studies have demonstrated a significant increase in the growth performance of Nile tilapia-fed diet supplemented with *B. subtilis* at either 5×10^6 cfu/g (Telli et al., 2014), 1×10^7 /g for 60 days (Aly et al., 2008), or 10^8 cfu/g for 60 days (Liu et al., 2017). In addition, Liu et al. (2017) exhibited an increase in intestinal probiotic recovery and digestive enzyme activities. A 10-week oral use of *B. subtilis* at $0.42–1.35 \times 10^7$ cfu/g in juvenile large yellow croaker (*Larimichthys crocea*) in floating sea cages revealed a better growth performance at the higher dosage of the probiotic than lower dosage, suggesting the significant role of the optimization of probiotic dosage (Ai et al., 2011). Further, a 60-day oral administration of *B. subtilis* in combination with *B. licheniformis* at different dosages in juvenile Asian sea bass (*Lates calcalifier*) revealed significantly higher growth and better performance in terms of digestive enzymes, that is, protease, lipase, and amylase, body composition, and total protein at 1×10^6 cfu/g probiotics than the higher or lower concentrations of these probiotics (Adorian et al., 2018). These results exhibited that the dosage optimization is a function of both the bacterial species used as a probiotic and the target host species.

Para-probiotics confer potential replacements for live probiotics due to their beneficial effects on the performance of aquatic animals including fish. For instance, in a study by Shawky et al. (2023), a two-month effect of heat-inactivated *B. subtilis* (strain HIB) at 0.5, 1, and 2 g/kg feed in striped catfish exhibited a significant improvement in growth performance and intestinal morphometry. The beneficial effect of *B. subtilis* as a probiotic and a para-probiotic in goldfish revealed an improvement in the growth performance of fish compared to control fish (Shahbazi et al., 2023). The juvenile catfish fed 2% *B. subtilis* combined with 0.05% inulin for 60 days exhibited a better productive performance, ensuring greater weight gain and growth (Oliveira et al., 2022).

Besides *B. subtilis*, studies exhibited the potential of other *Bacillus* species as fish growth promoters. In their study by Bandyopadhyay and Mohapatra (2009), a two-month oral administration of *B. circulans* PB7 obtained from the intestine of *Catla catla* in fingerlings of the same fish species at different concentrations, that is, 2×10^4, 2×10^5, and 2×10^6 cfu/100 g, exhibited a positive effect on growth, with the best results seen in fish fed with the probiotic at 2×10^5 cfu/100 g. In a two-month study by Sun et al. (2010), feeding orange-spotted grouper with a single dose of two species of *Bacillus* revealed no significant difference in fish growth performance. In a subsequent study,

TABLE 10.1
Efficacy of *Bacillus* Bacteria on Growth Performance, Immune Response, and Disease Resistance in Fish Culture

Bacillus Species	Fish Species	Water temp (°C)	Dosage/ Duration (day)	Route	Growth Factor	Disease/Stress Resistance	Immune Response	Reference
B. subtilis + *B. licheniformis* (BioPlus2B)	Trout (*O. mykiss*)		1:1 mixture of each *Bacillus* (total: 4 × 10^4 spores g^{-1})	Oral	Unknown	*Yersinia ruckeri* (+)	Antibody titre (+), lymphocyte (+), total protein (+)	(Raida et al., 2003)
B. subtilis	Rohu (*L. rohita*)	27	0.5, 1.0 and 1.5 ×10^7 CFU g^{-1} (15 days)	Oral	WG (+)	Unknown	Unknown	(Kumar et al., 2006)
B. subtilis AB1	Trout	14	10^7 g kg^{-1} (14 days)	Oral	Unknown	*Aeromonas* sp. (+)	Respiratory burst (+), serum and gut lysozyme (+), peroxidase (+), phagocytic killing (+), total a1-antiprotease (+), lymphocyte (+)	(Newaj-Fyzul et al., 2007)
Commercial *B. subtilis* and *B. licheniformis* (*Bio Plus 2B*)	Trout	13	4.8×10^8, 1.2×10^9, 2.01×10^9, 3.8×10^9, 6.1×10^9 CFU g^{-1}	Oral	WG (+), SGR (+), FCR (+)	Unknown	Unknown	(Bagheri et al., 2008)
B. circulans PB7	Catla (*C. catla*)	25–29	2×10^4, 2×10^5, 2×10^6 CFU 100 g^{-1} (60 days)	Oral	WG (+), SGR (+), PER (+), FCR (+)	*A. hydrophila* (+)	Unknown	(Bandyopadhyay & Mohapatra, 2009)
B. clausii	Grouper (*E. coioides*)	30	1.0 × 10^8 CFU g^{-1} (30-60 days)	Oral	NS	Unknown	Lysozyme (+), phagocytosis (+), complement C3 (+), complement C4 (NS), superoxide dismutase (NS), IgM (+)	(Sun et al., 2010)
B. pumilus	Grouper (*E. coioides*)	30	1.0 × 10^8 CFU g^{-1} (30-60 days)	Oral	NS	Unknown	Lysozyme (+), phagocytosis (+), complement C3 (+), complement C4 (NS), superoxide dismutase (NS), IgM (NS)	(Sun et al., 2010)

(Continued)

TABLE 10.1 (Continued)
Efficacy of *Bacillus* Bacteria on Growth Performance, Immune Response, and Disease Resistance in Fish Culture

Bacillus Species	Fish Species	Water temp (°C)	Dosage/ Duration (day)	Route	Growth Factor	Disease/Stress Resistance	Immune Response	Reference
B. subtilis	Yellow croaker (*L. crocea*)	22.5–31.5	1.35× 10^7 CFU g^{-1} (10 weeks)	Oral	SGR (+), FE (+)	*V. harveyi* (+)	Complement pathway (NS), superoxide dismutase (+), lysozyme activity (+), respiratory burst (NS)	(Ai et al., 2011)
B. subtilis E20	Grouper (*E. coioides*)	27	10^4, 10^6 and 10^8 CFU g^{-1} (28 days)	Oral	FE (+), WG (+)	*Streptococcus* sp. (+)	Lysozyme (+), phagocytosis (+), respiratory burst (+), complement activity (+), superoxide dismutase (+)	(Liu et al., 2012)
B. subtilis Ch9	Grass carp (*C. Idella*)	22–28	1.0×10^9, 3.0×10^9 and 5.0×10^9 CFU kg^{-1} (56 days)	Oral	WG (+), SGR (+), FCR (+)	Unknown	Unknown	(Wu et al., 2012)
B. amyloliquifaciens FPTB16	Catla (*C. catla*)	26	10^7, 10^8 and 10^9 CFU g^{-1} (30 days)	Oral	Unknown	*Edwardsiella tarda* (+)	Serum lysozyme (+), myeloperoxidase (+), respiratory burst (+), nitric oxide (+),mucus lysozyme (+), mucus myeloperoxidase (+)	(Das and Ghosh 2013)
B. subtilis, B. pumilus or B. licheniformis	Olive flounder (*P. olivaceus*)	N/A	1 × 10^{10} CFU g^{-1} (56 days)	Oral	FW (+), FCR (+), PER (+)	*Streptococcus iniae* (+)	respiratory burst (+), lysozyme (+), superoxide dismutase (+), myeloperoxidase (+)	(Cha et al., 2013)
B. subtilis	Tilapia (*O. niloticus*)	25	10^7 CFU ml^1 (40 days)	Bath	NS	Unknown	Lysozyme (NS), phagocytosis (+), myeloperoxidase (NS), superoxide dismutase (+), catalase (+)	(Zhou et al., 2010)
B. coagulans	Tilapia (*O. niloticus*)	25	10^7 CFUml^{-1} (40 days)	Bath	FW (+), SGR (+)	Unknown	Lysozyme (NS), phagocytosis (+), myeloperoxidase (+), superoxide dismutase (+), catalase (+)	(Zhou et al., 2010)

B. latrospores + *B. licheniformis*	Silver carp		Commercial (1-2 × 10^6 CFU mL^{-1} and 1-2 × 10^6 CFU mL^{-1} with or without rotifer for 14 days)	Bath	FCR (-), WG (+), SGR (+)	Unknown	Survival (+)	(Sahandi et al., 2012)
B. subtilis	Tilapia *(O. niloticus)*	N/A	5×10^6 CFU g^{-1} (84 days)	Oral	FW (+), WG (+), FW (+)	Unknown	Lysozyme (+), phagocytosis (+)	(Telli et al., 2014)
B. coagulans (MTCC 9872), B. licheniformis (MTCC 6824)	Carp *(C. carpio)*	26.8–29.7	Each probiotic at 10^9 CFU g^{-1} (80 days)	Oral	Unknown	*A. hydrophila* (+)	Lysozyme (+), myeloperoxidase (+), respiratory burst (NS)	(Gupta et al., 2014)
B. subtilis (VSG1) *B. subtilis* + *Lactobacillus plantarum* (VSG3) *B. subtilis*+ *Pseudomonas aeruginosa* (VSG2) *B. subtilis* + *L. plantarum* +*P. aeruginosa*	Rohu *(L. rohita)*	27	*B. subtilis* (1 × 10^8 CFUg-[1]) *L. plantarum* (0.5 × 10^8 CFUg^{-1}) *P. aeruginosa* (0.5 × 10^8 CFUg^{-1}) (60 days)	Oral	WG (+), SGR (+), FCR (+)	*A. hydrophila* (+)	Lysozyme (+), complement (+), phagocytic activity (+), respiratory burst (+), superoxide dismutase (+), IgM (+)	(Giri et al. 2014)
B. amyloliquefaciens	Tilapia *(O. niloticus)*	N/A	10^6 and10^6 CFU g^{-1} (30 days)	Oral	Unknown	Unknown	Lysozyme (+), phagocytosis (+), nitric oxide (+), bactericidal activity (+), immune genes expression (+)	(Reda & Selim, 2015)
B. amyloliquefaciens	Tilapia *(O. niloticus)*	24.35	1×10^6 CFU g^{-1}; 5×10^6 CFU g^{-1}; and 1×10^7 CFU g^{-1} (90 days)	Oral	NS	Unknown	Unknown	(Silva et al., 2015)

(Continued)

TABLE 10.1 (Continued)
Efficacy of *Bacillus* Bacteria on Growth Performance, Immune Response, and Disease Resistance in Fish Culture

Bacillus Species	Fish Species	Water temp (°C)	Dosage/ Duration (day)	Route	Growth Factor	Disease/Stress Resistance	Immune Response	Reference
B. licheniformis Dahb1	Catfish *(P. hypophthalmus)*	26.8–29.7	1×10^5 CFU mL^{-1} for 24 days	Bath	WG (+)	*V. parahaemolyticus* (+)	Myeloperoxidase (+), respiratory burst (+), natural complement (+) lysozyme (+), antioxidant including glutathione-S-transferase, reduced glutathione, total glutathione (+)	(Gobi et al., 2016)
P. polymyxa	Carp (*C. carpio*)	29.5	10^3–10^4 CFU mL^{-1} for 8 weeks	Bath	Unknown	*A. hydrophila* (+)	Lysozyme (+), respiratory burst (+), myeloperoxidase (+), catalase (+), superoxidase dismutase (+)	(Gupta et al., 2016)
B. amyloliquefaciens	Turbot *(S. maximus)*	Unknown	10^8 CFU g^{-1} for 42 days	Oral	WG (+), SGR (+), FCR (+)	*V. anguillarim* (+)	Protease (+), amylase (+) protease, lipase (+), superoxide dismutase (+) total protein (+)	(Chen et al., 2016)
B. subtilis HAINUP40	Tilapia *(O. niloticus)*	Unknown	10^8 cfu/g^{-1} (8 weeks)	Oral	FW (+), WG (+), SGR (+), FCR (+)	*Streptococcus agalactiae* (+)	total antioxidant capacity (+), serum superoxide dismutase (+) respiratory burst (+), lysozyme (+)	(Liu et al., 2017)
B. velezensis (CGMCC 10149)	Atlantic salmon (*S. salar*)	Unknown	5 × 10^6 CFU g^{-1} (62 days)	Oral	WG (+), FCR (+)	*A. salmonicida* (+)	Total superoxide dismutase (+), Glutathione peroxidase (+), catalase (+), total antioxidant capacity (+), malondialdehyde (-), nitric oxide (-), glutamic pyruvic transaminase (-), glutamic oxalacetic transaminase (-), lysozyme (NS), cortisol (+)	(Wang et al., 2019)

B. velezensis	Tilapia (*O. niloticus*)	22	10^7 CFU g^{-1} (30 days)	Oral	SGR (+), WG (+), FCR (+)	*S. agalactiae* (+)	skin mucus lysozyme (+), peroxidase (+), serum peroxidase (+), complement (+), respiratory burst (+)	(Doan et al., 2018)
B. subtilis	Tilapia (*O. niloticus*)	Unknown	1.07 ± 0.2 × 10^4 and 1.23 ± 0.3 × 10^3 CFUg^{-1} (51 days)	Oral	Unknown	Unknown	Respiratory burst (+), phagocytosis (+), lysozyme (+), cytokine (+), complement (+)	(Galagarza et al., 2018)
B. licheniformis + *B. subtilis*)	Sea bass (*L. calcalifer*)	27	1 × 10^3, 1× 10^6, 1×10^9 CFU g^{-1} (56 days)	Oral	FW (+), SGR (+), FCR(+), CF (+)	Unknown	Leukocytes (+), lysozyme (+)	(Adorian et al., 2018)
B. amyloliquefaciens	Zebrafish		2 × 10^6 CFU g^{-1} (30 days)	Oral	Unknown	*A. hydrophila* (+), *S. agalactiae* (+)	Liver oxidative stress-related genes (SOD, Gpx, NOS2, and Hsp70) (+), apoptotic gene (tp53) (+), anti-apoptotic gene (bcl67 2) (+) Innate immune-related genes (IL-1β, IL-6, IL-21, TNF-α, TLR-1, -3, -4) (+)	(Lin et al., 2019)
B. velezensis + *B. Amyloliquefaciens* + *B. Subtilis*	Nile tilapia	28 ± 2.00 °C	1 × 10^8 CFU/mL (4 weeks)	Oral	FW (+), SGR (+), FCR (+), CF (+)	S. agalactiae	↑GOT, ↓ MDA, ↑SOD, ↑CAT, lysozyme activity (+), Nitric oxide (+), ACP (+), AKP (+), TNF-α (+), TLR-2 (+), IgM (+)	(Kuebutornye et al., 2020)
B. cereus QSI-1	Crucian carp	25 ± 2	14 days	Oral		*A. hydrophila*	↓TNF- α, IL-8, and IL-10, ↑MyD 88, Occludin and ZO-1(+)	(Jiang et al., 2020)
B. amyloliquefaciens + *Spirulina platensis*	Nile tilapia	26.1 ± 1.2	1 × 10^9 CFU/g 60days	Oral	FW (+), WG (+), SGR (+), FCR		SOD, TNF, HSP70 (+)	(Al-Deriny et al., 2020)
B. subtilis, B. Megaterium & *B. licheniformis*	Nile tilapia	24 ± 2	1 × 10^6 cfu g^{-1} (1,3, 5gkg^{-1}) 120days	Oral	WG, ↑SGR, ↓FCR	*S. agalactiae*	Myeloperoxidase, lysozyme activity, IL-1β, TNF-α	(Van Doan et al., 2021)

(*Continued*)

TABLE 10.1 (Continued)
Efficacy of *Bacillus* Bacteria on Growth Performance, Immune Response, and Disease Resistance in Fish Culture

***Bacillus* Species**	**Fish Species**	**Water temp (°C)**	**Dosage/ Duration (day)**	**Route**	**Growth Factor**	**Disease/Stress Resistance**	**Immune Response**	**Reference**
Bacillus sp. +*L. plantarum*	Olive flounder (Paralichthys olivaceus)		10^8 CFU/g (Mixture 3:1 and 1:1) (8 weeks)	Oral	↑WG, ↑SGR, ↓FCR	*S. iniae*	respiratory burst activity, lysozyme, myeloperoxidase activities, ↑TNF-α, IL-1B, IL-6, IL-10 (+)	(Hasan et al., 2021)
B. coagulans + ginger (*Zingiber officinale*)	Catla catla	24 ± 1	3000CFU mL^{-1} + 10 gm ginger	Oral	↑WG, ↑SGR, ↓FCR, ↑PER		phagocytic activity & index, respiratory burst activity, serum protein, serum bactericidal activity (+)	(Bhatnagar and Saluja 2021)
B. subtilis (subcellular component)	Lebeo rohita	24 ± 1	4 µL 40.4 µg protein/mL	IP (1 month)		*A. hydrohila*	ACP, AKP, TAOC, TNFα, IL1β, MHC1, IgM (+)	(Mohanty et al., 2022)
B. subtilis CotC MCP spores	Red spotted grouper	23–26	10^6, 10^7, 10^8 $CFUg^{-1}$	Oral/bath		*RGNNV*	SOD, lysozyme, MPO, Respiratory burst, ↑TNF-α, NF-kB, IL-6, IL-8, GH, TLR (+)	(Mai, Yan, and Xin 2022)
Bacillus species (PM8313)+ 0.1% β-glucan	Red sea bream (Pagrus major)	18.3	10^8 CFU+ 0.1% β-glucan	Oral	WG, SGR, FCR, FI, CF, HIS, VSI, intestinal length index	*E. trada*	IL-1β, IL-6, IL8, IL10, TNF-α, MyD88, TGFB (+)	(Jang et al., 2023)
B. subtilis, B. velezensis, B. tequilensis	Hybrid grouper	28–30	1 × 10^9 CFU (6 weeks	Oral	No sig diff was detected in FCR, WG, SGR, FW		CAT & TAOC (oral application), SOD & ACP in both application methods, lysozyme	(Amoah et al., 2023)
B. subtilis + RNA	Asian Sea bass	28–30	1 × 10^8 CFU kg^{-1} (45 days)	Oral	Unknown	*A. hydrophila*	Innate & humoral immune parameters; phagocytic activity & index, lysozyme, respiratory burst, immune-related genes; HEPC1, A2M, C3, CC, LYS, HSP70, HSP90	(Saengrung et al., 2023)

Bacillus mojavensis B191 and *Bacillus subtilis* MRS11	Nile tilapia	25–27	10^8, 10^6 CFU/g (60 days)	Oral	WG, ↑SGR, ↓FCR	*S. iniae*	TGF-β, IL-10, TNF-α, IL-1β	(Etyemez Büyükdeveci et al., 2023)
Heat inactivated *B. subtilis*	Striped catfish	27 ± 0.5	1 × 10^8 cfu/gm (0, 0.5, 1, 2 g/kg) 60 days	Oral	WG, ↑SGR, ↓FCR		Lysozyme, phagocytosis, SOD, CAT, GSH, ↓MDA	(Shawky et al., 2023)
B. subtilis B0E9 + *E. Faecalis* ATS	Yellow drum	28 ± 2	10^8 cfu/g (6 weeks)	Oral	Not significant	*V. harveyi*	Improve liver health ↓ GOT, GPT, improve C3, IL6, IL10, and lysozyme expression pre and post-challenge.	(Ding et al., 2023)
B. subtilis 1-C-7	Chinese perch	23 ± 2	0, 10^8, 10^9, 10^{10} cfu/kg diet (10 weeks)	Oral	Not significant (high concentration affects the growth performance negatively)	A. *hydrophila*	AST (+), ALT (+), MDA (+)	(Ji et al., 2023)
B. licheniformis HGA8B+ *Paenibacillus polymyxa*	Nile tilapia	27 ± 2	10^6, 10^8 cfu g^{-1} 60 days	Oral	WG, ↑SGR, ↓FCR, GHR-1, GHR-2, IGF1, IGF2	A. *hydrophila*	TLR-2, IL10, TNF-α, Muc2, CAT, GPx, GR, SOD, MDA	(Jose et al., 2023)

(+) = positive effect, (-) = negative effect, NS: not significant; FW = final weight, WG = weight gain, CF = condition factor, FCR = food conversion ratio, SGR = specific growth rate, FE = feed efficiency, PER = protein efficiency ratio, (ALT) = alanine aminotransferase, (AST) = aspartate aminotransferase, MDA = malondialdehyde in the liver.

Sun et al. (2013) demonstrated that bio-encapsulated *B. clausii* DE5 and *B. pumilus* SE5 in the copepod *Pseudodiaptomus annandale* could improve the weight of grouper larvae and reduce the growth of pathogenic *Vibrio* including *V. fischeri, V. scophtham*, and *Vibrio* sp. These results could be in part due to the alkaline and acid phosphatases in the fish intestinal digestive systems, while improved fish resistance may be due to an enhancement in fish immune responses such as increased lysozyme and superoxide dismutase activities (Sun et al., 2013). When grouper was fed with inulin alone, no effect was seen on the growth status, but the fish growth was promoted up to 8.5% when inulin was used in combination with *B. licheniformis* probiotic (Wang et al., 2017), suggesting a synergistic effect. In addition, findings by He et al. (2017) proposed that the administration of *B. cereus* in combination with *C. butyricum* and *L. acidophilus* in hybrid grouper (*Epinephelus lanceolatus* ♂×*E. fuscoguttatus* ♀) exhibited a higher growth and revealed higher activity of digestive enzymes than the single application of the probiotic. When Nile tilapia was fed with *B. amyloliquefaciens* for two months, the highest dosage (1×10^6 cfu/g) resulted in an improved intestinal villi height, and more numbers of goblet cells and intraepithelial lymphocytes compared to lower dosages of bacilli or control fish (Reda & Selim, 2015). Furthermore, higher levels of lipid and protein contents were measured at the slaughter in fish fed higher levels of bacilli probiotics (Reda & Selim, 2015). In contrast, in a study by Gobi et al. (2016), the use of *B. licheniformis* Dahb1 at a lower dosage (10^5 cfu/mL) was more effective at enhancing the growth of catfish (*Pangasius hypophthalmus*) than a higher dosage (10^7 cfu/mL). Silva et al. (2015) demonstrated no significant change in the growth of Nile tilapia-fed *B. amyloliquifaciens* at $1{\times}10^6$, $5{\times}10^6$, and $1{\times}10^7$ cfu/g for three months. Feeding turbot with *B. amyloliquefaciens* for 42 days exhibited only a marginal improvement in fish growth, but digestive enzymes including amylase and protease in the hepato-pancreas, lipase in the stomach, and protease in the intestine of treated fish significantly increased compared to control fish (Chen et al., 2016). Silver carp larvae fed with *B. latrospores* and *B. licheniformis* via rotifers (*Brachionus plicatilis*) as a probiotic vehicle exhibited an increase in growth (Sahandi et al., 2012). Oral use of *B. velezensis* obtained from aquaculture rearing water enhanced the growth and antioxidant activity of juvenile Atlantic salmon under a recirculating bio-filtration system (Wang et al., 2019). Wang et al. (2019) also reported that a mixed administration of *B. velezensis* and *R. mucilaginosa* showed a better effect on growth performance, feed utilization, and antioxidant activity, and a decrease was seen in fish mortality. Such beneficial effects in terms of weight gain, SGR, and FCR were also shown by Giri et al. (2014) after feeding rohu with *B. subtilis* in a mixture with *P. aeruginosa* and *L. plantarum* compared to fish fed *B. subtilis* alone. However, in a study by Doan et al. (2018), no significant difference was seen in the growth of tilapia fed either *B. velezensis* alone or in combination with *L. plantarum* for one month. Dietary supplementation of *Spirulina platensis* and *B. amyloliquefaciens* showed a synergistic influence on the growth performance, intestinal histomorphology, and immune response of Nile tilapia (Al-Deriny et al., 2020). The good growth performance and intestinal morphometry also were recorded in tilapia-fed *B. subtilis* and *E. faecalis* and over 76% survival rate against *S. agalactiae* infection that may be in part due to a better gut micro-ecological environment provided by the probiotic (W. Liu et al., 2021). Today, non-digestible carbohydrates from corn, wheat, and soy are being used in most aquafeed formulations; these complex carbohydrates represent 50–70% of all ingredients. However, energy production throughout these carbohydrates is scarce in cultivated fish, therefore, carbohydrase overproduction by *Bacillus* strains could increase the energy intake from these grains. Hence, the administration of *Bacillus* strains with the capacity to overproduce alpha-galactosidase, alpha glycosidase, cellulase, and pectinase enzymes could solve one of the biggest problems of the aquaculture industry (Olmos Soto 2021). In addition, *Bacillus* species are reported to produce a variety of cell wall lytic enzymes with antimicrobial properties which make them important for the aquaculture industry and in the digestion of fish. The cell wall of pathogenic microbes is made up of proteins, cellulose, chitins, etc. that can be broken by proteases, cellulases, and chitinases, respectively. Probiotics are used to improve soybean delivery as a plant protein in the diet of red sea bream. However, consuming excessive plant protein hinders growth performance and overall health due to anti-nutritional substances and poorer nutritional values in these plant protein-based sources. *B. subtilis* C-3102 (BSC-3102) can decrease the impact of several anti-nutritional factors and enhance the palatability of food (Zaineldin et al., 2021). The soybean supplementation with *B. subtilis* efficiently formulated low fish meal diets for carnivorous marine fish like red sea bream. Seabream-fed *B. subtilis* supplemented diets had relatively high hepato-somatic indices, which might indicate increased liver lipid reserves, which would enhance overall fish health (Zaineldin et al., 2021). *B. subtilis* was also successful in increasing the digestibility of soybean meal. This outcome could be attributed to probiotics' involvement in the nutritional digesting process via enzyme production, intestinal health, and/or increased intestinal microbiota activity. In addition, inflammatory or degenerative alterations in the gut, which are signs of soy-induced enteritis, were not seen. Inclusion of *B. subtilis* into fish meal substituted diets improved anti-protease activity at 500 g/kg replacement level but was unable to do that at a higher replacement level (750 g/kg). Fermentation of soya bean meal (FSBM) by *B. subtilis* and *B. licheniformis* with the mixed ratio 1:2 and replacement of fish meal by 30% of FSBM improved the protease activity, immune activity, and antioxidant activity of Koi carp (Zhao et al., 2022). As the results show, FSBM used in aquaculture effectively reduced feed expenditure, improved economic and environmental benefits, and prevented the deterioration of water quality.

Ye et al. (2023) observed that dietary supplementation with *Bacillus* sp. improved protease and lipase activities, total protein, and fat contents of the proximate carcass of turbot. It also affected the contents of certain amino acids, such as sulfur amino acids. However, the effects of the dietary supplementation with *Bacillus* sp. on the contents of flesh and total essential amino acids were not significant. Moreover, the diet with *B. subtilis* SMF1 increased the contents of polyunsaturated fatty acids (PFAs) and docosahexaenoic acid (DHA) in the carcass of turbots. *B. subtilis* SMF1 significantly increased the relative abundances of *Mycoplasma* and *Chryseobacterium* in the intestines of fish. The effects of the dietary with *B. subtilis* SMF1 were positively related to PFAs or DHA levels. Meanwhile, *Bacillus* probiotic improved the nutritional composition of turbot carcass in turbot intestine (Ye et al., 2023) and increased the populations of *Acinetobacter, Lactobacillus*, and *Bifidobacterium*, but reduced the load of harmful bacteria, such as *Vibrio* and Bacteroides in turbot intestine (Ye et al., 2023).

A four-week feeding trial was carried out to study the special effects of the dietary probiotic, Sanolife® PRO-F (a combination of *B. subtilis* 3.25×10^9 cfu/g, *B. licheniformis* at 3.50×10^9 cfu/g, and *B. pumilus* 3.25×10^9 cfu/g) on growth performance, intestinal digestive enzymes activities, and intestinal histomorphometry of Nile tilapia (El-Son et al., 2022). The treated fish revealed a significant improvement in weight gain, specific growth rate, and a reduction in feed utilization compared to the control group. In addition, the specific activities of protease, amylase, and lipase enzymes were significantly increased upon *Bacillus* strains' incorporation in the fish fed the probiotics. The length of intestinal villi, absorptive epithelium, and goblet cell counts were also greater in the fish fed on *Bacillus* strains diets (El-Son et al., 2022). Also, mirror carp-fed *B. subtilis* and *B. indicus* supplementation had no significant effects on growth performance, but significantly influenced the gut microbial communities (Baumgärtner et al., 2024). Further, no effect on growth or the microbial community was observed in treated goldfish with the probiotics. The results highlighted substantial species-specific differences in response to probiotics, in two closely related cyprinid species.

Dietary inclusion of *Bacillus* PM8313 isolated from the intestine of wild red sea bream (*Pagrus major*) demonstrated an improvement in growth performance and feed utilization (Jang et al., 2023). A 60-day feeding trial of Nile tilapia with *B. mojavensis* B191 and *B. subtilis* MRS11 was carried out to assess the effect of the potential host-associated probiotics on the growth performance (Büyükdeveci et al., 2023). The treated fish exhibited a significant increase in growth performances, with the *B. subtilis* (10^8 CFU/g) group being the highest (Büyükdeveci et al., 2023). Feeding yellow drum (*Nibea Albiflora*) with *B. megaterium* B1M2, *B. subtilis* B0E9, *B. velezensis* DM5, and *B. siemens* B0E14 exhibited an increase in fish growth (Ding et al., 2023). Six weeks of dietary supplementation of three *Bacillus* species including *B. velezensis, B. Subtilis*, and *B. tequilensis* previously isolated from the gut of hybrid grouper exhibited a significant improvement in growth performance, intestinal morphometry, whole fish body proximate composition, and intestinal biochemical indexes (Amoah et al., 2023).

Although *Bacillus* probiotics are usually applied through oral administration, some studies have assessed the bacilli effect of bath applications on fish growth. For example, in a study with Nile tilapia, Zhou et al. (2010) applied *B. coagulans* to the rearing water at 10^7 cfu/ml per day for 40 days and the results exhibited significantly a greater final weight-specific growth rate compared to control fish. In addition, the application of *B. subtilis* at 10^7 cfu /ml with 2-day intervals to the rearing water for 56 days also caused a remarkable increase in final weight and specific growth rate (Zhou et al., 2010). Application of probiotics in the rearing water can act as a good bioremediation of water quality (see Chapter 8); however, the mode of action of a bath application of probiotics on fish growth warrants further research.

The combination of *B. subtilis* TISTR001, *B. megaterium* TISTR067, and *B. licheniformis* DF001 (1×10^6 cfu/g) at 3 g/kg diet improved weight gain, average daily gain, specific growth rate, and FCR in Nile tilapia (Van Doan et al., 2021). *B. subtilis* D1-2 could effectively promote the growth of juvenile sea cucumber (*Apostichopus japonicus*) and improve its digestion to a certain extent. Moreover, it could actively regulate the intestinal microflora including *Lactobacillus, Clostridium, Lactococcus, Bifidobacterium*, and *Streptococcus* (M. Wang et al., 2022). The results also revealed that the lipase activity of the *B. subtilis* supplemented groups was significantly increased, thereby improving the digestibility of sea cucumbers (M. Wang et al., 2022).

The application of probiotic *Bacillus* in combination with feed additives such as prebiotics has exhibited various results. For instance, dietary supplementation of 10 g ginger (*Zingiber officinale*) and 3000 cfu/mL *B. coagulans* in *Catla catla* for 90 days revealed significantly high values of weight gain and specific growth rate (Bhatnagar & Saluja, 2021). In contrast, the combination of benzoic acid and *Bacillus* spp. in the diet of Nile tilapias shows no synergistic effect on the growth performance and survival (Santos et al., 2023).

10.3 *BACILLUS* PROBIOTICS AS ENHANCERS OF IMMUNE STATUS AND DISEASE RESISTANCE IN FISH

Like growth assessment, most of the studies have been directed to evaluate the efficacy and potency of *B. subtilis* on the immune response and disease resistance of commercial fish species. In the first research work, Newaj-Fyzul et al. (2007) isolated *B. subtilis* AB1 from the intestine of rainbow trout, and its inclusion in the diet trout at 10^7 g /kg for two weeks significantly improved immune responses, including phagocytic activity, respiratory burst, lysozyme activity of serum and gut, and activities of peroxidase, and anti-protease. The treated fish with probiotics also exhibited a higher resistance to *Aeromonas* sp. infection. Application of *B. subtilis* E20 isolated from fermented boiled soybeans in grouper (*Epinephelus coioides)* induced a significant enhancement in the activity of phagocytosis, lysozyme, respiratory burst, and complement (Liu et al., 2012). The treated fish also revealed resistance in a challenge with *Streptococcus* sp. infection, with the best effects being seen in fish fed with higher concentrations of the probiotic. Similarly, olive flounder demonstrated a satisfied resistance to infection by *Streptococcus iniae* after fish were subjected to dietary administration of *B. subtilis* (Cha et al., 2013). This resistance was attributed to the *B. subtilis* effect on the fish immunity, as the authors detected a significant improvement in lysozyme, respiratory burst, and superoxide dismutase and myeloperoxidase activities in treated fish. In their studies by Zhou et al. (2010) and Telli et al. (2014), the efficacy of *B. subtilis* was evaluated on immune functions of Nile tilapia. The studies differed in terms of both dosage (10^7 cfu/mL or 5×10^6 cfu/g) and route of administration (oral or bath), but both studies exhibited a significant enhancement in the fish immune responses including activities of phagocytosis, lysozyme, catalase, and myeloperoxidase. Also, feeding Nile tilapia with *B. subtilis* revealed an increase in immune variables including neutrophil adherence, lysozyme activity, and nitroblue tetrazoliume assay compared to control fish (Aly et al., 2008). Yellow croaker fed with *B. subtilis* exhibited a significant improvement in immune responses and a higher survival in a challenge with *V. harveyi* infection; and the inclusion of fructo-oligosaccharide as a source of feed for the probiotic in the fish diet revealed no significant effect on the fish immune responses and disease resistance in probiotic-fed fish (Ai et al., 2011). *Administration of B. subtilis* LR1 isolated from the rohu intestine significantly decreased the bacterial pathogenicity in Indian major carp (Banerjee et al., 2017). Liu et al. (2017) demonstrated that feeding Nile tilapia with *B. subtilis* HAINUP40 enhanced the fish immune responses and increased the fish resistance in a challenge with *Streptococcus agalactiae* infection (Table 10.1). Galagarza et al. (2018) exhibited that strains of *B. subtilis* were able to stimulate immune responses of tilapia at both local and systemic levels (Table 10.1). *Bacillus* spores exhibited to be highly efficient as an oral vaccine delivery system because of their strong specialty, gene operability, safety, and adjuvant properties. For instance, in a study by Jian et al. (Jian et al., 2019), higher resistance, higher specific Ig M and IgZ titers, and an up-regulation of immune-related genes were observed in grass carp orally immunized with recombinant *B. subtilis* spores carrying grass carp *Reovirus* VP4 protein than control fish. The vaccinated fish demonstrated significantly higher survival in challenges with the virulent strain of the virus. Dietary supplementation of *B. subtilis* in Chinese perch increased malondialdehyde (MDA) and increased the abundance of probiotics in midguts such as Tenericutes and Bacteroides, whereas it reduced the abundance of pernicious bacteria such as *Proteobacteria, Actinobacteria, Thermophilia,* and *Spirochaetes* (Ji et al., 2023). In a challenge test, the treated fish also showed an increase in resistance against *A. hydrophila* infection (Ji et al., 2023). Further, feeding the fish at 0.85×10^8 CFU/kg *B. subtilis* could improve the intestinal microbiota, intestinal health, and disease resistance, but more or excessive supplementation could reduce growth performance and have negative effects on fish health (Ji et al., 2023).

Feeding rainbow trout with *Bacillus* sp. JB-1 obtained from rainbow trout at various doses (10^3, 10^6, 10^8, and 10^{10} cells/g feed) exhibited high survival up to 100% in a challenge with infections by *Lactococcus garvieae* and *Streptococcus iniae* compared to 20–28% survival rate in the control fish (Brunt et al., 2007). More protection was, however, obtained using higher doses of the *Bacillus* than lower ones, suggesting the effectiveness of probiotic dosage optimization in disease resistance against lactococcosis and streptococcosis in fish. This protection induced by the bacilli probiotic in trout may in part be due to the improvement of the enhancement of fish innate immune responses including, respiratory burst, lysozyme activity, and phagocytosis, leucocyte, and total protein in serum and mucus in the treated trout.

B. velezensis strain JW exhibited antimicrobial activity against various bacterial fish pathogens including *Aeromonas hydrophila, Aeromonas salmonicida, Lactococcus garvieae, Streptococcus agalactiae*, and *Vibrio Parahemolyticus* (Yi et al., 2018). Feeding carp (*Carassius auratus*) with the diets containing 10^7 and 10^9 cfu/g of this probiotic for 4 weeks revealed a significant increase in the activity of glutathione peroxidase, acid phosphatase, and alkaline phosphatase; improvement in the expression of interferon gamma gene (IFN- γ), tumor necrosis factor-α (TNF-α), interleukin-1 (IL-1), interleukin-4 (IL-4), interleukin-10 (IL-10), and interleukin-12 (IL-12); and downregulated the expression of IL-10 and IL-12 genes. In addition, the treated fish revealed a significant increase in survival rate in a challenge with *A. hydrophila* infection (Yi et al., 2018).

Dietary inclusion of *Bacillus* PM8313 isolated from the intestine of wild red sea bream (*Pagrus major*) demonstrated an improvement in nonspecific immune responses including superoxide dismutase, lysozyme, myeloperoxidase, and respiratory burst in red sea bream (Jang et al., 2023). An up-regulating of immune-related genes including IL-6, NF-kB, IL8, HSP70, TNF-α, GH, and TLR was also observed in the treated fish. In addition, the treated fish exhibited an increase in disease resistance toward *Edwardsiella tarda* infection (Jang et al., 2023).

Immune responses and disease resistance of Nile tilapia against *Streptococcus iniae* infection were assessed by Büyükdeveci et al. (2023) after fish were fed with *B. mojavensis* B191 and *B. subtilis* MRS11 for two months. The best protection against *S. iniae* was observed at *B. subtilis* 10^8 CFU/g followed by *B. subtilis* 10^6 CFU/g, *B. mojavensis* (10^6 CFU/g), and *B. mojavensis*

(10^8 CFU/g). A general trend of up-regulation of some immune-mediated cytokines including TGF-β, IL-10, TNF-α, and IL-1β was also observed in all probiotic-fed groups (Büyükdeveci et al., 2023).

Application of *Bacillus* probiotics in combination with RNA can be considered a beneficial feed additive and immunostimulant for fish culture. For instance, feeding Asian sea bass (*Lates calcarifer*) *B. subtilis* mixed with RNA for 14 days exhibited a significant increase in innate cellular and humoral immune parameters, including phagocytic activity, phagocytic index, respiratory burst, serum lysozyme, and bactericidal activities, as well as upregulated expression of immune-related genes, including HEPC1, A2M, C3, CC, CLEC, LYS, HSP70, and HSP90 (Saengrung et al., 2023). Furthermore, significant increases were observed in the ileal villus height and goblet cell numbers in the intestinal villi in treated fish. The combination treatment did not cause histopathological abnormalities in the intestine and liver, suggesting such a synbiotic application is safe for use in the fish. The treated Asian seabass also exhibited a significant increase in survival rate after the challenge with *Aeromonas hydrophila* infection (Saengrung et al., 2023).

Feeding goldfish with *B. subtilis* in the forms of probiotic and para-probiotic was able to reduce the density of the parasite, *Ichtiophetirios multifiliis*, on skin and gills and reduced the level of histopathological alterations in these tissues of treated fish (Shahbazi et al., 2023). The treated fish also exhibited a higher expression of lysozyme and tumor necrosis factor-α than the control fish (Shahbazi et al., 2023). The heat-inactive *B. subtilis* var. natto was reported to positively affect Nile tilapia intestine microbiota and could regulate the intestine and fecal metabolite production to improve the intestine immune network (Pan et al., 2023).

Recently some researchers have focused on the efficacy and potency of probiotics as adjuvants for various vaccines in aquaculture, and this topic has been reviewed by Soltani et al. (2018). However, most of the reports are related to *Lactobacillus* spp., and less data is available evaluating *Bacillus* as an adjuvant of vaccines. For instance, in a study by Aly et al. (2016), feeding a mixture of probiotics including *B. subtilis, Lactobacillus acidophilus, Saccharomyces cerevisiae*, and *Aspergillus oryzae* to *A. hydrophila*-vaccinated tilapia for five months significantly enhanced specific antibody level and survival of fish from the challenge infection.

Intraperitoneal injection of subcellular components of *B. subtilis* isolated from the mangrove forest in India at a dose of 40.4 μg protein/μL, in *Labeo rohita*, significantly revealed an enhancement in leucocyte and erythrocyte populations, bactericidal activity, total protein, albumin, and globulin compared to the control group one month post-immunization (Mohanty et al., 2022). The immunized fish also demonstrated a higher protection toward *A. hydrophila* infection. The authors suggested the subcellular components of *B. subtilis* as a candidate against motile *Aeromonas* septicemia caused by *A. hydrophila* in rohu aquaculture (Mohanty et al., 2022). Chinese perch (*Siniperca chuatsi*) fed with high dietary inclusion of *B. subtilis* (0.9×10^{10} cfu/kg diet) displayed the highest activity of ALT and AST (Ji et al., 2023).

B. subtilis ABP1 strain revealed a higher adherence capacity to rainbow trout epithelial cells than the ABP2 strain (Docando et al., 2022). It also showed an increased capacity to induce the transcription of immune genes in epithelial cells and gut explants. The potential adjuvant capacity of the ABP1 strain was further confirmed by establishing its capacity to activate splenic IgM and B cells and up-regulate the transcription of several immune genes, after a single oral administration in rainbow trout (Docando et al., 2022).

The inclusion of 2% *B. subtilis* mixed with 0.05% inulin in the diet of catfish inhibited the cortisol increase when fish were challenged with *A. hydrophila* infection (Oliveira et al., 2022). Further, Yang et al. (2022) demonstrated that *B. subtilis* CK3 could act as a water additive to improve the immune response of *Procambarus clarkii* against *A. veronii* challenge. In addition, after *B. subtilis* CK3 immersion, the antioxidant enzymes and immune-related enzyme activities in the fish hepatopancreas were significantly higher than in the control fish. The authors suggested that the application of *B. subtilis* CK3 into rearing water can decrease morbidity and mortality caused by *A. veronii* infection. Dietary inclusion of *B subtilis* JCL16 in the diet of largemouth bass significantly increased the survival rate from 21.67% to 68.3% in a challenge with nocardiosis caused by *Nocardia seriolae*

(X. Wang et al., 2022). This effective protection may be in part due to the synergistic effect of the bacilli antimicrobial production including surfactin, bacilysin, and subtilisin (X. Wang et al., 2022). Dietary inclusion of 10^8 cfu/g *B. subtilis* enhanced IgM, alkaline phosphatase, superoxide dismutase, and catalase and reduced lipid peroxidation and tissue damage in *Carassius auratus* exposed to saline-alkaline stress condition (Lu et al., 2022). Based on the characteristics observed so far, *B. subtilis* LSG2-1 could form potential probiotic candidates in the digestive tract of *R. lagowskii* to fight diseases in aquaculture (J. Wang et al., 2022). The dietary supplementation with *B. subtilis* C-3102 (CAISPORIN)© in rates of 1%, 2%, 3%, and 4% in juvenile *Pseudoplatystoma* caused significant improvement in the intestinal histomorphometry as well as non-specific immunity such as phagocytic activity and leukocytes populations and promoted fish survival in a challenge with *A. hydrophila* infection (Nunes et al., 2020). The highest survival percentage was also recorded in the fish fed 3% *B. subtilis* (Nunes et al., 2020).

B. subtilis spores could be used as an adjuvant as they can enhance the protection efficacy of the spring viremia of carp subunit vaccine in common carp (J. Liu et al., 2021). Oral vaccination was an effective and economical measure because of the advantages of non-invasion, no size limitation, lower cost, and ease of operation. Based on *B. subtilis* spores, Wang et al. (2023) successfully constructed a vaccine using one of the proteins of *B. subtilis* named CotC against the largemouth bass virus (LMBV), and the protective efficacy and immune responses of the constructed CotC-LMBV recombinant *B. subtilis* spores were evaluated in largemouth bass. After the challenge assay with the largemouth bass virus, the relative percent survival of largemouth bass orally vaccinated with CotC-LMBV spores was 45.0% which was significantly higher than the control group. In addition, the specific IgM level in serum in the vaccinated fish was significantly higher than in the control one. Further, the expression of the immune-related genes in the spleen of vaccinated fish exhibited an increasing trend in different degrees, suggesting a positive effect on both innate and adaptive immune responses of fish. Another study indicated that the oral administration of recombinant *B. subtilis* spores in grouper was effective in preventing Singapore grouper iridovirus infection (Liang et al., 2023). This study provided a feasible strategy for controlling fish virus diseases. The spores of *B. subtilis* WB600 were utilized as the vehicle to deliver major capsid protein (MCP) of red-spotted grouper nervous necrosis virus (RGNNV) partnered with CotC to the gastrointestinal tract of grouper (Mai et al., 2022). The grouper immunized with B·s-CotC-MCP vaccine at 1×10^6, 1×10^7, and 1×10^8 cfu/g via either oral or bath immunization revealed a significant improvement in antibody titer against RGNNV and the transcription of immune-related genes, that is, TNF-α, IL-1β, MHC1, and IgM as well as non-specific immune parameters including acid phosphatase, alkaline phosphatase, and total antioxidant capacity. However, only the highest dose of bath vaccination induced a higher production of specific antibodies and up-regulated transcriptions of IgM and MHC1. In addition, the vaccinated showed a higher protection against RGNNV infection, suggesting that the *B. subtilis* CotC MCP vaccine could protect fish against this viral infection. Interestingly, the relative percentage survival in the orally vaccinated fish (88.89%) was much higher than the bath group (18.89%), which was consistent with the production of specific serum antibodies, non-specific immune response, and immune-related gene expression. These data show that *subtilis* CotC MCP spores can trigger high levels of mucosal and humoral immunity and represent a promising candidate vaccine against nervous necrosis virus infection (Mai et al., 2022).

Apart from *B. subtilis*, the immune responses of fish treated with other bacilli have been also demonstrated. For instance, in a study by Sun et al. (2010), orange-spotted grouper fed with *B. clausii* and *B. pumilus* previously obtained from this fish at 10^8 cfu/g (a single dose) for two months revealed a significant enhancement in the activities of phagocytosis, lysozyme, and complement C3, but no significant effect was seen on the activity of superoxide dismutase. These results raised the beneficial effects of endogenous *Bacillus* spp. on fish immune responses, but still more detailed studies including species, strain, and virulent level of the pathogenic agents used for the challenge assays warranted further investigations to confirm the probiotic efficacy against the pathogenic agents in fish culture. The disease resistance and immunomodulatory effects of *B. amyloliquefaciens* were studied in Nile tilapia (Reda and Selim 2015) and catla (Das & Ghosh, 2013) and in both experiments, the probiotic improved fish immune responses including respiratory burst in serum, lysozyme and myeloperoxidase and bactericidal activities in serum and mucus, as well as nitric oxide and the expression of some genes related to the fish immunity. Interestingly, *the* bacilli probiotic improved the mucosal immunity of catla and increased the resistance of the fish toward *Edwardsiella tarda* infection (Das & Ghosh, 2013). Gupta et al. (2014) assessed the efficacy of *B. coagulans* MTCC 9872, *B. licheniformis* MTCC 6824, and *Paenibacillus polymyxa* MTCC 122) on the non-specific immune responses of common carp and the resistance toward *A. hydrophila* challenge. Feeding the fish with a single dose at 10^9 cfu/g for 80 days revealed a significant enhancement in levels of lysozyme, myeloperoxidase, and respiratory burst activities compared to control fish, but *P. polymyxa* was superior to two other probiotics. Also, this improvement in fish immune responses resulted in a remarkable resistance toward *A. hydrophila* challenge, and a better result was seen in fish fed with *P. polymyxa*. Grouper fed with *B. pumilus* SE5 (1.0×10^8 cells/g) for two months presented an upregulation in the expression of TLR1, TLR2, and IL-8 genes, and the populations of potential pathogens including *Psychroserpens burtonensis* and *Pantoea agglomerans* were decreased in the intestine of the treated fish (Yang et al., 2014).

When *B. amyloliquefaciens* G1 was orally used at 3×10^7 and 3×10^9 cfu/g in eel (*Anguilla anguilla*), significantly provided lower mortalities than control fish in a challenge with *A. hydrophila* infection (Lu et al., 2011). Inclusion of *B. amyloliquefaciens* to the diet of tilapia enhanced populations of immunocompetent cells and erythrocytes, and the content of hemoglobin and hematocrit. A higher dosage of the probiotic also exhibited a better effect on the values of total protein and globulin in the fish serum (Reda & Selim 2015). Feeding turbot with *B. amyloliquefaciens* significantly increased the activity of sera superoxide dismutase and total protein content and also induced a higher relative survival (62.7%) against *V. anguillarim* challenge (Chen et al., 2016). In a study by Lin et al. (2019), the use of xylanase-expressing *B. amyloliquefaciens* R8 in zebrafish exhibited enhanced nutrient metabolism and hepatic stress tolerance in fish and increased immune status and resistance of fish toward *A. hydrophila* and *S. agalactiae* infections.

Oral application of rainbow trout with *B. subtilis* and *B. licheniformis* spores (BioPlus2B) in the form of the mixture (1:1) increased fish immune responses including antibody titer, lymphocyte, and total protein and revealed an increase in survival in the challenge with *Yersinia ruckeri* infection (Raida et al., 2003), suggesting that *Bacillus* spp. and *Yersinia* spp. may share antigens, with B-lymphocyte or T-lymphocyte clones induced by the *Bacillus* spp. conferring some protection against *Yersinia* septicemia in susceptible fish. Feeding Asian sea bass (*Lates calcarifer*) with a mixture of *B. licheniformis* and *B. subtilis* exhibited an increase in the populations of leukocytes (Adorian et al., 2018). Bath immersion of a lower dosage of *B. licheniformis* (10^5 cfu/mL) was more effective on catfish immune responses and antioxidant activity and also provided a higher protection in fish challenged with *V. parahaemolyticus* infection than the higher dosage (10^7 cfu/mL) (Table 10.1) (Gobi et al., 2016). An enhancement was demonstrated in common carp immune responses including the activity of respiratory burst, lysozyme, myeloperoxidase, catalase, and superoxides dismutase after the addition of *Paenibacillus polymyxa* in the fish-rearing water for 8 weeks (Table 10.1) (Gupta et al., 2016). This improvement in fish immunity was confirmed by increasing the resistance of treated fish in a challenge with *A. hydrophia* infection. Feeding Atlantic salmon either with single *B. velezensis* alone or in a mixture with *R. mucilaginosa* revealed an improvement in immune parameters (Table 10.1) and increased fish survival in a challenge with *A. salmonicida* infection (Wang et al., 2019). Further, Nile tilapia fed with *B. velezensis* either in a single form or in a mixture form with *L. plantarum* demonstrated a significant improvement in innate immune responses and resistance toward infection by *S. agalactiae* infection compared to control fish (Doan et al., 2018). In addition, in a study by Giri et al. (2014), the use of *B. subtilis* as a single probiotic or in a mixed form with *P. aeruginosa* and *L. plantarum* in rohu promoted immune responses and increased fish survival toward *A. hydrophia* infection. Dietary four-week addition of *B. velezensis, B. amyloliquefaciens*, and *B. subtilis* alone or in a mixture form in Nile tilapia was effective and increased fish survival toward infection by *Streptococcus agalactiae* (Kuebutornye et al., 2020). Feeding hybrid grouper with *B. velezensis, B. subtilis*, and *B. tequilensis* significantly improved blood hematological parameters, liver health, and intestinal biochemical indexes (Amoah et al., 2023). A combination efficacy of *B. subtilis, B. licheniformis*, and *B. pumilus* (Sanolife® PRO-F) was studied on serum biochemical immune parameters, antioxidant activity, and the expression of pro-inflammatory cytokines and antioxidant-related genes of Nile tilapia (El-Son et al., 2022). There was a significant reduction in the serum cholesterol and triglyceride levels in the fish-fed *Bacillus* species. In addition, serum lysozyme activity significantly showed a greater value in treated fish compared to control, but no changes were seen in the levels of glucose, alanine aminotransferase (ALT), and aspartate aminotransferase (AST), and renal biomarkers (urea, creatinine, and uric acid) in the treated fish. Further, the antioxidant capacity of catalase and superoxide dismutase in the liver and spleen significantly increased with diminished malondialdehyde (El-Son et al., 2022). When striped catfish were orally subjected to a diet containing heat-inactivated *B. subtilis* (strain HIB), a significant enhancement was seen in the values of antioxidant activity and nonspecific immune response (Shawky et al., 2023) (Table 10.1).

Feeding *Catla* with *B. coagulans* in combination with ginger exhibited a significant increase in the values of immunocompetent cell populations, phagocytic activity, respiratory burst activity, serum protein, and serum bactericidal activity (Bhatnagar & Saluja, 2021). The feed additive administration of the probiotic *B. cereus* QSI-1 proved to enhance the expression levels of TNF-α IL-8, IL-10, and MyD88 in the intestine of crucian carp (Jiang et al., 2020). In addition, administration of the bacilli strain could prevent intestinal mucosal barrier damage and decrease inflammation induced by *A. hydrophilia*) infection (Table 10.1). Furthermore, the influence of dietary supplement of quorum quenching bacteria *B. cereus* QSI-1 on the intestinal mucosal barrier functions and nonspecific immune response in crucian carp was investigated. The permeability of the intestinal barrier is mainly dominated by tight junction proteins including ZO-1 and Occludin that maintain the integrity of the intestinal mucosal barrier. The results also showed that *B. cereus* QSI-1 can significantly increase the expression of these proteins in the intestinal mucosa of crucian carp.

A mixture of *Bacillus* species and *L. plantarum* in the diet of Olive flounder at ratios (3:1 and 1:1) improved the growth performance, respiratory burst, lysozyme, and myeloperoxidase activities. Similar modulation was observed in immune-related gene expressions such as TNF-α, IL-1β, IL-6, and IL-10. The group fed on the mixture in a ratio (3:1) showed superior resistance against *S. iniae* infection (Hasan et al., 2021). Immune parameters including myeloperoxidase, lysozyme, IL-1β, and TNF-α gene expressions were also significantly higher in Nile tilapia fed *B. subtilis* TISTR001, *B. megaterium* TISTR067, and *B. licheniformis* DF00 than a control group. In addition, the survival of treated fish was higher than control fish in a challenge with *Streptococcus agalactiae* infection (Van Doan et al., 2021).

B. velezensis has recently been proven as a potential probiotic in aquaculture with good antagonistic properties against *Aeromonas* and *Vibrio* pathogens (Sam-on et al., 2023). In addition to its antimicrobial metabolites encoding genes, has adhesion capability to the host intestine as well as the acid and bile salt tolerance genes (Sam-on et al., 2023). In a study by Ding et al. (2023) the effect of four bacilli candidates including *Bacillus megaterium* B1M2, *B. subtilis* B0E9, *B. velezensis* DM5, and *B. siemens* B0E14 were assessed on the serum and skin immunities, resistance to red-head disease and gut and, skin microbiota in the yellow drum. Autochthonous *B. subtilis* B0E9 also improved the up-regulation of lysozyme gene and inflammation-related gene expression, positively shaped gut, and skin mucosal microbiota, and enhanced resistance against red-head disease caused by *Vibrio harveyi* (Ding et al., 2023).

10.4 CONCLUSION

Rapid development in the aquaculture sector requires the employment of new biotechnology particularly in cases of aquafeed and disease protection. Research on the efficacy and potency of *Bacillus* probiotics on the growth performance and immunity of fish has raised a promising hope to reduce the use of chemotherapeutic agents in aquatic ecosystems. Diverse physiological functions of *Bacillus* probiotics can improve digestion and absorption of feed nutrients with a consequent enhancement in fish growth. *Bacillus* probiotics can also change the ecology of the gut microbiota of fish via bacterial competition resulting in the prevention of pathogen attachment to the intestine and increasing the health condition of fish. Further, *Bacillus* probiotics can modulate fish innate immunity and cause a change in fish cell physiology such as increasing neutrophil adherence capacity, neutrophil migration, and plasma bactericidal activity that, all can result in the enhancement of immune effector functions such as enhancement in complement activity, immunoglobulin production, and cell cytotoxicity and finally increase in disease resistance. However, further studies are required to elucidate the actual mechanisms involved in immune-stimulatory effects by *Bacillus* spp. in the gut-associated lymphoid tissue of fish.

BIBLIOGRAPHY

Adorian TJ, Jamali H, Ghafari Farsani H, Darvishi P, Hasanpour S, Bagheri T, Roozbehfar R (2018) Effects of probiotic bacteria *Bacillus* on growth performance, digestive enzyme activity, and hematological parameters of Asian sea bass, *Lates calcarifer* (Bloch). *Probiotics and Antimicrobial Proteins*. https://doi.org/10.1007/s12602-018-9393-z

Ai Q, Xu H, Mai K, Xu W, Wang J, Zhang W (2011) Effects of dietary supplementation of *Bacillus subtilis* and fructooligosaccharide on growth performance, survival, non-specific immune response and disease resistance of juvenile large yellow croaker, *Larimichthys crocea*. *Aquaculture* **317**, pp.155–161.

Al-Deriny SH, Dawood MA, Abou Zaid AA, Wael F, Paray BA, Van Doan H, Mohamed RA (2020). The synergistic effects of *Spirulina platensis* and *Bacillus amyloliquefaciens* on the growth performance, intestinal histomorphology, and immune response of Nile tilapia (*Oreochromis niloticus*). *Aquaculture Reports*, ***17***, p.100390. https://doi.org/10.1016/j.aqrep.2020.100390

Aly S, Mohamed AAZ, Rahmani AH, Nashwa MAA (2016) Trials to improve the response of *Orechromis niloticus* to *Aeromonas hydrophila* vaccine using immunostimulants (garlic, Echinacea) and probiotics (Organic Green TM and Vet-Yeast TM). *African Journal of Biotechnology* **15**, pp.989–994.

Aly SM, Ahmed YSG, Ghareeb AAA, Mohamed MF (2008) Studies on *Bacillus subtilis* and *Lactobacillus acidophilus*, as potential probiotics, on the immune response and resistance of Tilapia nilotica (*Oreochromis niloticus*) to challenge infections. *Fish and Shellfish Immunology* **25**, pp.128–136. https://doi.org/10.1016/j.fsi.2008.03.013

Amiri O, Kolangi Miandare H, Hoseinifar SH, Shabani A, Safari R (2017) Skin mucus protein profile, immune parameters, immune related genes expression and growth performance of rainbow trout (*Oncorhynchus mykiss*) fed white bottom mushroom (*Agaricus bisporus*) powder. *International Journal of Medicinal Mushrooms* **20**, pp.337–347.

Amoah K, Tan B, Zhang S, Chi S, Yang Q, Liu H, Dong X. (2023). Host gut-derived Bacillus probiotics supplementation improves growth performance, serum and liver immunity, gut health, and resistive capacity against Vibrio harveyi infection in hybrid grouper (*♀ Epinephelus fuscoguttatus×♂ Epinephelus lanceolatus*). *Animal Nutrition*, **14**, pp.163–184. https://doi.org/10.1016/j.aninu.2023.05.005

Bagheri T, Hedayati SA, Yavari V, Alizade M, Farzanfar A (2008) Growth, survival and gut microbial load of rainbow trout (*Onchorhynchus mykiss*) fry given diet supplemented with probiotic during the two months of first feeding. *Turkish Journal of Fisheries and Aquatic Sciences*, **8**, pp.43–48.

Bandyopadhyay P, Mohapatra PKD (2009) Effect of a probiotic bacterium *Bacillus circulans* PB7 in the formulated diets: On growth, nutritional quality and immunity of *Catla catla* (Ham.). *Fish Physiology and Biochemistry*, **35**, pp.467–478.

Banerjee G, Nandi A, Ray AK (2017) Assessment of hemolytic activity, enzyme production and bacteriocin characterization of *Bacillus subtilis* LR1 isolated from the gastrointestinal tract of fish. *Archives of Microbiology* **199**, pp.115–124.

Baumgärtner S, Creer S, Jones C, James J, Ellison A. (2024). Bacillus indicus and *Bacillus subtilis* as alternative health and colouration promoters to synthetic astaxanthin in cyprinid aquaculture species. *Aquaculture*, **578**, p.740016. https://doi.org/10.1016/j.aquaculture.2023.740016

Bhatnagar A, Saluja S. (2021). Role of Zingiber officinale and autochthonous probiotic *Bacillus coagulans* in feeds of *Catla catla* (Hamilton, 1822) for growth promotion, immunostimulation, histoprotection, and control of DNA damage. *Fish Physiology and Biochemistry*, **47**, pp.2081–2100. https://doi.org/10.1007/s10695-021-01030-8

Brunt J, Newaj-Fyzul A, Austin B (2007). The development of probiotics for the control of multiple bacterial diseases of rainbow trout, Oncorhynchus mykiss (Walbaum). *Journal of Fish Diseases*, **30**(10), pp.573–579. https://doi.org/10.1111/j.1365-2761.2007.00836.x

Büyükdeveci ME, Cengizler İ, Balcázar JL, Demirkale I (2023). Effects of two host-associated probiotics *Bacillus mojavensis* B191 and *Bacillus subtilis* MRS11 on growth performance, intestinal morphology, expression of immune-related genes and disease resistance of Nile tilapia (*Oreochromis niloticus*) against *Streptococcus iniae*. *Developmental & Comparative Immunology*, ***138***, p.104553. https://doi.org/10.1016/j.dci.2022.104553

Cha J-H, Rahimnejad S, Yang S-Y, Kim K-W, Lee K-J (2013) Evaluations of *Bacillus* spp. as dietary additives on growth performance, innate immunity and disease resistance of olive flounder (*Paralichthys olivaceus*) against *Streptococcus iniae* and as water additives. *Aquaculture*, **402**, pp.50–57.

Chen Y, Li J, Xiao P, Li GY, Yue S, Huang J, et al. (2016) Isolation and characterization of *Bacillus* spp. M001 for potential application in turbot (*Scophthalmus maximus* L) against *Vibrio anguillarum*. *Aquaculture Nutrition* **22**, pp.374–381.

Das P, Ghosh K (2013) Evaluation of phytase-producing ability by a fish gut bacterium, *Bacillus subtilis* subsp. *subtilis*. *Journal of Biological Sciences* **13**, pp.691–700.

Dawood MAO, Koshio S (2016) Recent advances in the role of probiotics and prebiotics in carp aquaculture: A review. *Aquaculture*, **454**, pp.243–251.

Dawood MAO, Koshio S, Esteban MÁ (2017) Beneficial roles of feed additives as immunostimulants in aquaculture: A review. *Reviews in Aquaculture*, n/a-n/a.

Ding XY, Wei CY, Liu ZY, Yang HL, Han F, Sun YZ (2023). Autochthonous *Bacillus subtilis* and *Enterococcus faecalis* improved liver health, immune response, mucosal microbiota, and red-head disease resistance of yellow drum (*Nibea albiflora*). *Fish & Shellfish Immunology*, **134**, p.108575. https://doi.org/10.1016/j.fsi.2023.108575

Doan HV, Hoseinifar SH, Khanongnuch C., Kanpiengjai A, Unban K, Kim VV, et al (2018) Host-associated probiotics boosted mucosal and serum immunity, disease resistance and growth performance of Nile tilapia (*Oreochromis niloticus*). *Aquaculture* **491**, pp.94–100.

Docando F, Nuñez-Ortiz N, Serra C R, Arense P, Enes P, Oliva-Teles A, ... Tafalla C. (2022). Mucosal and systemic immune effects of *Bacillus subtilis* in rainbow trout (*Oncorhynchus mykiss*). *Fish & Shellfish Immunology*, **124**, pp.142–155. https://doi.org/10.1016/j.fsi.2022.03.040

El-Son MA, Elshopakey GE, Rezk S, Eldessouki EA, Elbahnaswy S (2022). Dietary mixed Bacillus strains promoted the growth indices, enzymatic profile, intestinal immunity, and liver and intestinal histomorphology of Nile tilapia, *Oreochromis niloticus*. *Aquaculture Reports*, *27*, p.101385. https://doi.org/10.1016/j.aqrep.2022.101385

Fečkaninová A, Koščová J, Mudroňová D, Popelka P, Toropilová J (2017) The use of probiotic bacteria against *Aeromonas* infections in salmonid aquaculture. *Aquaculture*, **469**, pp.1–8.

Galagarza OA, Smith SA, Drahos DJ, Robert JD, Williams C, Kuhn DD (2018) Modulation of innate immunity in Nile tilapia (*Oreochromis niloticus*) by dietary supplementation of *Bacillus subtilis* endospores. *Fish & Shellfish Immunology* **83**, pp.171–179.

Giri SS, Sukumaran V, Sen SS, Jena PK (2014) Effects of dietary supplementation of potential probiotic *Bacillus subtilis* VSG1 singularly or in combination with *Lactobacillus plantarum* VSG3 or/and *Pseudomonas aeruginosa* VSG2 on the growth, immunity and disease resistance of *Labeo rohita*. *Aquaculture Nutrition*, **20**, pp.163–171.

Gobi N, Malaikozhundan B, Sekar V, Shanthi S, Vaseeharan B, Jayakumar R, Nazar AK (2016) GFP tagged *Vibrio parahaemolyticus* Dahv2 infection and the protective effects of probiotic *Bacillus licheniformis* Dahb1 on the growth, immune and antioxidant responses in *Pangasius hypophthalmus*. *Fish and Shellfish Immunology* **52**, pp.230–238.

Gupta A, Gupta P, Dhawan A (2014) Dietary supplementation of probiotics affects growth, immune response and disease resistance of *Cyprinus carpio* fry. *Fish & Shellfish Immunology*, **41**, pp.113–119.

Gupta A, Gupta P, Dhawan A (2016). *Paenibacillus polymyxa* as a water additive improved immune response of *Cyprinus carpio* and disease resistance against *Aeromonas hydrophila*. *Aquaculture Reports*, **4**, pp.86–92.

Hai NV (2015) Research findings from the use of probiotics in tilapia aquaculture: A review. *Fish & Shellfish Immunology*, **45**, pp.592–597.

Hasan MT, Jang WJ, Lee BJ, Hur SW, Lim SG, Kim KW, ... Kong IS (2021). Dietary supplementation of *Bacillus* sp. SJ-10 and *Lactobacillus plantarum* KCCM 11322 combinations enhance growth and cellular and humoral immunity in olive flounder (*Paralichthys olivaceus*). *Probiotics and Antimicrobial Proteins*, pp.1–15. https://doi.org/10.1007/s12602-021-09749-9

He RP, Feng J, Tian JX L, Dong S L, Wen B (2017) Effects of dietary supplementation of probiotics on the growth, activities of digestive and non-specific immune enzymes 8 in hybrid grouper (*Epinephelus lanceolatus♂× Epinephelus fuscoguttatus♀*). *Aquaculture Research* **48**, pp.5782–5790.

Hoseinifar SH, Dadar M, Ringø E (2017) Modulation of nutrient digestibility and digestive enzyme activities in aquatic animals: The functional feed additives scenario. *Aquaculture Research*, **48**, pp.3987–4000.

Hoseinifar SH, Ringø E, Shenavar Masouleh A, Esteban MÁ (2016) Probiotic, prebiotic and synbiotic supplements in sturgeon aquaculture: A review. *Reviews in Aquaculture* **8**, pp.89–102.

Hoseinifar SH, Sun Y-Z, Caipang CM (2017) Short chain fatty acids as feed supplements for sustainable aquaculture: An updated view. *Aquaculture Research*, **48**, pp.1380–1391.

Jang WJ, Lee KB, Jeon MH, Lee SJ, Hur SW, Lee S, … Lee EW (2023). Characteristics and biological control functions of *Bacillus* sp. PM8313 as a host-associated probiotic in red sea bream (Pagrus major) aquaculture. *Animal Nutrition*, **12**, pp.20–31. doi: 10.1016/j.aninu.2022.08.011

Jose M.S., Arun D, Neethu S, Radhakrishnan EK, Jyothis M. (2023) Probiotic *Paenibacillus polymyxa* HGA4C and *Bacillus licheniformis* HGA8B combination improved growth performance, enzymatic profile, gene expression and disease resistance in *Oreochromis niloticus*. *Microbial Pathogenesis*, 174, p.105951. doi: 10.1016/j.micpath.2022.105951

Ji Z, Zhu C, Zhu X, Ban S, Yu L, Tian J, … Jiang M (2023). Dietary host-associated *Bacillus subtilis* supplementation improves intestinal microbiota, health, and disease resistance in Chinese perch (*Siniperca chuatsi*). *Animal Nutrition*, **13**, pp.197–205. https://doi.org/10.1016/j.aninu.2023.01.001

Jian H, Bian Q, Zeng W, Ren P, Sun H, Lin Z, Huang Y (2019). Oral delivery of *Bacillus subtilis* spores expressing grass carp reovirus VP4 protein produces protection against grass carp reovirus infection. *Fish & Shellfish Immunology*, **84**, pp.768–780.

Jiang Y, Zhou S, Sarkodie EK, Chu W (2020). The effects of Bacillus cereus QSI-1 on intestinal barrier function and mucosal gene transcription in Crucian carp (*Carassius auratus* gibelio). *Aquaculture Reports*, **17**, p.100356. https://doi.org/10.1016/j.aqrep.2020.100356

Kuebutornye FK, Tang J, Cai J, Yu H, Wang Z, Abarike ED, Afriyie G (2020). In vivo assessment of the probiotic potentials of three host-associated *Bacillus* species on growth performance, health status and disease resistance of *Oreochromis* niloticus against *Streptococcus agalactiae*. *Aquaculture*, **527**, p.735440. https://doi.org/10.1016/j.aquaculture.2020.735440

Kumar R, Mukherjee SC, Prasad KP, Pal AK (2006) Evaluation of *Bacillus subtilis* as a probiotic to Indian major carp *Labeo rohita* (Ham.). *Aquaculture Research*, **37**, pp.1215–1221.

Liang X, Liang J, Cao J, Liu S, Wang Q, Ning Y, Zhou S (2023). Oral immunizations with *Bacillus subtilis* spores displaying VP19 protein provide protection against Singapore grouper iridovirus (SGIV) infection in grouper. *Fish & Shellfish Immunology*, **138**, p.108860. https://doi.org/10.1016/j.fsi.2023.108860

Lin Y-S, Saputra F, Chen Y-C, Hu S-Y (2019) Dietary administration of *Bacillus amyloliquefaciens* R8 reduces hepatic oxidative stress and enhances nutrient metabolism and immunity against *Aeromonas hydrophila* and *Streptococcus agalactiae* in zebrafish *(Danio rerio)*. *Fish and Shellfish Immunology*. https://doi.org/10.1016/j.fsi.2018.11.047

Liu C-H, Chiu C-H, Wang S-W, Cheng W (2012) Dietary administration of the probiotic, *Bacillus subtilis* E20, enhances the growth, innate immune responses, and disease resistance of the grouper, *Epinephelus coioides*. *Fish & Shellfish Immunology*, **33**, pp.699–706.

Liu H, Wang S, Cai Y, Guo X, Cao Z, Zhang Y, et al. (2017) Dietary administration of *Bacillus subtilis* enhances growth, digestive enzyme activities, innate immune responses and disease resistance of tilapia, *Oreochromis niloticus*. *Fish and Shellfish Immunology*, **60**, pp.326–33.

Liu J, Cheng Y, Lu Y, Xia C, Wang N, Li Y (2021). *Bacillus subtilis* spores as an adjuvant to enhance the protection efficacy of the SVCV subunit vaccine (SVCV-M protein) in German mirror carp (*Cyprirnus carpio songpa* Linnaeus Mirror). *Aquaculture Research*, ***52***(10), pp.4648–4660. https://doi.org10.1111/are.15299

Liu Q, Wen L, Pan X, Huang Y, Du X, Qin J, Lin Y (2021). Dietary supplementation of *Bacillus subtilis* and *Enterococcus faecalis* can effectively improve the growth performance, immunity, and resistance of tilapia against Streptococcus agalactiae. *Aquaculture Nutrition*, **27**(4), pp.1160–1172. https://doi.org/10.1111/anu.13256

Lu L, Cao H, He S, Wei R, Diong M (2011) *Bacillus amyloliquefaciens* G1: A potential antagonistic bacterium against eel-pathogenic *Aeromonas hydrophila*. *Evidence-Based Complementary and Alternative Medicine*. https://doi.org/10.1155/2011/824104

Lu Y, Zhang Y, Zhang P, Liu J, Wang B, Bu X, Li Y (2022). Effects of dietary supplementation with *Bacillus subtilis* on immune, antioxidant, and histopathological parameters of *Carassius auratus gibelio* juveniles exposed to acute saline-alkaline conditions. *Aquaculture International*, ***30***(5), pp.2295–2310. https://doi.org/10.1007/s10499-022-00902-x

Mai W, Yan B, Xin J (2022). Oral immunizations with *Bacillus subtilis* spores expressing MCP protein provide protection against red-spotted grouper nervous necrosis virus (RGNNV) infection in juvenile grouper, *Epinephelus coioides*. *Aquaculture*, ***552***, p.738008. https://doi.org/10.1016/j.aquaculture.2022.738008

Mohanty D, Roy P, Sahu A, Panda SP, Sahoo AK, Das BK (2022). Subcellular component of Bacillus subtilis (AN11) induces protective immunity against *Aeromonas hydrophila* in *Labeo rohita* (Ham.). *Aquaculture Research*, **53**(2), pp.367–376. https://doi.org/10.1111/are.15578

Nayak SK (2021). Multifaceted applications of probiotic *Bacillus* species in aquaculture with special reference to *Bacillus subtilis*. *Reviews in Aquaculture*, *13*(2), pp.862–906. https://doi.org/10.1111/raq.12503

Newaj-Fyzul A, Adesiyun AA, Mutani A, Ramsubhag A, Brunt J, Austin B (2007) *Bacillus subtilis* AB1 controls Aeromonas infection in rainbow trout (*Oncorhynchus mykiss*, Walbaum). *Journal of Applied Microbiology*, **103**, pp.1699–1706.

Ng W-K, Koh C-B (2016) The utilization and mode of action of organic acids in the feeds of cultured aquatic animals. *Reviews in Aquaculture*, https://doi.org/10.1111/raq.12141

Nunes AL, Owatari MS, Rodrigues RA, Fantini LE, Kasai RYD, Martins ML, de Campos CM (2020). Effects of *Bacillus subtilis* C-3102-supplemented diet on growth, non-specific immunity, intestinal morphometry, and resistance of hybrid juvenile P*seudoplatystoma* sp. challenged with *Aeromonas hydrophila*. *Aquaculture International*, **28**, pp.2345–2361. https://doi.org/10.1007/s10499-020-00586-1

Oliveira FC, Soares MP, Oliveira BPN, Pilarski F, de Campos CM (2022). Dietary administration of *Bacillus subtilis*, inulin, and its synbiotic combination improves growth and mitigates stress in experimentally infected *Pseudoplatystoma reticulatum*. *Aquaculture Research*, **53** (12), pp.4256–4265. https://doi.org/10.1111/are.15923

Olmos Soto, J (2021). Feed intake improvement, gut microbiota modulation and pathogens control by using *Bacillus* species in shrimp aquaculture. *World Journal of Microbiology and Biotechnology*, **37**(2), p.28.

Pan P K, Wang KT, Wu TM, Chen YY, Nan FH, Wu YS (2023). Heat inactive Bacillus subtilis var. natto regulate Nile tilapia (*Oreochromis niloticus*) intestine microbiota and metabolites involved in the intestine phagosome response. *Fish & Shellfish Immunology*, **134**, p.108567. https://doi.org10.1016/j.fsi.2023.108567

Raida MK, Larsen JL, Nielsen ME, Buchmann K (2003) Enhanced resistance of rainbow trout, *Oncorhynchus mykiss* (Walbaum), against *Yersinia ruckeri* challenge following oral administration of *Bacillus subtilis* and *B. licheniformis* (BioPlus2B). *Journal of Fish Diseases* **26** (**8**), pp.495–498.

Reda RM, Selim KM (2015) Evaluation of *Bacillus amyloliquefaciens* on the growth performance, intestinal morphology, hematology and body composition of Nile tilapia, *Oreochromis niloticus*. *Aquaculture International* **23**, pp.203–217.

Ringø E, Hoseinifar SH, Ghosh K, Doan HV, Beck BR, Song SK (2018). Lactic acid bacteria in finfish—An update. *Frontiers in Microbiology*, **9**, p.376234.

Ringø E, Van Doan H, Lee SH, Soltani M, Hoseinifar SH, Harikrishnan R, Song SK (2020). Probiotics, lactic acid bacteria and bacilli: Interesting supplementation for aquaculture. *Journal of Applied Microbiology*, **129**(1), pp.116–136.

Saengrung J, Bunnoy A, Du X, Huang L, An R, Liang X, Srisapoome P. (2023). Effects of ribonucleotide supplementation in modulating the growth of probiotic *Bacillus subtilis* and the synergistic benefits for improving the health performance of Asian seabass (Lates calcarifer). *Fish & Shellfish Immunology*, **140**, p.108983. https://doi.org/10.1016/j.fsi.2023.108983

Sahandi J, Jafaryan H, Roozbehfar R, Dehestani M (2012) The use of two enrichment forms (*Brachionus plicatilis* enrichment and rearing water enrichment) with probiotic bacilli spore on growth and survival of Silver carp (*Hypophthalmichthys molitrix*). *Iranian Journal of Veterinary Research* **13**(4), pp.289–295.

Sam-On MFS, Mustafa S, Hashim AM, Yusof MT, Zulkifly S, Malek AZA, ... Asrore MSM (2023). Mining the genome of *Bacillus velezensis* FS26 for probiotic markers and secondary metabolites with antimicrobial properties against aquaculture pathogens. *Microbial Pathogenesis*, **181**, p.106161. https://doi.org/10.1016/j.micpath.2023.106161

Santos GG, Libanori MCM, Pereira SA, Ferrarezi JVS, Ferreira MB, Soligo TA, Mouriño JLP (2023). Probiotic mix of *Bacillus* spp. and benzoic organic acid as growth promoter against Streptococcus agalactiae in Nile tilapia. *Aquaculture*, **566**, p.739212. https://doi.org/10.1016/j.aquaculture.2022.739212

Shahbazi P, Sheikhzadeh N, Siahtan MAN, Ghadimi A K, Soltani M, Nofouzi K, ... Firouzamandi M (2023). Efficacy of dietary live or heat-killed Bacillus subtilis in goldfish (*Carassius auratus*) infected with *Ichthyophthirius multifiliis*. *Veterinary Medicine and Science*, **9** (4), pp.1636–1645. https://doi.org/10.1002/vms3.1183

Silva TFA, Petrillo TR, Yunis-Aguinaga J, Marcusso PF, da Silva Claudiano G, de Moraes FR, de Moraes JRE (2015) Effects of the probiotic *Bacillus amyloliquefaciens* on growth performance, hematology and intestinal morphometry in cage-reared Nile tilapia. *Latin American Journal of Aquatic Research*, **43**, pp.963–971.

Shawky A, Abd El-Razek IM, El-Halawany RS, Zaineldin AI, Amer AA, Gewaily MS, Dawood MA (2023). Dietary effect of heat-inactivated *Bacillus subtilis* on the growth performance, blood biochemistry, immunity, and antioxidative response of striped catfish (*Pangasianodon hypophthalmus*). *Aquaculture*, **575**, p.739751. https://doi.org/10.1016/j.aquaculture.2023.739751

Soltani M, Lymbery A, Song SKand Hosseini Shekarabi P (2018) Adjuvant effects of medicinal herbs and probiotics for fish vaccines. *Reviews in Aquaculture*, https://doi.org/10.1111/raq.12295

Sumon TA, Hussain M A, Sumon MAA, Jang WJ, Abellan FG, Sharifuzzaman SM, Hasan MT (2022). Functionality and prophylactic role of probiotics in shellfish aquaculture. *Aquaculture Reports*, **25**, p.101220. https://doi.org/10.1016/j.aqrep.2022.101220

Sun Y, Wen Z, Li X, Meng N, Mi R, Li Y, Li S (2012) Dietary supplement of fructooligosaccharides and *Bacillus subtilis* enhances the growth rate and disease resistance of the sea cucumber *Apostichopus japonicus* (Selenka). *Aquaculture Research*, **43**, pp.1328–1334.

Sun Y-Z, Yang H-L, Ma R-L, Lin W-Y (2010) Probiotic applications of two dominant gut *Bacillus* strains with antagonistic activity improved the growth performance and immune responses of grouper *Epinephelus coioides*. *Fish & Shellfish Immunology*, **29**, pp.803–809.

Sun YZ, Yang HL, Huang KP, Ye JD, Zhang CX (2013) Application of autochthonous *Bacillus* bioencapsulated in copepod to grouper *Epinephelus coioides* larvae. *Aquaculture* **392**, pp.44–50.

Telli GS, Ranzani-Paiva MJT, de Carla Dias D, Sussel FR, Ishikawa CM, Tachibana L (2014) Dietary administration of *Bacillus subtilis* on hematology and non-specific immunity of Nile tilapia *Oreochromis niloticus* raised at different stocking densities. *Fish & shellfish immunology*, **39**, pp.305–311.

Van Doan H, Wangkahart E, Thaimuangphol W, Panase P, Sutthi N (2021). Effects of *Bacillus* spp. mixture on growth, immune responses, expression of immune-related genes, and resistance of Nile Tilapia against Streptococcus agalactiae infection. *Probiotics and Antimicrobial Proteins*, pp.1–16. https://doi.org/10.1007/s12602-021-09845-w

Wang C, Liu Y, Sun G, Li X, Liu Z (2019) Growth, immune response, antioxidant capability, and disease resistance of juvenile Atlantic salmon (*Salmo salar* L.) fed *Bacillus velezensis* V4 and *Rhodotorula mucilaginosa* compound. *Aquaculture*, **500**, pp.65–74.

Wang J, Wu Z, Wang S, Wang X, Zhang D, Wang Q, ... Chen Y (2022). Inhibitory effect of probiotic *Bacillus* spp. isolated from the digestive tract of *Rhynchocypris lagowskii* on the adhesion of common pathogenic bacteria in the intestinal model. *Microbial Pathogenesis*, **169**, p.105623. https://doi.org/10.1016/j.micpath.2022.105623

Wang M, Lv C, Chen Y, Bi X, Yang D, and Zhao J (2022). Effects of the potential probiotic *Bacillus subtilis* D1-2 on growth, digestion, immunity, and intestinal flora in juvenile sea cucumber, *Apostichopus japonicus*. *Fish & Shellfish Immunology*, **124**, pp.12–20. https://doi.org/10.1016/j.fsi.2022.03.043

Wang Q, Liang X, Ning Y, Liu S, Liang Z, Zhang Z, ... Zhou S (2023). Surface display of major capsid protein on *Bacillus subtilis* spores against largemouth bass virus (LMBV) for oral administration. *Fish & Shellfish Immunology*, **135**, p.108627. https://doi.org/10.1016/j.fsi.2023.108627

Wang T, Cheng Y, Chen X, Liu Z, Long X (2017). Effects of small peptides, probiotics, prebiotics, and synbiotics on growth performance, digestive enzymes, and oxidative stress in orange-spotted grouper, *Epinephelus coioides*, juveniles reared in artificial seawater. *Chinese Journal of Oceanology and Limnology*, **35** (1), pp.89–97.

Wang X, Onchari MM, Yang X, Xu L, Yin X, Wan F, ... Luo C (2022). Genome analysis of *Bacillus subtilis* JCL16 and the synergistic relationship among its metabolites reveals its potential for biocontrol of Nocardia seriolae. *Biological Control*, **167**, p.104855. https://doi.org/10.1016/j.biocontrol.2022.104855

Wu Z, Feng X, Xie L, Peng X, Yuan J, Chen X (2012) Effect of probiotic *Bacillus subtilis* Ch9 for grass carp, *Ctenopharyngodon idella* (Valenciennes, 1844), on growth performance, digestive enzyme activities and intestinal microflora. *Journal of Applied Ichthyology*, **28**, pp.721–727.

Yang H, Zhang M, Ji T, Zhang Y, Wei W, Liu Q (2022). Bacillus subtilis CK3 used as an aquatic additive probiotics enhanced the immune response of crayfish *Procambarus clarkii* against newly identified *Aeromonas veronii* pathogen. *Aquaculture Research*, **53**(1), pp.255–264. https://doi.org/10.1111/are.15571

Yang H L, Xia H Q, Ye Y D, Zou W C, Sun Y Z (2014) Probiotic *Bacillus pumilus* SE5 shapes the intestinal microbiota and mucosal immunity in grouper *Epinephelus coioides*. *Diseases of Aquatic Organisms* **111**, pp.119–127.

Ye Y, Yu D, Liu Q, Ma S, Zhang M, Zhao M, ... Yu J (2023). Nutritional composition of fresh carcass of turbot fed with *Bacillus subtilis* SMF1, B. licheniformis LMF1, and B. siamensis DL3 and its relationship with intestinal flora. *Animal Feed Science and Technology*, **300**, p.115627. https://doi.org/10.1016/j.anifeedsci.2023.115627

Yi Y, Zhang Z, Zhao F, Liu H, Yu L, Zha J, Wang G (2018). Probiotic potential of *Bacillus velezensis* JW: Antimicrobial activity against fish pathogenic bacteria and immune enhancement effects on *Carassius auratus*. *Fish & Shellfish Immunology*, **78**, pp.322–330. https://doi.org/10.1016/j.fsi.2018.04.055

Zaineldin A I, Hegazi S, Koshio S, Ishikawa M, El Basuini MF, Dossou S, Dawood MA (2021). The influences of Bacillus subtilis C-3102 inclusion in the red sea bream diet containing high levels of soybean meal on growth performance, gut morphology, blood health, immune response, digestibility, digestive enzymes, and stress resistance. *Aquaculture Nutrition*, **27**(6), pp.2612–2628. https://doi.org/10.1111/anu.13389

Zhang Q, Yu H, Tong T, Tong W, Dong L, Xu M, Wang Z (2014) Dietary supplementation of *Bacillus subtilis* and fructooligosaccharide enhance the growth, non-specific immunity of juvenile ovate pompano, *Trachinotus ovatus* and its disease resistance against *Vibrio vulnificus*. *Fish & Shellfish Immunology*, **38**, pp.7–14.

Zhao M, Yu D, Liu Q, Ma S, Xu J, Yu J (2022). Co-fermentation of *Bacillus subtilis* and *Bacillus licheniformis* and its application in the feeding of Koi. *Aquaculture Research*, **53**(17), pp.6056–6068. https://doi.org/10.1111/are.16077

Zhou X, Tian Z, Wang Y, Li W (2010) Effect of treatment with probiotics as water additives on tilapia (*Oreochromis niloticus*) growth performance and immune response. *Fish Physiology and Biochemistry*, **36**, pp.501–509.

11 *Bacillus* as Probiotics in Shellfish Culture

Mehdi Soltani
Murdoch University, Perth, Australia, University of Tehran, Tehran, Iran

Sohrab Ahmadivand
Ludwig-Maximilians University Munich, Munich, Germany

Einar Ringø
UiT The Arctic University of Norway, Tromsø, Norway

11.1 INTRODUCTION

Shrimp aquaculture is fast growing industry with a great vision. The global shrimp production achieved above 4 million tons in 2019 and has been increased in 9.4 million tons in the year 2022 (Fletcher 2021; FAO, 2023). Countries of Southeast Asia, the United States of America, the European Union, and Japan are the most demanded markets for shrimp products (Geetha et al., 2020). Various novelties have been employed to improve shrimp culture particularly using white leg shrimp (*Litopenaeus vannamei*) that is the main cultured shrimp species (Davis et al., 2022). However, many infectious diseases such as vibriosis, photobactriosis, aeromonasis, tenacibaculosis, shewanellasis, and acute hepatopancreatic necrosis disease (AHPND) are frequently occurring in shrimp aquaculture worldwide. Application of immunomodulators such as probiotics and medicinal herbs has received a high promising attention for increasing the shrimp product and reducing the side effects of chemotherapy (Jahangiri et al., 2018, Soltani et al., 2024). For instance, probiotics as a part of beneficial bacteria available in the host intestine can enhance the growth and animal immunity resulting in increasing animal resistance toward diseases that are one of the most serious obstacles to the sustainability on shrimp aquaculture development (Kerry et al., 2018; Guo et al., 2022). Studies have shown that probiotics are able to improve the shrimp production through the enhancement of growth performance and resistance to infectious diseases (Table 11.1). For instance, data analysis of the available literature of shrimp aquaculture showed that probiotics can increase the survival of shrimp by up to 95% against some infectious diseases in comparison with controls (Toledo et al., 2019). This chapter addressed the efficacy and potency of *Bacillus* probiotics in shrimp aquaculture particularly on performance, immunity, and disease resistance of species of *L. vannamei*.

11.2 *BACILLUS* PROBIOTICS AS GROWTH ENHANCER IN SHELLFISH AQUACULTURE

One of the benefits of *Bacillus* probiotics in shrimp aquaculture is a direct growth-promoting effect through production of various digestive enzymes, including lipase, protease, and amylase, that can enhance the natural digestive functions of the target animal (Liu et al., 2009; Wang, 2007). *Bacillus* probiotics are able to support host nutrition, in particular by supplying vitamins and fatty acids and increasing growth and survival of *P. monodon* at postlarvae stage with no water exchange (Devaraja et al., 2013; NavinChandran et al., 2014; Kumar et al., 2016). In a study by Zokaeifar et al. (2012b, 2014), the use of *B. subtilis* in *L. vannamei* exhibited the activity of digestive enzymes including protease and amylase resulting in an increase in the shrimp growth (Zokaeifar et al., 2012b;

DOI: 10.1201/9781003503811-11

TABLE 11.1
Effect of *Bacillus* Probiotics on Growth Performance, Immune Response, and Disease Resistance in Shrimp Culture

Bacillus Species	Shrimp Species	Water Temp (°C)	Dosage/ Duration (day)	Route	Growth Factor	Disease/Stress Resistance	Immune Response	Reference
Bacillus S11	Tiger shrimp (*P. monodon*)	30	~10^{10} CFU g^{-1} feed (100 days)	Oral	MSW (+)	*Vibrio harveyi* (+)	Unknown	Rengpipat et al. (1998b)
Bacillus S11	Tiger shrimp	30	~10^{10} CFU g^{-1} feed (90 days)	Oral	MSW (+)	*V. harveyi* (+)	Phagocytosis (+) and phenoloxidase (+)	Rengpipat et al. (2000)
Bacillus S11	Tiger shrimp	26–31	~10^{10} CFU g^{-1} mixed with 3 kg of feed (100 days)	Oral	AGR (+)	*V. harveyi* (+)	Unknown	Rengpipat et al. (2003)
B. subtilis BT23	Tiger shrimp	Unknown	10^6–10^8 CFU ml^{-1} (6 days)	Bath	Unknown	*V. harveyi* (+)	Unknown	Vaseeharan et al. (2003)
Bacillus spp.	Tiger shrimp	28	1 × 10^{11}/ 0.8 ha pond	Bath	AGR (+)	*Vibrio* sp. (+)	Unknown	Dalmin et al. (2001)
B. subtilis E20	White leg shrimp (*L. vannamei*)	23	10^6, 10^7, 10^8 CFU kg^{-1} feed (108 days)	Oral	Unknown	*V. alginolyticus* (+)	Phenoloxidase (+) and phagocytosis (+)	Tseng et al. (2009)
Bacillus OJ	White leg shrimp	26	10^8 and 10^{10} CFU g^{-1} feed (28 days)	Oral	Unknown	White spot viral disease (+)	Phenoloxidase (+), phagocytosis (+), respiratory burst (+), and acid phosphatase (+)	Li et al. (2009)
B. subtilis E20	White leg shrimp	26	10^6–10^8 CFU kg^{-1} (7 days)	Oral	FCR (+) WG (+)	Survival (+)	Unknown	Liu et al. (2009)
		30	10^8–10^9 CFU L^{-1} (14 days)	Bath	Unknown	Low temperature (+), low salinity (+), high salinity (+), and high nitrite-N (+)	Lysozyme (+) and prophenoloxidase I and II (+)	Liu et al. (2010)
Bacillus sp. NL110	Giant freshwater prawn (*Macrobrachium rosenbergii*)	Unknown	4.73 ± 2.87 ×10^9 CFU g^{-1} (twice in a day for 60 days) 1.15 ±0.56 106 CFU ml^{-1} (once in a week for 60 days)	Oral or bath	SGR (+) WG (+)	Survival (+)	Total hemocyte count (+), phenoloxidase activity (+), and respiratory burst (+)	Mujeeb Rahiman et al. (2010)

(Continued)

TABLE 11.1 (Continued)
Effect of *Bacillus* Probiotics on Growth Performance, Immune Response, and Disease Resistance in Shrimp Culture

Bacillus Species	Shrimp Species	Water Temp (°C)	Dosage/ Duration (day)	Route	Growth Factor	Disease/Stress Resistance	Immune Response	Reference
B. subtilis	White leg shrimp	28	1.2×10^4 CFU g^{-1} (28 days)	Oral	WG (+) FCR (+)	Ammonia (+), oxygen deficiency (+), and survival (+)	Unknown	Olmos et al. (2011)
B. endophyticus and *B. tequilensis*	White leg shrimp	Unknown	1×10^5 CFU mL^{-1} (9 days)	Bath	Unknown	*V. campbellii (+), V. vulnificus (+), V. parahaemolyticus(+)*, and *V. alginolyticus* (+)	Unknown	Luis-Villaseñor et al. (2012)
Lactobacillus plantarum	White leg shrimp	30	2-4 $\times 10^8$ CFU g^{-1} feed (42 days)	Oral	RGW (+) RGR (+) FCR (+)	*V. harveyi* (+)	Unknown	Kongnum et al. (2012)
Bacillus subtilis L10 and G1	White leg shrimp	28	10^5 and 10^8 CFU mL^{-1} (8 weeks)	Oral	WG (+) SGR (+)	*V. harveyi* (+)	Prophenoloxidase (+), peroxinectin (+), lipopolysaccharide (+), β-1,3-glucan binding protein (+), and serine protein (+)	Zokaeifar et al. (2012a, 2012b)
Mixed *Bacillus* (*B. thuringiensis, B. megaterium, B. polymyxa, B. licheniformis, B. subtilis*, and *B. circulans*	White leg shrimp	31-32	10^9 CFU g^{-1} in the form of enriched *Artemia* (4 d for larvae and 14 and 22 d for post-larvae) in	Oral	LG (+) WG (+) ADG (+) SGR (+)	Unknown	Unknown	Nimrat et al. (2012)
B. subtilis and *B. licheniformis*	*M. japonicus*	30	1.0×10^8 CFU g^{-1} feed (60 days)	Oral	FW (+) SGR (+)	High temperature (+)	Phenoloxidase (+), lysozyme (+), superoxide dismutase (+), and hemocyanin subunit L (+)	Dong et al. (2014)
B. cereus	Tiger shrimp	28	0.1–0.4%/100 g feed (90 days)	Oral	FCR (+) SGR (+) FCE (+) AGR (+)	*V. harveyi* (+)	Phenoloxidase (+), lysozyme (+), respiratory burst (+), and bactericidal activity (+)	Chandran et al. (2014)

B. subtilis L10 and G1	White leg shrimp	28	10^5 and 10^8 CFU mL^{-1} (8 weeks)	Bath	WG (+) SGR (+) FCR (+)	*V. harveyi* (+)	Prophenoloxidase (+), peroxinectin (+), lipopolysaccharide (+), β-1,3-glucan binding protein (+), and serine protein (+)	Zokaeifar et al. (2014)
Bacillus sp. JL47 and *Bacillus* sp. JL1	Tiger shrimp	30	4 g wet bacterial weight/kg feed (30 days)	Oral	ABW (+) ABL (+)	Ammonia (+) *V. campbellii* (+)	Unknown	Laranja et al. (2014)
Bacillus probiotic PC465	White leg shrimp	30	10^7 and 10^9 CFU g^{-1} (20 days)	Oral	GR (+) WG (+)	White spot viral disease (+)	Prophenoloxidase (+), peroxinectin (+), penaeidin (+), thioredoxin (+), lectins (+), hemocyanin (+), and crustin (+)	Chai et al. (2016)
Bacillus sp. JL47	Tiger shrimp	30	0.5 g wet bacterial weight/L water (6 h)	Oral	Unknown	*V. campbellii* (+)	Prophenoloxidase (+), Tgase (+), and Hsp70 (+)	Laranja et al. (2017)
B. subtilis and *B. licheniformis*	White leg shrimp	30	10^4 and 10^8 CFU g^{-1} (60 days)	Oral	WG (+) SGR (+) FCR (+)	Unknown	Lysozyme (+), total hemocyte (+), granular cells (+), semi-granular cell (+), and hyaline cells (+)	Sadat Hoseini Madani et al. (2018)
B. licheniformis and *B. flexus*	White leg shrimp	28	*B. licheniformis* (2.0×10^9 CFU/g^{-1}), *B. flexus* (2.0×10^9 CFU/g^{-1}), in combination with each probiotic at 1.0×10^9 CFU/g^{-1}	Oral	WG (+), SGR (+)	Fresh water (+) and *V. harveyi* (+)	Alkaline phosphatase (+) and peroxidase (+)	Cai et al. (2019)
B. aryabhattai	White leg shrimp		1.0×10^8 CFU g^{-1} (6 weeks)			*V. harveyi* (+)	Upregulation of antioxidant enzymes (C-type lec, pen3a, hsp60, trx, and fer) (+), phenoloxidase (+), total antioxidant activity (+), total hemocyte counts (NS), and superoxide dismutase (NS)	Tepaamorndech et al. (2019)
B. subtilis	White leg shrimp	28–29	1.0×10^6 CFU g^{-1} (60 days)		FW (+)	*V. parahaemolyticus* (+)	Prophenoloxidase gene (+), lipopolysaccharide, β-1,3-glucan-binding protein gene (+), and hemocyanin gene (+)	Interaminensea et al. (2019)

(Continued)

TABLE 11.1 (Continued)
Effect of *Bacillus* Probiotics on Growth Performance, Immune Response, and Disease Resistance in Shrimp Culture

Bacillus Species	Shrimp Species	Water Temp (°C)	Dosage/ Duration (day)	Route	Growth Factor	Disease/Stress Resistance	Immune Response	Reference
Bacillus licheniformis	*L. vannamei*	–	10^8 CFU/g (35 days)	Oral		–	Glutathione peroxidase, peroxidase, total antioxidant capacity, and superoxide dismutase (+)	Fan et al. (2021)
B. subtilis (Aqua grow)®	*L. vannamei*	28C	0.5, 1, 1.5 g/kg (56 days)	Oral	WG, SGR, FCR	*A. hydrophila, V. parahemolytica,* and *V. alginolyticus*	Phagocytosis, lysozyme activity, phenol oxidase, and SOD activity (+)	Eissa et al. (2022)
Encapsulated B. Subtilis E20 in alginate/chitosan coat	*L. vannamei*		10^7, 10^8, 10^9 cfu mL^{-1} (56 days)	Oral	↑WG, ↑SGR, ↓FCR, ↑PER	*V. parahaemolyticus*	WBC, NBT, bactericidal activity, and serum biochemical activity (+)	Adilah et al. (2022)
B. subtilis +B. Amyloliquefaciens coated with chitosan	*L. vannamei*	17.5 ± 1	5 weeks	Oral	↑WG, ↑SGR, ↓FCR, ↑PER		Phagocytic activity and immune-related genes ALF & prpPO (+)	Kewcharoen and Srisapoome (2022)
B. subtilis	*L. vannamei*	28–30	10^6 CFU/mL	Oral/ water	No effect	*V. harveyi*	SOD (oral supplementation), CAT, T-AOC (both groups), lysozyme, ACP, and AIP not affected	He et al. (2023)

(+) = positive effect, (-) = negative effect.

MSW = mean shrimp weight, RGW = relative gain weight, RGR = relative gain rate, FCR = food conversion ratio, SGR = specific growth rate, FCE = feed conversion efficiency, AGR = average growth rate, WG = weight gain, GR = growth rate, ABW = average body weight, ABL = average body length, ADG = average daily growth, LG = length gain, PL = post-larva.

Source: Adapted from Soltani et al. (2019).

Zokaeifar et al., 2014). Administration of commercial *Bacillus* probiotics in *L. vannamei* significantly improved weight gain, specific growth rate (SGR), and reduced food conversion ratio (FCR) compared to controls (Nimrat et al., 2012; Sadat Hoseini Madani et al., 2018). Inclusion of *Bacillus* sp. in the form of probiotics into the rearing water of *L. vannamei* significantly enhanced weight gain and decreased FCR of the treated animals (Bachruddin et al., 2018). Feed administration of *Bacillus* bacteria isolated from the intestine of healthy, wild shrimps in the growth of *L. vannamei* exhibited an increase in total weight, SGR, improved digestion and nutrient absorption of feed, and reduced FCR value (Kongnum et al., 2012; Chai et al., 2016). A 100-day feeding trial of *P. monodon* with *Bacillus* sp. also exhibited a higher growth rate than control shrimp (Rengpipat et al., 2003). In addition, survival and growth of giant freshwater prawns were increased after the prawns were fed *Bacillus* sp. either orally or as a bath immersion (Rahiman et al., 2010). Olmos et al. (2011) demonstrated a higher weight gain and a lower FCR in *L. vannamei* juvenile orally fed *B. subtilis* than both shrimps fed *B. megaterium* and basal diet (control), suggesting a selection of probiotic species should be considered as an important factor in probiotic therapy. Feeding *L. vannamei* with *B. flexus* (strain LD-1) or *B. licheniformis* (strain LS-1) alone or in a combination form increased weight gain and SGR as well as an improvement in water quality parameters, suggesting a beneficiary effect of bacilli probiotics to both host and the quality of rearing water (Cai et al., 2019). *B. subtilis* alone or in a mixture with *B. pumilus* (10^7 CFU/g feed) at a level of about 0.2% in the feed could improve the digestibility, growth performance, and innate immunity of *L. vannamei* and reduce FCR as well as a decrease in ammonia content of the culture water (Lee et al., 2021). For this shrimp species, the same amount of *B. velezensis* BV007 has been recommended as a dietary supplement for growth promotion and modulation of the shrimp gut microbiota (Chen et al., 2021). Synergistic positive effects of *B. subtilis* (ATCC 6633) and *Lactobacillus plantarum* (RITCC 1273) on the growth performance of *L. vannamei* postlarvae without influencing water quality features such as pH, ammonia nitrogen, nitrate nitrogen, and nitrite nitrogen have also been reported (Azhdari et al., 2023). Comparing four different *Bacillus* species including *B. subtilis, B. megaterium, B. cereus,* and *B. infantis* in *L. vannamei* showed that *B. subtilis* was superior in improving growth performance and digestive enzyme activities (Tamilarasu et al., 2021). This strain can also reduce dietary fishmeal requirements, as Tao et al. (2022) recently reported an improved growth performance of Pacific white shrimp fed 0.3% Gutcare® (*B. subtilis* DSM 32315), regardless of low or high fishmeal diets (5% or 20%). Cao et al. (2022) have reported that the 0.5% concentration of *B. subtilis* supplement after 30 days could significantly increase the weight gain, crude protein content of muscle, and a variety of enzyme indexes of *L. vannamei*, and it could also improve the composition of intestinal flora and adjust the number of dominant bacterial phyla. These benefits can be further enhanced by adding 0.5% yeast. In a study by Amoah et al. (2019), an eight-week feeding *L. vannamei* with *B. coagulans* exhibited an increase in the growth and intestine morphology. Dietary supplementation of *Paenibacillus polymyxa* also known as *Bacillus polymyxa* with an optimal dosage of 10^8 cfu/g feed in *L. vannamei* for 8 weeks has significantly increased final weight, SGR, protein efficiency rate (PER), and decreased FCR, as well as improved mid-intestinal morphological structures including villi height, villi width, and muscle thickness. The bacilli probiotic could also enhance the shrimp digestive enzyme activities including amylase, trypsin and lipase (Amoah et al., 2020). Probiotic *B. licheniformis* administrated at 6.0 mL/kg feed exhibited an improvement in the growth and moulting of blue swimming crablets (Boonyapakdee and Bhujel, 2020). In a study conducted by Huang et al. (2023), the use of two *Bacillus* strains, *B. subtilis* NT9, and *B. cereus* YB3 in the rearing water of a biofloc culture of *L. vannamei* improved growth performance of the shrimp and water quality parameters. However, the authors' results exhibited a better effect by *B. subtilis* on shrimp growth than *B. cereus*, while *B. cereus* was more effective in improving the water quality factors, that is, dissolved oxygen and total bacterial count. The bacilli bacteria were used for a 7-day biofloc pre-formation trial and during a 63-day shrimp culture. The addition of the commercial probiotic SANOLIFE®PRO-W containing *B. subtilis* and *B. licheniformis* in the rearing water of white leg shrimps not only improved the water quality parameters but also enhanced the growth performance, and the animal digestive enzymes including chymotrypsin, trypsin, protease, lipase, and amylase (Monier et al., 2023). *L. vannamei* fed *B. licheniformis* at 10^8 CFU/g feed for 35 days revealed a significant increase in weight gain, SGR, and reduced FCR (Fan et al., 2021). There was also an increase in amylase, lipase, trypsin, glutamine synthase, hexokinase, and malate dehydrogenase. In addition, values of glutathione peroxidase, peroxidase, total antioxidant capacity, and superoxide dismutase were improved in the treated shrimp. Further, exposing the treated shrimp to nitrite stress revealed a higher survival rate compared to control shrimp. In addition, 16S rRNA sequencing showed that *B. licheniformis* optimized the intestinal micro-ecological composition in particular, a decrease in populations of *Proteobacteria* and *Planctomycetes*, while an increase in densities of *Firmicutes* and *Bacteroidetes* was seen (Fan et al., 2021). The higher dose (1.5 g/kg feed) of dietary *B. subtilis* improved the growth performance, feed utilization, survival rate, and immune parameters of *L. vannamei* (Eissa et al., 2022).

Poly-β-hydroxybutyrate (PHB) is a bacterial storage compound deposited intracellularly in an amorphous state, and acts as a cellular energy and carbon reservation (Borah et al., 2002). Alongside the beneficiary effect of bacilli probiotics in shrimp growth performance, there are some reports suggesting an advantage of PHB as a bio-control agent for crustacean culture and demonstrated a promising effect against pathogenic bacteria such as *Vibrio* spp. (Jiang et al., 2008; Borah et al., 2002; Defoirdt et al., 2007; Wang et al., 2012; Sui et al., 2012). For instance, *B. megaterium* and *B. pasteurii* recovered from shrimp intestines accumulated PHB up to 79% on cell dry weight under an optimized condition (Singh et al., 2009; Kaynar & Beyatli, 2009). Laranja et al. (2014) exhibited that the PHB accumulating *Bacillus* species isolated from Philippine shrimp culture ponds were able to improve growth and survival of *Penaeus monodon* and also the level of ammonia stress (Laranja et al., 2014).

An eight-week dietary administration of encapsulated *B. subtilis* E20 in alginate/chitosan coat at 10^7 cfu/kg feed improved the growth of *L. vannamei* but the best results of growth performance were recorded in the shrimp fed with encapsulated probiotic at >10^8 cfu/kg feed (Adilah et al., 2022). The synbiotic mixture of *B. subtilis* and *B. amyloliquefaciens* coated with prebiotic chitosan and fed to Pacific white shrimp for 5 weeks upregulated the weight gain, length average daily gain, specific growth rate, and reduced feed conversion ratio (Kewcharoen and Srisapoome 2022). Such growth enhancement could be associated with two growth-related genes (Rap-a and GF-II) in the hepatopancreas and intestines of treated shrimp. Additionally, it significantly enhanced the hepatopancreas characteristics and epithelial and intestinal wall thickness.

Probiotics have been used to augment the biofloc technology system to achieve maximum benefits. Inclusion of *B. subtilis* in feed of *L. vannamei* in a biofloc system demonstrated a better result on shrimp performance, intestinal digestive enzyme activity, and non-specific immune enzyme activities than its inclusion in the rearing water, suggesting a better mode of action of *B. subtilis* when it is included in the feed than water column (He et al., 2023). Concerning water quality, the addition of *B. subtilis* did not also significantly affect the dynamic and mean concentrations of total ammonia nitrogen and total nitrogen. However, NO2-N was higher in the *B. subtilis* water-exposed group as this bacterium was found to be active in nitrification (He et al., 2023).

11.3 *BACILLUS* PROBIOTICS AS A PROMOTER OF SHELLFISH IMMUNITY AND DISEASE RESISTANCE

Advantages of probiotics on the immune responses and resistance to infectious diseases in shrimp have received great attention during the last two decades (Tseng et al., 2009; Kumar et al., 2016). Probiotic *Bacillus* have demonstrated a significant role in improving the health conditions of shrimp culture by stimulating the animal's innate immune responses that result in increasing in the survival rate toward infections agents (Rengpipat et al., 2000; Sanchez-Ortiz et al., 2016; Laranja et al., 2017).

In the earlier studies, a higher survival was seen in *Penaus monodon* received *Bacillus* probiotics and challenged with *Vibrio harveyi* infection (Rengpipat et al., 1998a; Rengpipat et al., 1998b). Parts of the mechanisms of disease resistance were understood following researches on the shrimp immune responses (Tseng et al., 2009; Li et al., 2009; Chandran et al., 2014 Zokaeifar et al., 2014; Sadat Hoseini Madani et al., 2018). In a study by Rengpipat et al. (2000), probiotic *Bacillus* S11 enhanced *P. monodon* immune responses through the activation of phenoloxidase, phagocytosis, and antimicrobial activity in the hemolymph and increased the shrimp resistance against *V. harveyi* infection (Rengpipat et al., 2000). Feeding *P. monodon* with *Bacillus* sp. caused a higher resistance against *V. harveyi* challenge (Rengpipat et al., 2003). In their study by Rahiman et al. (2010), an increase was seen in total hemocyte count, phenoloxidase, and respiratory burst activities of giant freshwater prawn fed a *Bacillus* sp. obtained from the same prawn species, suggesting a probiotic species-specific effect in this animal species. Li et al. (2009) demonstrated a decrease in total viable counts of bacteria and *Vibrio* bacteria in the intestine of *L. vannamei* fed with *Bacillus* probiotics as a supplemented diet. Guo et al. (2009) exhibited bath immersion of *L. vannamei* with *B. fusiformis* either daily or each day interval improved the animal survival during the larvi-culture stage. Various immune responses including phenoloxidase, lysozyme, respiratory burst, and bactericidal activity were enhanced in *P. monodon*-fed *B. cereus* in the form of probiotic and the treated shrimp exhibited higher survival after challenging with *V. harveyi* infection (Chandran et al., 2014). Some immune responses including lysozyme and pro-phenoloxidase I and II, survival rate, and stress tolerance to water temperature, salinity, and nitrite-N in *L. vannamei* were improved after administrating the shrimp with *B. subtilis* E20 isolated from fermented soya beans (Liu et al., 2010). Also, *L. vannamei* orally fed *B. subtilis* exhibited a higher survival and a higher stress tolerance to ammonia oxygen deficiency than control shrimps (Olmos et al., 2011).

In their study by Dong et al. (2014), the efficiency of *Bacillus* probiotic was assessed in kuruma shrimp (*Marsupenaeus japonicus*) juveniles against temperature stress, and results exhibited an increase in growth performance and immune responses as well as a reduction in the generated free radicals induced by insufficient oxygen metabolism. The use of *Bacillus* spp. including *B. subtilis* and *B. licheniformis* in the rearing water of *L. vannamei* culture enhanced shrimp immune responses including prophenoloxidase (ProPO), peroxinectin, lysozyme, lipopolysaccharide, serine protein, and β-1,3-glucan-binding protein (Zokaeifar et al., 2012a; Zokaeifar et al., 2014; Sadat Hoseini Madani et al., 2018). These probiotics also increased *L. vannamei* resistance against *V. harveyi* infection (Zokaeifar et al., 2012b; Zokaeifar et al., 2014). Administration of *B. subtilis* strain E20 isolated from fermented soyabeans increased the survival of *L. vannamei* against *V. alginolyticus* by enhancing the level of phenoloxidase and phagocytic activities, and *Bacillus* strain OJ exhibited an enhancement in shrimp immunity including phagocytosis and phenoloxidase, and a protected the animals toward white spot viral disease (Tseng et al., 2009; Li et al., 2009). *P. monodon* treated with cell-free extracts of *B. subtilis* strain BT23 significantly exhibited a decrease in mortality rate after a challenge with *V. harveyi* infection (Vaseeharan and Ramasamy, 2003). Growth performance and genes' expression of prophenoloxidase, β-1,3-glucan-binding protein, lipopolysaccharide, and hemocyanin were enhanced in *L. vannamei* treated with *B. subtilis* before and after challenging with *V. parahaemolyticus* infection (Interaminensea et al. (2019). In addition, *L. vannamei* fed with *B. coagulans* demonstrated an enhancement in immune responses and resistance to *V. parahemolyticus* infection (Amoah et al., 2019). In a study by Tepaamorndech et al. (2019), the administration of *B. aryabhattai* isolated from shrimp environment (sediment) in *L. vannamei* reduced the population of *Vibrio* bacteria in the shrimp gastrointestinal tract. Further, the treated shrimp revealed an increase in immunity, antioxidant activity, and resistance to *V. harveyi* challenge.

Studies have shown that *Bacillus* bacteria obtained from the same environment can enhance the cellular and humoral components of innate immunity in shrimp species. For example, immunological variables including lysozyme, respiratory burst, phenoloxidase, and bactericidal activity were enhanced in *P. monodon* treated with *B. cereus* (Chandran et al., 2014). The increase of such immunological variables was confirmed by demonstrating a higher survival in treated shrimp than control after challenging with *V. harveyi* infection (Chandran et al., 2014). In a study by Chai et al. (2016), *Bacillus* probiotic strain PC465, isolated from the gut of *Fenneropenaeus chinensis*, enhanced immune responses including penaeidin, thioredoxin, lectins, hemocyanin, crustin, ProPO, and peroxinectin and exhibited good protection toward infection by white spot viral disease in *L. vannamei* (Chai et al., 2016).

Mirbakhsh et al. (2021) showed a reasonable correlation between increased biological activities, growth performance, digestive enzyme activity, and immune response against the pathogenic bacteria and higher survival rate in the shrimp exposed to the probiotic *B. subtilis* IS02 challenged with the pathogen *V. harveyi*. A simultaneous upregulation of the immune-related gene of prophenoloxidase was also demonstrated. The results suggest that adding the probiotic during the hatchery stages and continuing its administration throughout the farming stages are necessary to maximize survival and growth in the shrimp culture.

As mentioned earlier, some *Bacillus* species such as *B. thuringiensis* carry poly-β-hydroxybutyrate (PHB) that is able to stimulate specific and non-specific immune functions in aquatic organisms (Suguna et al., 2014; Defoirdt et al., 2007). Such a beneficial role of PHB-accumulating has been reported by Laranja et al. (2014) who demonstrated its positive effect on immunity and survival of shrimp species against pathogenic microorganisms (Laranja et al., 2014). In addition, Laranja et al. (2017) exhibited an increased immune stimulation effect (relative expression of innate immune genes for ProPO, transglutaminase, and heat shock protein) by *Bacillus* strains with PHB in *P. mondon* postlarvae; feeding *P. mondon* postlarvae with *Bacillus* sp. JL47 enriched *Artemia* larvae; the treated shrimp showed resistance against *V. campbellii* challenge. In addition, the activities of some innate immune enzymes, digestive enzymes, stress tolerance, and disease resistance to *V. harveyi* infection have been shown in *L. vannamei* fed with *B. licheniformis* and *B. flexus* either alone or in a mixture form for three weeks at 28°C (Cai et al., 2019).

Application of *B. amyloliquefaciens* (up to 3.79×10^5 cfu/mL) in the rearing water in a biofloc system enhanced the immune system of *L. vannamei* by increasing the percentage of granular hemocytes and the concentration of total protein in the hemolymph and by decreasing the number of cells with apoptosis (Llario et al., 2020). Dietary administration of *B. subtilis* DSM 32315 (Gutcare®) significantly enhanced immunity and antioxidant capacity and improved intestinal function of *L. vannamei* (Tao et al., 2022). Also, a 12-week feeding white leg shrimp with *B. subtilis* reduced the load of *Vibrio* bacteria and improved the immunity, that is, lysozyme and phagocytic activity, of the animals (Lee et al., 2022*)*. Water application of a commercial *Bacillus* probiotic (SANOLIFE®PRO-W) containing *B. subtilis* and *B. licheniformis* up to 0.03 g/m^3 improved the innate immunity of white leg shrimp manifested by increasing activities of lysozyme, respiratory burst, prophenoloxidase, and the total hemocyte count, resulting in an enhancement of shrimp resistance toward *Fusarium solani* infection (Monier et al., 2023).

Encapsulation is an advanced technique used to improve probiotic viability and minimize sensitivity during processing, storage, and gastrointestinal exposure. For instance, a five-week dietary administration of *B. subtilis* and *B. amyloliquefaciens* coated with prebiotic chitosan in Pacific white shrimp elevated phagocytic activity and expression of immune-related genes of anti-lipopolysaccharide (ALF) and prophenoloxidase (prpPO) in the intestine. The treated group also significantly exhibited stronger resistance against *V. parahaemolyticus* (AHPND) with higher survivabilities of 41.7 to 52.8% versus 22.5% in control shrimp (Kewcharoen and Srisapoome 2022). An extended shelf-life and higher encapsulation survival of *B. subtilis* E20 has been demonstrated by Adilah et al. (2022) when exposed to adverse conditions. In addition, *L. vannamei* fed this bacilli probiotic showed a higher resistance to *Vibrio* infection at a dose of 10^7 cfu/kg feed in comparison to a higher dose of an unencapsulated probiotic (10^9 cfu/kg feed), which was required to increase the protective capacity (Adilah et al., 2022). Further, administration of encapsulated *B. subtilis* E20 increased population of beneficial bacteria such as *Bacillus*, while reduced harmful bacteria including *Vibrio* species in white leg shrimp, *L. vannamei* at the end of a 60-feeding trial (Cheng et al., 2023).

Dietary supplementation of certain strains of *Bacillus* spp. reduced mortality caused by bacterial infections, with stimulation of the host immune system and/or antagonism between bacteria as possible mechanisms. For instance, antimicrobial peptides (AMPs) including bacillomycin, fengycin, iturin, surfactin, bacilysin, and subtilin secreted by *Bacillus* spp. could inhibit the growth of other microorganisms, particularly bacteria and fungi (Sumi et al., 2015). The recent study by Proespraiwong et al. (2023) has shown that *Bacillus* spp. without AMP-related genes were unable to inhibit *V. parahaemolyticus* under *in vitro* condition, while other *Bacillus* spp. with at least two AMP-related genes exhibit various inhibitory activities. Strain K3 [B. subtilis (srfAA+ and bacA+)] isolated from shrimp remarkably inhibited *V. parahaemolyticus* infection in Pacific white shrimp (80% survival) and reduced shrimp mortality in different salinity ranges (75–95% survival). In addition, *B. subtilis* (K3) provided excellent protection against several strains of *V. parahaemolyticus* and shrimp survival remained stable in the groups tested (80-95% survival). *B. velezensis* (BV007, CGMCC No. 20039) isolated from the gut of *L. vannamei* can enhance the immune responses of this shrimp fed at a concentration of 10^7 CFU/g diet for 42 days and after a two-week exposure to *V. parahaemolyticus*, especially in terms of respiratory bursts and the activities of superoxide dismutase, catalase, and alkaline phosphatase (Chen et al., 2021). *B. licheniformis* ATCC 11946 at 10^8 CFU/g in feed can increase the survival rate of pacific white leg shrimp which was infected with *V. parahaemolyticus* (Amoah et al., 2020). Also, supplementing the diet of white leg shrimp with 1×10^8 CFU/g feed of *B. subtilis* WB60 reduced mortality in shrimp infected with *V. parahaemolyticus* (Won et al., 2020). The strain *B. velezensis* DH82 reduced the biomass accumulation of *V. parahaemolyticus* 17SZ, including the number of planktonic bacteria and biofilm formation in *in vitro* assays. It could inhibit quorum sensing regulation gene of *V. parahaemolyticus* to reduce pathogenicity by down-regulating the primary regulator AhpA, OpaR, and the virulence factor tlh. Dietary DH82 could also enrich their abundance, richness, and quantity in both the aquatic system and the intestine of *L. vannamei* and also inhibit the biomass of *Vibrio* bacteria, thereby reducing the damage to the non-specific immune system of alkaline phosphatase and superoxide dismutase activities and assisting the host in antibacterial activity, which may reduce the damage to the intestine, infected muscles, and survival rate in the challenge with *V. parahaemolyticus* infection (Sun et al., 2022). Strains of *B. subtilis* and *B. pumilus* showed inhibitory activity against 11 strains of *V. parahaemolyticus* and could effectively control the growth of *V. parahaemolyticus* in simulated aquaculture waste water when their concentration reached 1×10^7 CFU/mL (Jiang et al., 2023).

In contrast to bacterial infections such as vibriosis, for which the effects of probiotics are well documented, research is still needed to identify the right strains for viral diseases. Nevertheless, there are some evidence that probiotics may inhibit, reduce, or promote white spot virus, the cause of white spot viral disease resistance in shellfish. Dietary supplement of *B. amyloliquefaciens* significantly reduced the mortality of white spot viral disease in crayfish and reduced copy numbers of the virus (Lai et al., 2020). The dietary also increased the expression of several immune-related genes, including Toll-like receptor, C-type-lectin, and NF-κB, as well as effected total hemocyte count, phenoloxidase activity, and superoxide dismutase activity, and decreased hemocyte apoptosis in both infected and uninfected crayfish. In their study, Xu et al. (2021) isolated *B. amyloliquefaciens* A23 from the intestines of healthy red swamp crayfish and feeding the animal with a diet containing the isolated strain at 10^7 and 10^8 CFU/g feed for four weeks exhibited a successful ameliorated activity on innate immune enzymes and conferred protection against white spot viral disease. Further, the probiotic *B. subtilis* has also been engineered to produce dsRNAs against the cause of white spot viral disease using a gene transcript assay (Riet et al., 2021). The dsRNAs produced by the probiotic were efficient in inducing the RNAi mechanism in *L. vannamei* hemocytes, and the immunized shrimps showed a higher survival rate when they were challenged with the virus, with a significant reduction in viral load. Histological analysis also revealed no viral inclusions in the cuticle, gills, and gastric epithelium of the survivors.

11.4 CONCLUSION AND FUTURE REMARKS

Over last decades, the disease outbreaks especially by bacterial and viral agents have been remarkably increased in shrimp aquaculture, and administration of antibiotics in some regions where such diseases are prevalent has been largely increased. Hence, the misuse of antibiotics has developed some constraints including the development of bacterial resistance and antibiotic residuals in the carcas which is harmful to the both environment and human health. Therefore, in recent years, studies for friendly alternatives such as probiotics have been developed. Probiotics such as *Bacillus* as a dietary supplement can exclude the harmful bacteria by a bacterial competition action. Also, probiotic *Bacillus* are able to enhance the immune responses of shrimp without affecting its health status. In addition, studies have clearly exhibited a positive effect of probiotic *Bacillus* supplementation in shrimp feed that can significantly decrease disease outbreaks and enhance the activity of digestive enzymes, growth, and shrimp survival. However, bacilli probiotics must be applied in an appropriate way, that is, appropriate dosage for an appropriate period. In addition, studies of genetic, transcriptomic, and proteomic profiles of probiotic *Bacillus* are necessary to increase methods and comprehensive field applications of *Bacillus* use in shrimp farming. These findings, however, still require additional studies in the molecular pathways that regulate the mechanisms of bacilli probiotics that affect shrimp metabolism. Further, more studies are required to assess the efficacy and mode of actions of other naturally safe alternatives including para-probiotics, postbiotic, algae, and prebiotic in shrimp aquaculture.

BIBLIOGRAPHY

Adilah RN, Chiu ST, Hu SY, Ballantyne R, Happy N, Cheng AC, Liu CH (2022). Improvement in the probiotic efficacy of *Bacillus subtilis* E20-stimulates growth and health status of white shrimp, *Litopenaeus vannamei* via encapsulation in alginate and coated with chitosan. *Fish & Shellfish Immunology* **125**, pp.74–83.

Amoah K, Huang QC, Dong XH, Tan BP, Zhang S, Chi SY, Yang QH, Liu HY, Yang YZ (2020) *Paenibacillus polymyxa* improves the growth, immune and antioxidant activity, intestinal health, and disease resistance in *Litopenaeus vannamei* challenged with *Vibrio parahaemolyticus*. *Aquaculture* **518**, p.734563.

Amoah K, Huang QC, Tan BP, Zhang S, Chi SY, Yang QH, Liu HY, Dong XH (2019) Dietary sup-plementation of probiotic *Bacillus coagulans* ATCC 7050, improves the growth performance, intestinal morphology, microflora, immune response, and disease confrontation of Pacific white shrimp, *Litopenaeus vannamei*. *Fish & Shellfish Immunology* **87**, pp.796–808.

Azhdari S, Rezaei Tavabe K, Kazemzadeh Pournaki S, Hosseni S, Bagheri D, Javanmardi S, Azhdari A, Frinsko M. (2023). Effects of *Bacillus subtilis* and *Lactobacillus plantarum* probiotics on the *Litopenaeus vannamei* growth performance, hemolymph factors, and physicochemical parameters. *Aquaculture Reports*, **33**, p.101873.

Bachruddin M, Sholichah M, Istiqomah S, Supriyanto A (2018) Effect of probiotic culture water on growth, mortality, and feed conversion ratio of Vaname shrimp (*Litopenaeus vannamei* Boone). 7th ASEAN-FEN Int. Fish. Symposium IFS 2017 137. https://doi.org/10.1088/1755-1315/137/1/012036

Balcazar JL, de Blas I, Ruiz-Zaruela I, Cunningham D, Vendrell D, Muzquiz JL (2006) The role of probiotics in aquaculture. *Veterinary Microbiology* **114**, pp.173–186.

Banerjee S, Khatoon H, Shariff M, Yusoff FM (2010) Enhancement of *Penaeus monodon* shrimp post-larvae growth and survival without water exchange using marine *Bacillus pumilus* and periphytic microalgae. *Fisheries Science*, **76(3)**: pp.481–487.

Boonyapakdee A, Bhujel R, (2020). Determining the Dose of Dietary Probiotic (*Bacillus licheniformis*) for the Nursing of Blue Swimming Crablets (Portunus pelagicus, L, 1758). *Turkish Journal of Fisheries and Aquatic Sciences* **20**, pp.889–899. https://doi.org/10.4194/1303-2712-v20_12_05

Borah B, Thakur PS, Nigam JN (2002) The influence of nutritional and environmental conditions on the accumulation of poly-β-hydroxybutyrate in *Bacillus mycoides* RLJ B-017. *Journal of Applied Microbiology* **92**, pp.776–783.

Cai Y, Yuan W, Wang S, Guo W, Li A, Wu Y, et al. (2019) *In vitro* screening of putative probiotics and their dual beneficial effects: To white shrimp (*Litopenaeus vannamei*) postlarvae and to the rearing water. *Aquaculture*. doi:10.1016/j.aquaculture.2018.08.024

Cao H, Chen D, Guo L, Jv R, Xin Y, Mo W, Wang C, Li P, Wang, H. (2022). Effects of *Bacillus subtilis* on growth performance and intestinal flora of *Penaeus vannamei*. *Aquaculture Reports* **23**, p.101070.

Chai PC, Song XL, Chen GF, Xu H, Huang J (2016) Dietary supplementation of probiotic Bacillus PC465 isolated from the gut of *Fenneropenaeus chinensis* improves the health status and resistance of *Litopenaeus vannamei* against white spot syndrome virus. *Fish and Shellfish Immunology* **54**, pp.602–611. https://doi.org/10.1016/j.fsi.2016.05.011

Chandran MN, Iyapparaj P, Moovendhan Ramasubburayan R, Prakash S, Immanuel G, et al. (2014) Influence of probiotic bacterium *Bacillus cereus* isolated from the gut of wild shrimp *Penaeus monodon* in turn as a potent growth promoter and immune enhancer in *P. monodon. Fish & Shellfish Immunology* **36**, pp.38–45.

Chen L, Lv C, Li B, Zhang H, Ren L, Zhang Q, Zhang X, Gao J, Sun C, Hu S (2021) Effects of *Bacillus velezensis* supplementation on the growth performance, immune responses, and intestine microbiota of *Litopenaeus vannamei. Frontiers in Marine Science* **8**, p.744281.

Cheng A-C, Ballantyne R, Chiu S, Liu C. (2023). Microencapsulation of *Bacillus subtilis* E20 probiotic, a promising approach for the enrichment of intestinal microbiome in white shrimp, *Penaeus vannamei. Fishes* **8**(5): p.264. https://doi.org/10.3390/fishes8050264

Cutting SM (2011) *Bacillus* probiotics. Food Microbiol 28:214–220 Dalmin G, Kathiresan K, Purushothaman A (2001) Effect of probiotics on bacterial population and health status of shrimp in culture pond ecosystem. *Indian Journal of Experimental Biology* **39**, pp.939–942.

Daniels CL, Merrifield DL, Boothroyd DP, Davies SJ, Factor JP, Arnold KE (2010) Effect of dietary *Bacillus* spp. and mannan oligosaccharides (MOS) on European lobster (*Homarus gammarus* L.) larvae growth performance, gut morphology and gut microbiota. *Aquaculture* **304**, pp.49–57.

Davis RP, Boyd CE, Godumala R, Mohan ABC, Gonzalez A, Duy NP, Sasmita JPG, Ahyani N, Shatova O, Wakefield J, Harris B, McNevin AA, Davis DA (2022) Assessing the variability and discriminatory power of elemental fingerprints in whiteleg shrimp *Litopenaeus vannamei* from major shrimp production countries. *Food Control* **133**, p.108589.

Decamp O, Moriarty DJW, Lavens P (2008) Probiotics for shrimp larviculture: Review of field data from Asia and Latin America. *Aquaculture Research* **39**, pp.334–338. doi:10.1111/j.1365-2109.2007.01664.x

Defoirdt, T., Boon, N., Sorgeloos, P., Verstraete, W., and Bossier, P. (2007) Alternatives to antibiotics to control bacterial infections: Luminescent vibriosis in aquaculture as an example. *Trends in Biotechnology* **25**, pp.472–479.

Dong HB, Su YG, Mao Y, You XX, Ding SX, Wang J (2014) Dietary supplementation with *Bacillus* can improve the growth and survival of the kuruma shrimp *Marsupenaeus japonicus* in high-temperature environments. *Aquaculture International* **22**, pp.607–617.

Devaraja, T., Banerjee, S., Yusoff, F., Shariff, M., and Khatoon, H. A. 2013. A holistic approach for selection of *Bacillus* spp. as a bioremediator for shrimp post-larvae culture. *Turkish Journal of Biology*, **37**, pp.92–100 (2013). https://doi.org/10.3906/biy-1203-19

Eissa, E-S. H. Ahmed, El-Badawi, Munir, M.B., Abd Al-Kareem, O.M., Eissa, M.E.H., Hussien, E.H.M., Sakr, E.S.S. 2022. Assessing the influence of the inclusion of *Bacillus subtilis* AQUA-GROW® as feed additive on the growth performance, feed utilization, immunological responses and body composition of the Pacific white shrimp, *Litopenaeus vannamei*. Aquaculture Research, https://doi.org/10.1111/are.16129

Fan Y, Wang X, Wang Y, Ye H, Yu X, Wang S, Diao J, et al. (2021) Effect of dietary *Bacillus Licheniformis* on growth, intestinal health, and resistance to nitrite stress in Pacific white shrimp *Litopenaeus Vannamei. Aquaculture International* **29**(6): pp.2555–73, https://doi.org/10.1007/s10499-021-00764-9

FAO (2023) Information and analysis on markets and trade in fisheries and aquaculture products. www.fao.org/in-action/globefish/news-events/trade-and-market-news/q1-2023-janmar/en/#:~:text=World%20shrimp%20production%20reached%20a,9.4%20million%20tonnes%20in%202022

Fletcher R. (2021) Global Shrimp Production Sees Significant Growth in 2021. www.thefishsite.com/articles/global-shrimp-production-sees-significant-growth-in-2021-gorjan-nikolik-rabobank

Gao S, Pan L, Huang F, Song M, Tian C, Zhang M (2019) Metagenomic insights into the structure and function of intestinal microbiota of the farmed Pacific white shrimp (*Litopenaeus vannamei*). *Aquaculture* **499**, pp.109–118.

Geetha R, Ravisankar T, Patil PK, Avunje S, Vinoth S, Sairam CV, Vijayan KK (2020) Trends, causes, and indices of import rejections in international shrimp trade with special reference to India: A 15-year longitudinal analysis. *Aquaculture International* **28**(3): pp.1341–1369.

Gómez GD, Balcázar JL (2008) A review on the interactions between gut microbiota and innate immunity of fish. *FEMS Immunology and Medical Microbiology* **52**(2): pp.145–54.

Guo JJ, Liu KF, Cheng SH, Chang CI, Lay JJ, et al. (2009) Selection of probiotic bacteria for use in shrimp larviculture. *Aquaculture Research* **40**, pp.609–618.

Guo Q, Yao Z, Cai Z, Bai S, Zhang H (2022) Gut fungal community and its probiotic effect on *Bactrocera dorsalis*. *Journal of Insect Science* **29**(4): pp.1145–1158.

He X, Abakari G, Tan H, Liu W, Luo G (2023) Effects of different probiotics (*Bacillus Subtilis*) addition strategies on a culture of *Litopenaeus vannamei* in biofloc technology (BFT) aquaculture system. *Aquaculture* 566 (March). https://doi.org/10.1016/j.aquaculture.2022.739216

Hemdan EES, Ahmed NH, El-Badawi AA, Munir MB, Abd Al-Kareem OM, Eissa MEH, Hussien EHM, Sakr SES (2022) Assessing the influence of the inclusion of *Bacillus Subtilis* AQUA-GROW® as feed additive on the growth performance, feed utilization, immunological responses and body composition of the Pacific white shrimp, *Litopenaeus Vannamei. Aquaculture Research* **53**(18): pp.6606–15. https://doi.org/10.1111/are.16129

Hindu SV, Chandrasekaran N, Mukherjee A, Thomas J (2018) Effect of dietary supplementation of novel probiotic bacteria *Bacillus vireti* 01 on antioxidant defence system of freshwater prawn challenge with *Pseudomonas aeruginosa. Probiotics & Antimicrobial Proteins* **10**, pp.356–366.

Hu X, Xu Y, Su H, Xu W, Wen G, Xu C, Yang K, Zhang S, Cao Y (2023) Effect of a *Bacillus* probiotic compound on *Penaeus vannamei* survival, water quality, and microbial communities. *Fishes* **8**, p.362. https://doi.org/10.3390/fishes8070362

Hu XJ, Su HC, Xu Y, Xu WJ, Li SS, Huang XS, Cao YC, Wen GL (2020). Algicidal properties of fermentation products from *Bacillus cereus* strain JZBC1 dissolving dominant dinoflagellate species *Scrippsiella trochoidea, Prorocentrum micans*, and *Peridinium umbonatum. Biologia* **75**, pp.2015–2024.

Hu XJ, Wen GL, Xu WJ, Xu Y, Su HC, Yang K, Xu Y, Li ZJ, Cao YC (2019) Effects of the algicidal bacterium CZBC1 on microalgal and bacterial communities in shrimp culture. *Aquaculture Environment Interactions* **11**, pp.279–290.

Huang H, Li C, Lei Y, Zhou B, Kuang W, Zou W, Yang P (2023). Effects of *Bacillus* strain added as initial indigenous species into the biofloc system rearing *Litopenaeus vannamei* juveniles on biofloc preformation, water quality and shrimp growth. *Aquaculture* **569**, p.739375.

Interaminensea JA, Vogeleyb JL, Gouveiaa CK, Portelaa RS, Oliveirac JP, Silvad SMBC, et al. (2019) Effects of dietary *Bacillus subtilis* and *Shewanella algae* in expression profile of immune-related genes from hemolymph of *Litopenaeus vannamei* challenged with *Vibrio parahaemolyticus. Fish and Shellfish Immunology* **86**, pp.253–259.

Jiang, Y., Song, X., Gong, L., Li, P., Dai, C., and Shao, W. (2000) High poly(b-hydroxybutyrate) production by *Pseudomonas fluorescens* A2a5 from inexpensive substrates. *Enzyme and Microbial Technology*, **42**, pp.167–172.

Jiang N, Hong B, Luo K, Li Y, Fu H, Wang J. (2023). Isolation of *Bacillus subtilis* and *Bacillus pumilus* with Anti-*Vibrio parahaemolyticus* activity and identification of the anti-*Vibrio parahaemolyticus* substance. *Microorganisms*, **11**, p.1667. https://doi.org/10.3390/microorganisms11071667.

Jahangiri L, Esteban MÁ (2018) Administration of probiotics in the water in finfish aquaculture systems: A review. *Fishes* **3**(3): p.33.

Kerry RG, Patra JK, Gouda S, Park Y, Shin HS, Das G (2018) Benefaction of probiotics for human health: A review. *Journal of Food and Drug Analysis* **26**(3): pp.927–939.

Kewcharoen W, Srisapoome P (2022) Potential synbiotic effects of a *Bacillus mixture* and chitosan on growth, immune responses and VP(AHPND) resistance in Pacific white Shrimp (*Litopenaeus vannamei*, Boone, 1931). *Fish and Shellfish Immunology* **127**, pp.715–29. https://doi.org/10.1016/j.fsi.2022.07.017

Keysami MA, Mohammadpour M, Saad CR (2012) Probiotic activity of *Bacillus subtilis* in juvenile freshwater prawn, *Macrobrachium rosenbergii* (de Man) at different methods of administration to the feed. *Aquatic International* **20**, pp.499–511. https://doi.org/10.1007/s10499-011-9481-5

Kongnum K, Hongpattarakere T (2012) Effect of *Lactobacillus plantarum* isolated from digestive tract of wild shrimp on growth and survival of white shrimp (*Litopenaeus vannamei*) challenged with *Vibrio harveyi. Fish and Shellfish Immunology* **32**, pp.170–177.

Kaynar P, Beyatli Y Determination of poly-b-hydrox-ybutyrate production by *Bacillus* spp. isolated from the intestines of various fishes. *Fisheries Science*, **75**, pp.439–443.

Kumar V, Roy S, Meena DK, Sarkar UK (2016) Application of probiotics in shrimp aquaculture: Importance, mechanisms of action, and methods of administration. *Reviews in Fishery Science and Aquaculture* **24**, pp.342–368.

Lai Y, Luo M, Zhu F (2020). Dietary *Bacillus amyloliquefaciens* enhance survival of white spot syndrome virus infected crayfish. *Fish & Shellfish Immunology* **102**, pp.161–168.

Lara-Anguiano GF, Esparza-Leal HM, Sainz-Hernandez JC, Ponce-Palafox JT, Valenzuela-Quinonez W, Apun-Molina JP, Klanian MG (2013) Effects of inorganic and organic fertilization on physicochemical barameters, bacterial concentrations, and shrimp growth in *Litopenaeus vannamei* cultures with zero water exchange. *Journal of the World Aquaculture Society* **44**(4): pp.499–510.

Laranja JLQ, Amar EC, Ludevese-Pascual GL, Niu Y, Geaga MJ, De Schryver P, Bossier P, (2017) A probiotic *Bacillus* strain containing amorphous poly-beta-hydroxybutyrate (PHB) stimulates the innate immune response of *Penaeus monodon* postlarvae. *Fish and Shellfish Immunology* **68**, pp.202–210. https://doi.org/10.1016/j.fsi.2017.07.023

Laranja JLQ, Ludevese-Pascual GL, Amar EC, Sorgeloos P, Bossier P, De Schryver P (2014) Poly-β-hydroxybutyrate (PHB) accumulating Bacillus spp. improve the survival, growth and robustness of *Penaeus monodon* (Fabricius, 1798) postlarvae. *Veterinary Microbiology* **173**, pp.310–317. https://doi.org/10.1016/j.vetmic.2014.08.011

Lee C, Kim S, Shin J, Kim MG, Gunathilaka BE, Kim SH, Kim JE, Ji SC, Han JE, Lee KJ (2021). Dietary supplementations of *Bacillus* probiotic improve digestibility, growth performance, innate immunity, and water ammonia level for Pacific white shrimp, *Litopenaeus vannamei. Aquaculture International* **29**, pp.2463–2475.

Lee JW, Chiu ST, Wang ST, Liao YC, Chang HT, Ballantyne R, Lin JS, Liu CH (2022) Dietary SYNSEA probiotic improves the growth of white shrimp, *Litopenaeus vannamei* and reduces the risk of Vibrio infection via improving immunity and intestinal microbiota of shrimp. *Fish and Shellfish Immunology* **127**, pp.482–491.

Li E, Xu C, Wang X, Wang S, Zhao Q, Zhang M et al. (2018) Gut microbiota and its modulation for healthy farming of Pacific white shrimp *Litopenaeus vannamei. Reviews in Fisheries Science & Aquaculture* **26**, pp.381–399.

Li J, Tan B, Mai K (2009) Dietary probiotic *Bacillus* OJ and isomaltooligosaccharides influence the intestine microbial populations, immune responses and resistance to white spot syndrome virus in shrimp (*Litopenaeus vannamei*). *Aquaculture* **291**, pp.35–40. https://doi.org/10.1016/j.aquaculture.2009.03.005

Liu CH, Chiu CS, Ho PL, Wang SW (2009) Improvement in the growth performance of white shrimp, *Litopenaeus vannamei*, by a protease-producing probiotic, *Bacillus subtilis* E20, from natto. *Journal of Applied Microbiology* **107**, pp.1031–1041. https://doi.org/10.1111/j.1365-2672.2009.04284.x

Liu KF, Chiu CH, Shiu YL, Cheng W, Liu CH (2010) Effects of the probiotic, *Bacillus subtilis* E20, on the survival, development, stress tolerance, and immune status of white shrimp, *Litopenaeus vannamei* larvae. *Fish and Shellfish Immunology* **28**, pp.837–844.

Llario F, Romano LA, Rodilla M, Sebastia-Frasquet MT, Poersch LH. (2020). Application of *Bacillus amyloliquefaciens* as probiotic for *Litopenaeus vannamei* (Boone, 1931) cultivated in biofloc system. *Iranian Journal of Fisheries Sciences* **19**, pp.904–920.

Mirbakhsh M, Mahjoub M, Afsharnasab M, Kakoolaki S, Sayyadi M, Hosseinzadeh H (2021) Effects of *Bacillus subtilis* on the water quality, stress tolerance, digestive enzymes, growth performance, immune gene expression, and disease resistance of white shrimp (*Litopenaeus vannamei*) during the early hatchery period. *Aquaculture International* **29**(6): pp.2489–2506. https://doi.org/10.1007/s10499-021-00758-7

Mongkol P, Bunphimpapha P, Rungrassamee W, Arayamethakorn S, Klinbunga S, Menasveta P et al. (2018) Bacterial community composition and distribution in different segments of the gastrointestinal tract of wild-caught adult *Penaeus monodon. Aquaculture Research* **49**, pp.378–392.

Monier M, Kabary H, Elfeky A, Saadony S, El-Hamed N, Eissa M, Eissa E. 2023. The effects of Bacillus species probiotics (*Bacillus subtilis* and *B. licheniformis*) on the water quality, immune responses, and resistance of whiteleg shrimp (*Litopenaeus vannamei*) against Fusarium solani infection. *Aquaculture International* **31**, pp.3437–3455. https://doi.org/10.1007/s10499-023-01136-1

Moriarty DJW (1998) Control of luminous *Vibrio* species in penaeid aquaculture ponds. *Aquaculture* **164**, pp.351–358.

Moriarty DJW (1999) Disease control in shrimp aquaculture with probiotic bacteria. In: *Proceedings of the Eighth International Symposium on Microbial Ecology* (ed. Bell CR, Brylinsky M, Johnson-Green P) pp.237–244. Halifax, NS, Canada: Atlantic Canada Society for Microbial Ecology.

Ng W-K, Koh C-B (2016) The utilization and mode of action of organic acids in the feeds of cultured aquatic animals. *Reviews in Aquaculture*. https://doi.org/10.1111/raq.12141

Nimrat S, Suksawat S, Boonthai T, Vuthiphandchai V (2012) Potential *Bacillus* probiotics enhance bacterial numbers, water quality and growth during early development of white shrimp (*Litopenaeus vannamei*). *Veterinary Microbiology* **159**, pp.443–450.

Olmos J, Ochoa L, Paniagua-Michel J, Contreras R (2011) Functional feed assessment on *Litopenaeus vannamei* using 100% fish meal replacement by soybean meal, high levels of complex carbohydrates and *Bacillus* probiotic strains. *Marine Drugs* **9**, pp.1119–1132.

Olmos SJ (2021) Feed intake improvement, gut microbiota modulation and pathogens control by using *Bacillus* species in shrimp aquaculture. *World Journal of Microbiology and Biotechnology*. Springer Science and Business Media B.V. https://doi.org/10.1007/s11274-020-02987-z

Pérez T, Balcázar JL, Ruiz-Zarzuela I, Halaihel N, Vendrell D, De Blas I et al. (2010) Host-microbiota interactions within the fish intestinal ecosystem. *Mucosal Immunology* **3**, pp.355–360.

Proespraiwong P, Mavichak R, Imaizumi K, Hirono I, Unajak S (2023) Evaluation of *Bacillus* spp. as potent probiotics with reduction in AHPND-Related mortality and facilitating growth performance of Pacific white shrimp (*Litopenaeus vannamei*) farms. *Microorganisms* **11**, p.2176. https://doi.org/10.3390/microorganisms11092176

Rahiman KMM, Jesmi Y, Thomas AP, Hatha AAM (2010) Probiotic effect of *Bacillus* NL110 and *Vibrio* NE17 on the survival, growth performance and immune response of *Macrobrachium rosenbergii* (de Man). *Aquaculture Research* **41**, pp.120–134 doi:10.1111/j.1365-2109.2009.02473.x

Rengpipat S, Phianphak W, Piyatiratitivorakul S, Menasveta P (1998b) Effects of a probiotic bacterium on black tiger shrimp *Penaeus monodon* survival and growth. *Aquaculture* **167**, pp.301–313. https://doi.org/10.1016/S0044-8486(98)00305-6

Rengpipat S, Rukpratanporn S, Piyatiratitivorakul S, Menasaveta P (2000) Immunity enhancement in black tiger shrimp (*Penaeus monodon*) by a probiont bacterium (*Bacillus* S11). *Aquaculture* **191**, pp.271–288.

Rengpipat S, Rukpratanporn S, Piyatiratitivorakul S, Menasveta P (1998a) Probiotics in aquaculture: A case study of probiotics for larvae of the black tiger shrimp (*Penaeus monodon*). In: Flegel TW (ed) *Advances in shrimp biotechnology*, National Center for Genetic Engineering and Biotechnology, Bangkok, pp.177–181.

Rengpipat S, Tunyanun A, Fast AW, Piyatiratitivorakul S, Menasveta, P (2003) Enhanced growth and resistance to *Vibrio* challenge in pond-reared black tiger shrimp *Penaeus monodon* fed a *Bacillus* probiotic. *Diseases of Aquatic Organisms* **55**, pp.169–173.

Riet J, Costa-Filho J, Dall'Agno L, Medeiros L, Azevedo R, Nogueira LF, Marins LF (2021). Bacillus subtilis expressing double-strand RNAs (dsRNAs) induces RNA interference mechanism (RNAi) and increases survival of WSSV-challenged *Litopenaeus vannamei. Aquaculture* **541**, p.736834. https://doi.org/10.1016/j.aquaculture.2021

Sanchez-Ortiz AC, Angulo C, Luna-Gonzalez A, Alvarez-Ruiz P, Mazon-Suastegui JM, and Campa-Cordova AI Effect of mixed-*Bacillus* spp isolated from pustulose ark *Anadara tuberculosa* on growth, survival, viral prevalence and immune-related gene expression in shrimp *Litopenaeus vannamei. Fish & Shellfish Immunology* **59**, pp.95–102 (2016).

Singh, M., Patel, S. K., and Kalia, V. C. Bacillus subtilis as potential producer for polyhydroxyalkanoates. *Microbial Cell Factories* **8**, p.38 (2009). https://doi.org/10.1186/1475-2859-8-38

Sadat Hoseini Madani N, Adorian TJ, Ghafari Farsani H, Hoseinifar SH (2018) The effects of dietary probiotic bacilli (*Bacillus subtilis* and *Bacillus licheniformis*) on growth performance, feed efficiency, body composition and immune parameters of whiteleg shrimp (*Litopenaeus vannamei*) postlarvae. *Aquaculture Research* **49**, pp.1926–1933. https://doi.org/10.1111/are.13648

Soltani M, Ghosh K, Dutta D, Ringø E (2024) Prebiotics and probiotics as effective immunomodulators in aquaculture. In: Elumalai P, Soltani M, Lakshmi S (Eds.), *Immunomodulators in Aquaculture and Fish Health.* CRC Press, pp.136–168.

Sreenivasulu P, Suman Joshi DSD, Narendra K, Venkata Rao G, Krishna Satya A (2016) *Bacillus pumilus* as a potential probiotic for shrimp culture. *International Journal of Fisheries and Aquatic Studies* **4**(1): pp.107–110.

Suguna P, Binuramesh C, Abirami P, Saranya V, Poornima K, Rajeswari V, Shenbagarathai R (2014) Immunostimulation by poly-b hydroxybutyrate-hydroxyvalerate (PHB-HV) from *Bacillus thuringiensis* in *Oreochromis mossambicus. Fish & Shellfish Immunology* **36**, pp.90–97. https://doi.org/10.1016/j.fsi.2013.10.012

Sumi CD, Yang BW, Yeo In-Cheol Hahm YT (2015). Antimicrobial peptides of the genus *Bacillus*: A new era for antibiotics. *Canadian Journal of Microbiology* **61**, pp.93–103.

Sui L, Cai J, Sun H, Wille M, and Bossier P (2012) Effect of poly-b-hydroxybutyrate on Chinese mitten crab, *Eriocheir sinensis*, larvae challenged with pathogenic *Vibrio anguillarum. The Journal of Fish Disease* **35**, pp.359–364 (2012). https://doi.org/10.1111/j.1365-2761.2012.01351

Sun X, Liu J, Deng S, Li R, Lv W, Zhou S, Tang X, Sun Y-z, Ke M, Wang K (2022) Quorum quenching bacteria *Bacillus velezensis* DH82 on biological control of *Vibrio parahaemolyticus* for sustainable aquaculture of *Litopenaeus vannamei. Frontiers in Marine Science* **9**, p.780055. https://doi.org/10.3389/fmars.2022.780055

Sun Y, Wen Z, Li X, Meng N, Mi R, Li Y, Li S (2012) Dietary supplement of fructooligosaccharides and *Bacillus subtilis* enhances the growth rate and disease resistance of the sea cucumber *Apostichopus japonicus* (Selenka). *Aquaculture Research* **43**, pp.1328–1334.

Tamilarasu A, Ahilan B, Gopalakannan A, Somu Sunder Lingam R. (2021). Evaluation of probiotic potential of Bacillus strains on growth performance and physiological responses in *Penaeus vannamei. Aquaculture Research* **52**, pp.3124–3136. https://doi.org/10.1111/are.15159

Toledo A, Frizzo L, Signorini M, Bossier P, Arenal A. 2019. Impact of probiotics on growth performance and shrimp survival: A meta-analysis. *Aquaculture* **500**(1–4): pp.196–205.

Tao X, He J, Lu J, Chen Z, Jin M, Jiao L, Masagounder K, Liu W, Zhou Q (2022) Effects of *Bacillus subtilis* DSM 32315 (Gutcare®) on the growth performance, antioxidant status, immune ability and intestinal function for juvenile *Litopenaeus vannamei* fed with high/low-fishmeal diets. *Aquaculture Reports* **26**, p.101282.

Tepaamorndech S, Chantarasakha K, Kingcha Y, Chaiyapechara S, Phromson M, Sriariyanun M, Kirschke CP, Huang L, Visessanguan W (2019) Effects of *Bacillus aryabhattai* TBRC8450 on vibriosis resistance and immune enhancement in Pacific white shrimp, *Litopenaeus vannamei. Fish and Shellfish Immunology* **86**, pp.4–13. https://doi.org/10.1016/j.fsi.2018.11.010

Tseng DY, Ho PL, Huang SY, Cheng SC, Shiu YL, Chiu CS, Liu CH (2009) Enhancement of immunity and disease resistance in the white shrimp, *Litopenaeus vannamei*, by the probiotic, *Bacillus subtilis* E20. *Fish Shellfish Immunology* **26**, pp.339–344. https://doi.org/10.1016/j.fsi.2008.12.003

Tzeng T-D, Pao Y-Y, Chen P-C, Weng FC-H, Jean WD, Wang D (2015) Effects of host phylogeny and habitats on gut microbiomes of oriental river prawn (*Macrobrachium nipponense*). *PLOS One* **10**, p.e0132860.

Vargas-Albores F, Porchas-Cornejo MA, Martinez-Porshas M, Villalpando E, Gollas-Galván T, Martinez-Córdova LR (2017) Bacterial biota of shrimp intestine is significantly modified by the use of a probiotic mixture: A high throughput sequencing approach. *Helgoland Marine Research* **71**, p.5.

Vaseeharan BARP, Ramasamy P (2003) Control of pathogenic *Vibrio* spp. by *Bacillus subtilis* BT23, a possible probiotic treatment for black tiger shrimp *Penaeus monodon. Letters in Applied Microbiology* **36**, pp.83–87.

Velmurugan S, Palanikumar P, Velayuthani P, Donio MBS, Michael Babu M, Lelin C, et al. (2015) Bacterial white patch disease caused by *Bacillus cereus*, a new emerging disease in semi-intensive culture of *Litopenaeus vannamei. Aquaculture* **444**, pp.49–54.

Wang B, Sharma-Shivappa RR, Olson JW, Khan SA (2012) Upstream process optimization of polyhydroxybutyrate (PHB) by Alcaligenes latus using two-stage batch and fed-batch fermentation strategies. *Bioprocess and Biosystems Engineering* **35**, pp.1591–1602.

Wang YB (2007) Effect of probiotics on growth performance and digestive enzyme activity of the shrimp *Penaeus vannamei. Aquaculture* **269**, pp.259–264.

Wang YG, Lee KL, Najiah M, Shariff M, Hassan MD (2000) A new bacterial white spot syndrome (BWSS) in cultured tiger shrimp *Penaeus monodon* and its comparison with white spot syndrome (WSS) caused by virus. *Diseases of Aquatic Organisms* **41**, pp.9–18.

Won S, Hamidoghli A, Choi W, Bae JH, Jang WJ, Lee S, Bai SC (2020) Evaluation of Potential Probiotics *Bacillus subtilis* WB60, *Pediococcus pentosaceus*, and *Lactococcus lactis* on Growth Performance, Immune Response, Gut Histology and Immune-Related Genes in Whiteleg Shrimp, *Litopenaeus vannamei. Microorganisms* **8**, p.15. doi: 10.3390/microorganisms8020281

Wu DX, Zhao SM, Peng N, Xu CP, Wang J, Liang YX (2016) Effects of a probiotic (*Bacillus subtilis* FY99-01) on the bacterial community structure and composition of shrimp (*Litopenaeus vannamei*, Boone) culture water assessed by denaturing gradient gel electrophoresis and high-throughput sequencing. *Aquaculture Research* **47**, pp.857–869 doi:10.1111/are.12545

Xiong J, Zhu J, Dai W, Dong C, Qiu Q, Li C (2017) Integrating gut microbiota immunity and disease-discriminatory taxa to diagnose the initiation and severity of shrimp disease. *Environmental Microbiology* **19**, pp.1490–1501.

Xu L, Yuan J, Chen X, Zhang S, Xie M, Chen C, Wu Z (2021) Screening of intestinal probiotics and the effects of feeding probiotics on the digestive enzyme activity, immune, intestinal flora and WSSV resistance of *Procambarus clarkii. Aquaculture* **540**, p.736748.

Zhu J, Dai W, Qiu Q, Dong C, Zhang J, Xiong J (2016) Contrasting ecological processes and functional compositions between intestinal bacterial community in healthy and diseased shrimp. *Microbial Ecology* **72**, pp.975–985.

Zokaeifar H, Babaei N, Saad CR, Kamarudin MS, Sijam K, Balcazar JL (2014) Administration of *Bacillus subtilis* strains in the rearing water enhances the water quality, growth performance, immune response, and resistance against *Vibrio harveyi* infection in juvenile white shrimp, *Litopenaeus vannamei. Fish and Shellfish Immunology* **36**, pp.68–74.

Zokaeifar H, Balcázar JL, Kamarudin MS, Sijam K, Arshad A, Saad CR (2012a) Selection and identification of non-pathogenic bacteria isolated from fermented pickles with antagonistic properties against two shrimp pathogens. *Journal of Antibiotics* **65**, pp.289–294.

Zokaeifar H, Balcázar JL, Saad CR, Kamarudin MS, Sijam K, Arshad A, Nejat N (2012b) Effects of *Bacillus subtilis* on the growth performance, digestive enzymes, immune gene expression and disease resistance of white shrimp, *Litopenaeus vannamei. Fish and Shellfish Immunology* **33**, pp.683–689.

12 *Bacillus* as Paraprobiotics in Aquaculture

Dibyendu Kamilya
Indian Institute of Technology Kharagpur, Kharagpur, India

Tanmoy Gon Choudhury
Central Agricultural University, Lembucherra, West Tripura, India

12.1 INTRODUCTION

The use of probiotics as an efficient and environment-friendly solution to control infectious diseases in aquaculture is well recognized and has been demonstrated in a plethora of scientific researches (Merrifield et al. 2010; Nayak 2010; Prado et al. 2010; Aguirre-Guzmán et al. 2012; Mohapatra et al. 2013; Dawood et al. 2019). The application of probiotics in aquaculture has gained substantial momentum, particularly after apprehending the deleterious effects of drugs and chemicals, and subsequent regulatory restrictions on their use in the aquaculture sector. Although several definitions have been proposed for probiotics, the most commonly accepted definition is 'live microorganisms that when administered in adequate amounts confer health benefits on the host' (FAO/WHO 2001). This definition has been slightly modified (grammatically) in a recommendation by a consensus panel of the 'International Scientific Association for Probiotics and Prebiotics (ISAPP)' as 'live microorganisms that, when administered in adequate amounts, confer a health benefit on the host' (Hill et al. 2014). In all the definitions proposed so far, the provision that the probiotic microorganisms must be in 'live' condition to obtain health benefits has been approved by the scientific community (Gobbetti et al. 2010).

Although the beneficial effects of probiotic microbes in food, animal, and the aquaculture sector have been well researched and documented, the use of such microbes in 'live' conditions is not without problems and concerns. Some of the important apprehensions include the viability of the probiotic strain in feed or product, varying degrees of colonization and persistence in the intestinal milieu, and the probability of acquiring virulence genes from pathogenic bacteria via horizontal gene transfer (Choudhury & Kamilya 2019). Alongside, recent research outcomes suggest that the probiotics in the non-viable state are also beneficial to the host similar to their live counterparts (Kataria et al. 2009; Adams 2010; Almada et al. 2016; Choudhury & Kamilya 2019). These situations have incited using non-viable probiotic preparations as an alternative to using 'live' probiotic microorganisms. Such non-viable probiotics, termed paraprobiotics, are gaining increasing importance in food production sectors to obtain beneficial effects (Kataria et al. 2009; Adams 2010; de Almada et al. 2016).

Given the above, this chapter delineates an overview of the scientific studies in which paraprobiotics have been investigated to realize beneficial effects in the aquaculture sector. The concept of paraprobiotics, methods used for their preparation, and the effects of paraprobiotics on immune response, disease resistance, and growth performance of reared animals in aquaculture are discussed with special reference to one of the most extensively studied probiotics in aquaculture, *Bacillus* spp.

DOI: 10.1201/9781003503811-12

12.2 CONCEPT OF PARAPROBIOTICS IN AQUACULTURE

The term 'paraprobiotic' (or 'ghost probiotics') was proposed by Taverniti and Guglielmetti (2011) to indicate the use of inactivated (dead) microbial preparations as health-promoting agents and it was defined as 'non-viable microbial cells (intact or broken) or crude cell extracts (i.e., with complex chemical composition), which, when administered (orally or topically) in adequate amounts, confer a benefit on the human or animal consumer'. The prefix 'para' was chosen by the authors because of the meaning of this term as 'alongside of' or 'atypical' (in Greek), which can simultaneously indicate similarities to and differences from the classical definition of probiotics. Moreover, 'purified molecule of the microbial origin or pure microbial cell products' were omitted from the paraprobiotics concept, since their use is more pertinent in general pharmaceutical concepts. Additionally, after the demonstration of the health benefits of an inactivated microbial preparation, the assignation of that product into the paraprobiotic category should not be influenced by the methods used for inactivation (Taverniti & Guglielmetti 2011).

The health-promoting beneficial effects of paraprobiotic have been relatively well demonstrated in higher vertebrate models (Kataria et al. 2009; Adams 2010; de Almada et al. 2016). However, the concept of paraprobiotic application in the aquaculture sector is still in its infancy. In light of the emerging concerns about the administration of live microbial cells and recent scientific evidence indicating the potential advantages of non-viable microbes, the concept of paraprobiotics has also been gaining momentum in the aquaculture sector (Singh et al. 2017; Choudhury & Kamilya 2019).

12.3 PROBLEMS ASSOCIATED WITH THE ADMINISTRATION OF LIVE BENEFICIAL PROBIOTICS

Concerns are being raised regarding the safety and other problems associated with the extensive use of live microbial cells. For instance, the administration of viable probiotics to individuals with weaker immune systems was found to enhance inflammatory responses and/or to compromise mucosal barrier functions, thereby turning the harmless probiotic bacteria into detrimental microorganisms (Besselink et al. 2008). Another safety-related problem is the probability of acquiring virulence genes by probiotics from pathogenic bacteria through horizontal gene transfer (Newaj-Fyzul et al. 2014). Some practical applicability and functionality issues are also associated with live probiotic use. In the food production sector, the viability of probiotic bacteria is impacted by storage conditions and many probiotic bacteria are prone to lose the required viability during storage (Nayak 2010; Das et al. 2013). After the fresh probiotic preparation, the proportion of non-viable bacteria generally increases over time and any beneficial effect which is supposed to be derived from the viable probiotic bacteria may underestimate the effects that have been derived from the presence of the non-viable bacterial portion (Adams 2010; Das et al. 2013; Dash et al. 2015). To function optimally, the probiotic bacteria should be able to colonize and persist in the hostile environment of the intestine of the host (Ottesen & Olafsen 2000). To achieve this, the probiotic bacteria ideally should belong to a suitable strain specific to the host which is difficult to accomplish in practice (Adams 2010). Besides, the timing of probiotic administration (Ringo & Gatesoupe 1998; Balcázar et al. 2006) and temporary colonization by certain probiotic microbes (Nikoskelainen et al. 2003; Vieira et al. 2008; Son et al. 2009; Dash et al. 2014) are some of the constraints of the live probiotic application.

12.4 INACTIVATION METHODS

The commonly used methods for inactivating viable probiotic microorganisms are thermal treatment, irradiation, ultraviolet (UV) rays, high pressure, and sonication (de Almada et al. 2016). Published literature on paraprobiotic application in aquaculture indicates the use of heat, UV rays, formalin, and sonication as the usual methods for the inactivation of viable probiotics. These methods can kill microorganisms by several mechanisms. However, the inactivation method should be capable of retaining the beneficial attributes of the probiotic microorganisms in paraprobiotics (Raz & Rachmilewitz 2005).

12.4.1 Heat

Thermal treatment comprises of application of heat for a certain period of time to kill microorganisms. The effectiveness of heat inactivation is influenced by several factors including the type of microorganism, growth medium, stage of growth, the form of microorganism (vegetative or spore), pH, water activity, and mode of heating, among others (Juneja & Sofos 2002). The mechanisms by which microorganisms are inactivated include membrane damage, loss of nutrients and ions, ribosome aggregation, rupture of DNA filaments, inactivation of enzymes, and protein coagulation (Gould 1989; de Almada et al. 2016). Heat-treated bacterial suspension is a source of different soluble remnant molecules including parts of the cell wall components and bacterial DNA (Ivec et al. 2007). It has been observed that the CpG motif of bacterial DNA remains preserved even in dead bacteria after heat treatment (Jozsef et al. 2004). Many cell wall constituents including bacterial CpG motif have been reported to stimulate the immune response (Tassakka et al. 2002; Lee et al. 2003; Jozsef et al. 2004; Ivec et al. 2007; Taverniti & Guglielmetti 2011). Heat-inactivated probiotics, thus, can stimulate immune responses via such cellular components. However, the correlation between the inactivation condition and the expected beneficial outcomes should be critically evaluated before the production of paraprobiotics. For instance, bacterial cell coarseness, roughness, and adhesion capacities, which are important attributes in influencing beneficial effects, can become subdued if the inactivation temperature increases (Ou et al. 2011; Bermudez-Brito et al. 2012). In aquaculture research, *Bacillus* paraprobiotics have been prepared using different thermal conditions including 60°C for 1 h (Salinas et al. 2006, 2008), 60°C for 1 h (Kamilya et al. 2015; Singh et al. 2017, Midhun et al. 2018), autoclaving the culture at 121°C for 15 min (Hasan et al. 2019), and 95°C for 60 min (Yan et al. 2016; Wang et al. 2008).

12.4.2 Ultraviolet (UV) Rays

UV rays are part of the electromagnetic spectrum (200–400 nm) and can effectively inactivate a wide range of bacterial cells and spores (Mukhopadhyay & Ramaswamy 2012; Gayan et al. 2013). Exposure to UV rays causes bacterial proteins to denature and formation of DNA photoproducts. Among these mechanisms, the most significant is the formation of pyrimidine dimer on the same DNA strand that interrupts the transcription and translation processes leading to mutagenesis and cell death (Franz et al. 2009). There are limited works that have reported the inactivation of probiotic microorganisms by UV exposition to produce paraprobiotics. In one study, the inactivation of *Bacillus* probiotics was accomplished with a UV exposition time of 2.5 h (Kamilya et al. 2015).

12.4.3 Formalin

Formalin (37% aqueous solution of formaldehyde gas) inactivates microorganisms by alkylating the amino and sulfhydral groups of proteins and ring nitrogen atoms of purine bases (Favero & Bond 1991). Different concentrations of formalin with varying duration and temperature have been used to inactivate probiotics for use in aquaculture. In one study, *Bacillus* probiotics were killed by adding 2.0% (v/v) formalin for 48 h (Newaj-Fyzul et al. 2007). *Bacillus* probiotics were also inactivated by adding formalin to a final concentration of 1.0% (v/v) and incubating for 24 h at 4°C (Kamilya et al. 2015).

12.4.4 Sonication

Sonication is a physical method that involves the use of ultrasound waves (>16 kHz) to disrupt intermolecular interactions (Gibson et al. 2008). Sonication inactivates microbes by cell wall rupture, perturbations and thinning of the cell membranes, and DNA damage (Ross et al. 2003; Birmpa et al. 2013; Singh 2013). In one study, *B. subtilis* was inactivated by sonication to prepare the paraprobiotic (Newaj-Fyzul et al. 2007). This method has not been used much for the preparation of paraprobiotic.

12.5 APPLICATION OF *BACILLUS* AS PARAPROBIOTICS IN AQUACULTURE

The paraprobiotic concept is relatively new to the aquaculture sector. Long before the term 'paraprobiotic' was coined, heat-killed *Lactococcus lactis* was tested in turbot to investigate the immunostimulatory potential of the inactivated preparation (Villamil et al. 2002). Since then, several inactivated probiotic preparations (produced by different inactivation methods) have been investigated in fish and shellfish. An overview of the information on the application of paraprobiotics in aquaculture comprising the target probiotic microbes and host species, inactivation methods, and principal observations have been summarized in a recent review (Choudhury & Kamilya 2019). Probiotics, belonging to a diverse genus of Gram-positive and negative bacteria, have been used to prepare paraprobiotics; and immune response, growth, and disease resistance of fish and shellfish were investigated to delineate the paraprobiotic effects. Among the different genera, *Bacillus* constitutes the mostly studied genus to investigate the paraprobiotic effect. While some studies indicate almost equal effectiveness of paraprobiotics compared to their viable forms in terms of immunomodulation and disease resistance, other studies show better health benefits of probiotics compared to their non-viable counterparts. The work on *Bacillus* paraprobiotic in the aquaculture sector is summarized in Table 1, illustrating the target probiotic species, inactivation methods used to obtain paraprobiotic from them, target fish species, mode of the assay, and principal observations.

12.5.1 Effect on the Growth

Probiotics have been suggested to positively influence the digestive processes of aquatic animals through nutritional and/or enzymatic contribution (Balcázar et al. 2006). This attribute of probiotics cannot be expected from paraprobiotics, being the non-viable form. However, paraprobiotics, including the *Bacillus* paraprobiotics, have also been found to improve the growth performance and feed utilization of aquatic animals in certain instances. The mechanisms by which paraprobiotics improve the growth performance of fish are not clear. Dietary administration of heat-inactivated *B. pumilus* significantly improved the final weight, weight gain (WG), and specific growth rate (SGR) of juvenile grouper (*Epinephelus coioides*) at day 60 and significantly decreased the feed conversion ratio (FCR) at days 30 and 60 (Yan et al. 2016). The viable form of the probiotic *B. clausii* DE5 did not affect the final weight, WG, and SGR of *E. coioides* at days 30 and 60. However, both the heat-inactivated and the live forms significantly decreased the feed intake and feed FCR at day 60 (Wang et al. 2008). Heat-killed *Bacillus* sp. SJ-10 significantly enhanced the WG and PER of olive flounder (*Paralichthys olivaceus*) after 8 weeks of feeding (Hasan et al. 2019). The SGR and FCR of Nile tilapia (*Oreochromis niloticus*) fed with heat-killed *Bacillus* sp. NP5 were significantly enhanced after 30 days of feeding (Mulyadin et al. 2021). On the contrary, heat-killed dietary *B. subtilis* did not impart any significant changes in growth, PER, nutrient retention, digestibility, FCR, and gut colonization of rohu (*Labeo rohita*) (Mohapatra et al. 2012). Overall, the impact of *Bacillus* paraprobiotics in positively impacting the growth performance of fish is contradictory.

12.5.2 Effect on the Immune System and Disease Resistance

The health-promoting effect of probiotics and paraprobiotics is attributed to their ability to modulate the host's immune responses. In fact, the components of dead cells in the paraprobiotic preparation are believed to be responsible for immunostimulation (Taverniti & Guglielmetti 2011). Like the mammalian system, studies have also been performed in fish with *Bacillus* paraprobiotic and their ability to modulate the immune responses both under *in vitro* and *in vivo* conditions is noteworthy (Table 12.1). The published literature indicates that several *Bacillus* paraprobiotics can enhance both cellular immunity and immune-related gene expression in fish.

Several cellular immune parameters including respiratory burst, phagocytosis, myeloperoxidase content, and lysozyme activity were observed to be significantly enhanced when fish head-kidney leukocytes were exposed to *Bacillus* paraprobiotic preparations *in vitro* (Salinas et al. 2006;

TABLE 12.1
***Bacillus* Paraprobiotic Application in Aquaculture**

Test Organism (s)	Inactivation Method	Fish Species Under Study	Mode of Assay	Principal Observations[#]	References
B. subtilis	Heat	*Sparus aurata*	*In vitro*	↑ Leucocyte peroxidase content ↑ Respiratory burst activity ↑ Phagocytosis ↑ Cytotoxicity	Salinas et al. (2006)
Bacillus subtilis AB1	Formalin and sonication	*Oncorhynchus mykiss*	*In vivo*	↑ Respiratory burst ↑ Serum and gut lysozyme ↑ Peroxidase ↑ Phagocytic killing ↑ Total and α1-antiprotease ↑ Lymphocyte populations ↑ Fish survival	Newaj-Fyzul et al. (2007)
B. subtilis	Heat	*S. aurata*	*In vivo*	↑ Natural complement ↑ Serum peroxidase ↑ Phagocytic activities ↑ Total serum IgM ↑ Numbers of gut IgM⁺ cells ↑ Acidophilic granulocytes	Salinas et al. (2008)
B. subtilis	Heat	*Labeo rohita*	*In-vivo*	↔ Growth, protein efficiency ratio (PER), nutrient retention, digestibility, and food conversion ratio (FCR) ↔ Gut colonization	Mohapatra et al. (2012)
B. amyloliquefaciens (FPTB16) and *B. subtilis* (FPTB13)	Heat, UV, and formalin	*Catla catla*	*In vitro*	↑ Superoxide anion ↑ Nitric oxide ↑ Myeloperoxidase content ↑ Proliferative response	Kamilya et al. (2015)
B. pumilus SE5	Heat	*Epinephelus coioides*	*In vivo*	↑ Final weight, weight gain (WG), SGR, and FCR ↑ Phagocytic activity ↑ Serum complement C3 ↑ IgM levels ↑ SOD ↑ Expression of TLR2, IL-8, and IL-1b	Yan et al. (2016)
B. amyloliquefaciens FPTB16	Heat	*C. catla*	*In vivo*	↑ Oxygen radical production ↑ Serum lysozyme activity ↑ Total serum protein content ↑ Myeloperoxidase activity ↑ Alkaline phosphatase activity ↔ GPT, GOT, and glucose content ↑ Expression of IL-1β, TNF-α, C3, and iNOS ↓ Expression of IFN-γ gene	Singh et al. (2017)
B. coagulans (MTCC-9872)	Heat-killed	*Oreochromis niloticus*	*In Vitro*	↑ Inhibitory activity against fish pathogens *A. hydrophila, A. sobria, A. caviae, P. fluorescens*, and *P. aeruginosa*	Midhun et al. (2018)

(*Continued*)

TABLE 12.1 (Continued)
***Bacillus* Paraprobiotic Application in Aquaculture**

Test Organism (s)	Inactivation Method	Fish Species Under Study	Mode of Assay	Principal Observations#	References
B. clausii DE5	Heat-killed	*E. coioides*	*In vivo*	↔ Final weight, WG, and SGR ↓ Feed intake and FCR ↑ TLR5, IL-8, IL-1β, and TGF-β1 expression	Wang et al. (2008)
Bacillus sp. SJ-10	Heat-killed	*Paralichthys olivaceus*	*In Vivo*	↑ WG and PER ↑ Lysozyme and superoxide dismutase ↑ TNF-α, IL-1β, IL-6 expression in liver and gill ↑ Protection against *Streptococcus iniae*	Hasan et al. (2019)
Bacillus sp. NP5	Heat-killed	*O. niloticus*		↑ SGR and FCR ↑ Lysozyme activity ↑ Phagocytic activity ↑ Respiratory burst ↑ Total serum protein content ↑ Protection against *Streptococcus agalactiae*	Mulyadin et al. (2021)

'↑' increase; '↓' decrease; '↔' no change.

Kamilya et al. 2015). *In vivo*, studies by dietary administration of paraprobiotics have been reported to modulate several immune parameters of fish. Administration of paraprobiotics through feed has been demonstrated to significantly increase the serum peroxidase content, lysozyme activity, respiratory burst, alkaline phosphatase activity, antiprotease, immunoglobulin levels, complement, and several other immune-related activities (Newaj-Fyzul et al. 2007; Salinas et al. 2008, Singh et al. 2017, Yan et al. 2016, Hasan et al. 2019; Mulyadin et al. 2021). As regard the mRNA expression of immune genes, dietary administration of *Bacillus* paraprobiotics has been reported to induce the expression of pro-inflammatory cytokines (IL-1b, IL-6, and TNF-α), cell-mediated immune regulators (IL-12p35, IL-12p40, and IL-18), antiviral cytokines (IFN- γ), regulatory cytokines (IL-8 and TGF-b1), and other immune components (TLR2, TLR5, C3, and iNOS) (Yan et al. 2016; Singh et al. 2017; Wang et al. 2008; Hasan et al. 2019).

These findings indicate a potential health-promoting effect of *Bacillus* paraprobiotics by stimulating the immune system of fish. It can be safely assumed that the immunostimulatory activity of *Bacillus* paraprobiotics is due to different structural components of the bacterial cells present in the paraprobiotic preparation. In fact, the involvement of several cellular constituents of bacterial cells has been demonstrated to stimulate immune responses in animal models including fish (Kataria et al. 2009; Adams 2010; Taverniti & Guglielmetti 2011; de Almada et al. 2016, Choudhury & Kamilya 2019).

The increase in disease resistance capability of the host against infectious microorganisms is one of the beneficial effects of paraprobiotics. Generally, there is a positive correlation between immunostimulation and enhanced resistance of fish against pathogenic microorganisms. Thus, it is assumed that immunostimulation by paraprobiotics is the principal mode of action responsible for the increased resistance of fish against infectious diseases. Heat-killed whole-cell product of *B. coagulans* showed significant *in vitro* antagonistic activity against fish pathogenic bacteria including *Aeromonas hydrophila, A. sobria, A. caviae, Pseudomonas fluorescens*, and *P. aeruginosa* indicating the potential disease-resistant capability of the inactivated preparation (Midhun et al.

2018). Administration of formalin-killed and sonicated *B. subtilis* AB1 through feed significantly reduced the mortality of rainbow trout infected with pathogenic *Aeromonas* sp. (Newaj-Fyzul et al. 2007). Feeding *P. olivaceus* and *O. niloticus* with heat-killed *Bacillus* sp. SJ-10 and *Bacillus* sp. NP5, respectively, resulted in a significant increase in protection against streptococcosis (Hasan et al. 2019; Mulyadin et al. 2021). These dietary administration studies also reported enhanced immunity in paraprobiotic-fed fish, delineating immune correlates of protection during infection.

12.6 CONCLUSION

Published literature on *Bacillus* paraprobiotics indicates the health-promoting effect of inactivated *Bacillus* preparations in fish with stimulation of the immune responses being the mode of action. The potential beneficial effects of paraprobiotics can underscore the reliance on live probiotics as the use of the latter is associated with potential safety problems, besides functionality and applicability issues. Due to the fact that paraprobiotics can exert beneficial effects in a manner comparable to their viable counterparts, their application holds a promising new avenue for producing healthy fish in aquaculture. Nevertheless, future studies should be directed to elucidate the mechanism of action of the paraprobiotics, develop protocols for assessing their effectiveness, and gain knowledge about the fate and quality of the paraprobiotics, in the entire production process. As a final note, there is growing interest in paraprobiotics among the scientific community, and it holds a promising future application in aquaculture.

BIBLIOGRAPHY

Adams CA (2010) The probiotic paradox: Live and dead cells are biological response modifiers. *Nutrition Research Reviews* **23**, pp.37–46.

Aguirre-Guzmán G, Lara-Flores M, Sánchez-Martínez JG, Campa-Córdova AI, Luna-González A (2012) The use of probiotics in aquatic organisms: A review. *African Journal of Microbiology Research* **6**, pp.4845–4857.

Balcázar JL, Blas Id, Ruiz-Zarzuela I, Cunningham D, Vendrell D, Múzquiz JL (2006) The role of probiotics in aquaculture. *Veterinary Microbiology* **114**, pp.173–186.

Bermudez-Brito M, Plaza-Díaz J, Munoz-Quezada S, Gomez-Llorente C, Gil A (2012) Probiotic mechanisms of action. *Annals of Nutrition and Metabolism* **61**, pp.160–174.

Besselink MG, van Santvoort HC, Buskens E, Boermeester MA, van Goor H, Timmerman HM, Nieuwenhuijs VB, Bollen TL, van Ramshorst B, Witteman BJ, Rosman C (2008) Probiotic prophylaxis in predicted severe acute pancreatitis: A randomised, double-blind, placebo-controlled trial. *The Lancet* **371**, pp.651–659.

Birmpa A, Sfika V, Vantarakis A (2013) Ultraviolet light and ultrasound as nonthermal treatments for the inactivation of microorganisms in fresh ready-to-eat foods. *International Journal of Food Microbiology* **167**, pp.96–102.

Choudhury TG, Kamilya D (2019) Paraprobiotics: An aquaculture perspective. *Reviews in Aquaculture* **11**, pp.1258–1270.

Das A, Nakhro K, Chowdhury S, Kamilya D (2013) Effects of potential probiotic *Bacillus amyloliquifaciens* FPTB16 on systemic and cutaneous mucosal immune responses and disease resistance of catla (*Catla catla*). *Fish and Shellfish Immunology* **35**, pp.1547–1553.

Dash G, Raman RP, PaniPrasad K, Makesh M, Pradeep MA, Sen S (2014) Evaluation of *Lactobacillus plantarum* as feed supplement on host associated microflora, growth, feed efficiency, carcass biochemical composition and immune response of giant freshwater prawn, *Macrobrachium rosenbergii* (de Man, 1879). *Aquaculture* **432**, pp.225–236.

Dash G, Raman RP, PaniPrasad K, Makesh M, Pradeep MA, Sen S (2015) Evaluation of paraprobiotic applicability of *Lactobacillus plantarum* in improving the immune response and disease protection in giant freshwater prawn, *Macrobrachium rosenbergii* (de Man, 1879). *Fish and Shellfish Immunology* **43**, pp.167–174.

de Almada CN, Almada CN, Martinez RC, Sant'Ana AS (2016) Paraprobiotics: Evidences on their ability to modify biological responses, inactivation methods and perspectives on their application in foods. *Trends in Food Science and Technology* **58**, pp.96–114.

Dawood MA, Koshio S, Abdel-Daim MM, Van Doan H (2019) Probiotic application for sustainable aquaculture. *Reviews in Aquaculture* 11(3): pp.907–24.

FAO/WHO (2001) Evaluation of Health and Nutritional Properties of Powder Milk and Live Lactic Acid Bacteria. Food and Agriculture Organization of the United Nations and World Health Organization Expert Consultation Report. FAO, Rome, Italy.

Favero MS, Bond WW (1991) Chemical disinfection of medical and surgical materials. In: Block SS, ed. *Disinfection, sterilization, and preservation*. Philadelphia: Lea and Febiger, **1991**: pp.617–41.

Franz C, Specht I, Cho GS, Graef V, Stahl M (2009) UV-C inactivation of microorganisms in naturally cloudy apple juice using novel inactivation equipment based on Dean vortex technology. *Food Control* **20**, pp.1103–1.

Gayan E, Alvarez I, Condon S (2013) Inactivation of bacterial spores by UV-C light. *Innovative Food Science & Emerging Technologies* **19**, pp.140–145.

Gibson JH, Hai D, Yong N, Farnood RR, Seto P (2008) A literature review of ultrasound technology and its application in wastewater disinfection. *Water Quality Research Journal of Canada* **43**, pp.23–35.

Gobbetti M, Cagno RD, De Angelis M (2010) Functional microorganisms for functional food quality. *Critical Reviews in Food Science and Nutrition* **50**, pp.716–727.

Gould GW (1989) Heat induced injury and inactivation. In GW Gould (Ed.), *Mechanisms of action of food preservation procedures* (pp.11–42). London: Elsevier Applied Science.

Hill C, Guarner F, Reid G, Gibson GR, Merenstein DJ, Pot B, Morelli L, Canani RB, Flint HJ, Salminen S, Calder PC (2014) Expert consensus document: the international scientific association for probiotics and prebiotics consensus statement on the scope and appropriate use of the term probiotic. *Nature reviews Gastroenterology and hepatology* **11(8)**: pp.506–14.

Hasan MT, Jang WJ, Lee BJ, Kim KW, Hur SW, Lim SG, Bai SC, Kong IS (2019) Heat-killed Bacillus sp. SJ-10 probiotic acts as a growth and humoral innate immunity response enhancer in olive flounder (*Paralichthys olivaceus*). *Fish and shellfish Immunology* **88**, pp.424–31.

Ivec M, Botic T, Koren S, Jakobsen M, Weingartl, H, Cencic A (2007) Interactions of macrophages with probiotic bacteria lead to increased antiviral response against vesicular stomatitis virus. *Antiviral Research* **75**, pp.266–274.

Jozsef L, Khreiss T, Filep JG (2004) CpG motifs in bacterial DNA delay apoptosis of neutrophil granulocytes. *The FASEB Journal* **18**, pp.1776–1778.

Juneja VK, Sofos JN (2002) Thermal inactivation of microorganisms. In: M. Dekker (Ed.), *Control of food borne microorganisms* (pp.13–53). New York: Food Science and Technology.

Kamilya D, Baruah A, Sangma T, Chowdhury S, Pal P (2015) Inactivated probiotic bacteria stimulate cellular immune responses of catla, *Catla catla* (Hamilton) *in vitro*. *Probiotics and Antimicrobial Proteins* **7**, pp.101–106.

Kataria J, Li N, Wynn JL, Neu J (2009) Probiotic microbes: Do they need to be alive to be beneficial? *Nutrition Reviews* **67**, pp.546–550.

Lee CH, Jeong HDO, Chung JK, Lee HH, Kim KH (2003) CpG motif in synthetic ODN primes respiratory burst of olive flounder *Paralichthys olivaceus* phagocytes and enhances protection against *Edwardsiella tarda*. *Diseases of Aquatic Organisms* **56**, pp.43–48.

Merrifield DL, Harper GM, Dimitroglou A, Ringo E, Davies SJ (2010) Possible influence of probiotic adhesion to intestinal mucosa on the activity and morphology of rainbow trout (*Oncorhyncus mykiss*) enterocytes. *Aquaculture Research* **41**, pp.1268–1272.

Midhun SJ, Neethu S, Vysakh A, Radhakrishnan EK, Jyothis M. (2018) Antagonism Against Fish Pathogens by Cellular Components/Preparations of *Bacillus coagulans* (MTCC-9872) and It's In Vitro Probiotic Characterisation. *Current Microbiology* **75(9)**: pp.1174–81.

Mohapatra S, Chakraborty T, Kumar V, Deboeck G, Mohanta KN (2013) Aquaculture and stress management: A review of probiotic intervention. *Journal of Animal Physiology and Animal Nutrition* **97**, pp.405–430.

Mohapatra S, Chakraborty T, Prusty AK, Das P, Paniprasad K, Mohanta KN (2012) Use of different microbial probiotics in the diet of rohu, (*Labeo rohita*) fingerlings: Effects on growth, nutrient digestibility and retention, digestive enzyme activities and intestinal microflora. *Aquaculture Nutrition* **18**, pp.1–11.

Mukhopadhyay S, Ramaswamy R (2012) Application of emerging technologies to control Salmonella in foods: A review. *Food Research International* **45**, pp.666–677.

Mulyadin A, Widanarni W, Yuhana M, Wahjuningrum D (2021) Growth performance, immune response, and resistance of Nile tilapia fed paraprobiotic *Bacillus* sp. NP5 against *Streptococcus agalactiae* infection. Jurnal Akuakultur Indonesia **20**, pp.34–46.

Nayak SK (2010) Probiotics and immunity: A fish perspective. *Fish and Shellfish Immunology* **29**, pp.2–14.

Newaj-Fyzul A, Adesiyun AA, Mutani A, Ramsubhag A, Brunt J, Austin B (2007) *Bacillus subtilis* AB1 controls *Aeromonas* infection in rainbow trout (*Oncorhynchus mykiss*, Walbaum). *Journal of Applied Microbiology* **103**, pp.1699–1706.

Newaj-Fyzul A, Al-Harbi AH, Austin B (2014) Review: Developments in the use of probiotics for disease control in aquaculture. *Aquaculture* **431**, pp.1–11.

Nikoskelainen S, Ouwehand AC, Bylund G, Salminen S, Lilius EM (2003) Immune enhancement in rainbow trout (*Oncorhynchus mykiss*) by potential probiotic bacteria (*Lactobacillus rhamnosus*). *Fish and Shellfish Immunology* **15**, pp.443– 452.

Ottesen OH, Olafsen JA (2000) Effects on survival and mucous cell proliferation of Atlantic halibut, *Hippoglossus hippoglossus* L., larvae following microflora manipulation. *Aquaculture* **187**, pp.225–38.

Ou CC, Lin SL, Tsai JJ, Lin MY (2011) Heat-killed lactic acid bacteria enhance immunomodulatory potential by skewing the immune response toward Th1 polarization. *Journal of Food Science* **76**, pp.M260–M267.

Prado S, Romalde JL, Barja JL (2010) Review of probiotics for use in bivalve hatcheries. *Veterinary Microbiology* **145**, pp.187–197.

Raz E, Rachmilewitz D (2005) Inactivated probiotic bacteria and methods of use thereof. Patent n.US20050180962 A1, U.S. Patent Application 10/742,052.

Ringo E, Gatesoupe FJ (1998) Lactic acid bacteria in fish: A review. *Aquaculture* **160**, pp.177–203.

Ross AIV, Griffiths MW, Mittal GS, Deeth HC (2003) Combining nonthermal technologies to control foodborne microorganisms. *International Journal of Food Microbiology* **89**, pp.125–138.

Salinas I, Díaz-Rosales P, Cuesta A, Meseguer J, Chabrillón M, Morinigo MA, Esteban MA (2006) Effect of heat-inactivated fish and non-fish derived probiotics on the innate immune parameters of a teleost fish (*Sparus aurata* L.). *Veterinary immunology and immunopathology*. **111**, pp.279–286.

Salinas I. Myklebust R, Esteban MA, Olsen RE, Meseguer J, Ringø E (2008) In vitro studies of *Lactobacillus delbrueckii* subsp. *lactis* in Atlantic salmon (*Salmo salar* L.) foregut: Tissue responses and evidence of protection against *Aeromonas salmonicida* subsp. *salmonicida* epithelial damage. *Veterinary Microbiology* **128**, pp.167–177.

Singh RS (2013) A comparative study on cell disruption methods for release of aspartase from *E. coli* K-12. *Indian Journal of Experimental Biology* **51**, pp.997–1003.

Singh ST, Kamilya D, Kheti B, Bordoloi B, Parhi J (2017) Paraprobiotic preparation from *Bacillus amyloliquefaciens* FPTB16 modulates immune response and immune relevant gene expression in *Catla catla* (Hamilton, 1822). *Fish and Shellfish Immunology* **66**, pp.35–42.

Son VM, Chang CC, Wu MC, Guu YK, Chiu CH, Cheng W (2009) Dietary administration of the probiotic, *Lactobacillus plantarum*, enhanced the growth, innate immune responses, and disease resistance of the grouper *Epinephelus coioides*. *Fish and Shellfish Immunology* **26**, pp.691–698.

Tassakka ACMAR, Sakai, M (2002) CpG oligodeoxynucleotides enhance the non-specific immune responses on carp, *Cyprinus carpio*. *Aquaculture* **209**, pp.1–10.

Taverniti V, Guglielmetti S (2011) The immunomodulatory properties of pro- biotic microorganisms beyond their viability (ghost probiotics: Proposal of paraprobiotic concept). *Genes & Nutrition* **6**, pp.261–274.

Vieira FN, Buglione CC, Mourino JLP, Jatoba A, Ramirez C, Martins ML et al. (2008) Time-related action of *Lactobacillus plantarum* in the bacterial microbiota of shrimp digestive tract and its action as immunostimulant. *Pesquisa Agropecuária Brasileira* **43**, pp.763–739.

Villamil L, Tafalla C, Figueras A, Novoa B (2002) Evaluation of immunomodulatory effects of lactic acid bacteria in turbot (*Scophthalmus maximus*). *Clinical and Diagnostic Laboratory Immunology* **9**, pp.1318–1323.

Wang YB, Li JR, Lin J (2008) Probiotics in aquaculture: Challenges and outlook. *Aquaculture* **281**, pp.1–4.

Yan YY, Xia HQ, Yang HL, Hoseinifar SH, Sun YZ (2016) Effects of dietary live or heat-inactivated autochthonous *Bacillus pumilus* SE5 on growth performance, immune responses and immune gene expression in grouper *Epinephelus coioides*. *Aquaculture Nutrition* **22**, pp.698–707.

13 Application of Gut-Associated *Bacilli* in Feed-Biotechnology

Fawole Femi John
University of Ilorin, Ilorin, Nigeria

Shamna Nazeemashahul and L. Manjusha
ICAR-Central Institute of Fisheries Education, Mumbai
V. Vidhya and A.M. Babitha Rani
ICAR-Central Institute of Fisheries Education, Mumbai

S. Ferosekhan
ICAR-Central Institute of Freshwater Aquaculture, Bhubaneswar, Odissa

Karthireddy Syamala and Muralidhar P. Ande
ICAR-Central Institute of Fisheries Education, Kakinada

Upasana Sahoo
ICAR-Central Institute of Fisheries Education, Mumbai

R. Vidhya
ICAR-Central Institute of Brackishwater Aquaculture, Chennai, India

INTRODUCTION

The physiological well-being of fish is associated with the physical environment, its chemical characteristics, diet, nutrition, and immunological balance. The microbiota associated with fish play a crucial role in development, physiology, immune status, disease resistance, etc. (Yan et al., 2016). Of many factors affecting microbiota associated with fish, it is found that the prominent factors are the age of the fish, dietary preferences, husbandry practices, and trophic levels which they occupy (Michl et al., 2017). *Bacillus* bacteria are Gram-positive, rod-shaped, endospore-forming, mesophilic, facultative anaerobic microorganisms belonging to the phylum Firmicutes in the family *Bacillaceae*. Although originally an inhabitant of soil, *Bacillus* spp. have been isolated from various sources such as vegetables, meat products, air, water, and human and animal gut, in addition to fermented foods (Rao et al., 2015). This genus is known to harbor several bioactive substances of potential use in food, pharmaceutical, and health sectors (Ramirez-Olea et al., 2022). They can survive in different types of food products, as well as tolerate the acidity of the human intestinal tract, being endospore formers, when compared to other bacterial probiotic species (Ehling-Schultz et al., 2017). The environment in which the animal is raised has an important effect on the composition of the gut microbiome (Angthong et al., 2020; Pérez et al., 2010). Some researches, however, have found that changes in water microbial load have no influence on the gut microbiome (Zheng et al., 2017). The presence of *Bacillus* in the gut of fish is already reported by several authors (Silva-Brito et al., 2021). The biotechnological application of *Bacillus* in aquaculture is vast, as it acts as a probiotic and helps in disease control by strengthening immune response and promoting growth. *Bacillus* also acts as a bioremediator and is used best as a bio-augmentor.

DOI: 10.1201/9781003503811-13

Bacillus' method of action in aquaculture is enzyme synthesis, competitive exclusion, antimicrobial chemical generation, and immunological activation. *Bacillus* spores are particularly immune to extrinsic physical, chemical, and environmental factors due to their hierarchical structure. It may so survive for an extended period of time without losing viability, which is regarded as an advantage for utilizing the species as a probiotic in aquafeed compositions. In this context, the purpose of this chapter is to concentrate on the application of the bacteria *Bacillus* sp. as a biotechnology agent in aquaculture environments and fish feed.

GUT-ASSOCIATED MICROBES

Microbiota in the gut mainly represents the group of microbes which colonize the gut and contribute to the biochemical activities in the gut which are related to digestion and assimilation and also impart defense mechanism against pathogens (Gómez and Balcázar, 2008). Various investigations on gut microbiota indicated that diet is a major factor that determines the diversity of microbiota and it initiates from first feeding onward (Lauzon et al., 2010). It is reported that fish gut is colonized by more than 500 different species which constitutes both aerobic bacteria and facultative anaerobes (Romero and Navarrete, 2006). The gut microbiota is reported to be stabilized within the first 50 days of life for most of the fish species, even though the community changes with stages of life, environment especially temperature, and salinity (Macfarlane and Englyst, 1986), variation in diet (Larsen et al., 2014), and developmental stage (Romero and Navarrete, 2006).

Aeromonas, Bacteroidaceae, and *Clostridium* species of bacteria colonize the stomach of fish and are associated with amylase synthesis and, hence, play a significant role in starch digestion in the gastrointestinal tract (GIT) (Sugita et al., 1997). Some other bacteria belonging to genus *Enterobacter*, *Pichia*, and *Candida* were reported to produce enzymes which can neutralize antinutritional factors in feed ingredients (Mandal and Ghosh, 2013) and also certain bacteria belonging to *Actinobacteriacea* can produce biologically active compounds which are playing a role in gut health, digestion, etc. (Jami et al., 2015). Herbivorous marine fish microbiota is involved in the conversion of algal food elements into fatty acids with a short chain. The short-chain fatty acids (SCFAs) are the final products of the anerobic metabolism of bacteria in the gut's posterior area, indicating that microbial digestion occurs in that area (Willmott et al., 2005).

The microbiota of herbivores and fish has been found to be predominantly engaged in the breakdown of carbohydrates into SCFAs that are absorbed by the epithelium (Clements et al., 2007) and meet the energy requirement of fish (Mountfort et al., 2002). The extent of microbial fermentation is high in the hindgut region compared to the anterior gut of fish as indicated by various studies (Fidopiastis et al., 2006). In the hindgut region, the dominant bacteria described were *Proteobacteria, Enterovibrio*, and *Desulfovibrio* followed by *Bacteroidetes* and *Firmicutes*, but a dominance of *Clostridium* belonging to *Firmicutes* was described in the hindgut region by Moran et al. (2005). Smriga et al. (2010) proposed that *Proteobacteria*, *Bacteroidetes*, *Firmicutes*, and *Fusobacteria* are involved in food digestion in the gut by supplementing a variety of digestive enzymes. *Actinobacteria* is another phylum which is linked to metabolic process-related production of enzymes and other metabolites. The diversity of various communities among microbiota of the gut depends on the dietary content such as type, amount of protein and its quality, dietary lipid, and feed additives like prebiotics, probiotics, immunostimulants, antibiotics, and supplements of herbal origin (Merrifield et al., 2010; Dimitroglou et al., 2011).

Ray et al. (2012) discovered a diverse variety of microbiota involved in enzyme production in the GIT of fish, which has numerous functional metabolic activities relevant to digestion. The Clostridia group is mostly involved in the degradation of polymers and they use polysaccharides and proteins as substrates to produce SCFAs and alcohols (Mountfort et al., 2002). Lactic acid bacteria was proved to have probiotic effect which enhances the immunity of fish and shellfish (Merrifield et al., 2010; Soltani et al., 2019). It was also illustrated that the absence of microbiota in the intestine had led to a reduction in the assimilation of protein molecules in the hindgut (Bates et al., 2006).

The stomach of various fish species was reported to be dominated by *Proteobacteria, Actinobacteria,* and *Firmicutes* and the dominant family as Vibrionaceae. The *Vibrio* species were observed to have the ability to break down dietary compounds along with some strains that were reported to produce digestive enzymes such as amylase, lipase (Gatesoupe et al., 2007; Henderson and Millar, 1998), cellulase (Itoi et al., 2006; Sugita and Ito, 2006), and chitinase (MacDonald et al., 1986) (Ray et al., 2012). The gut microbiota can be altered through the diet supplementation of probiotics in fish and shrimp to enhance growth, nutrient utilization, and immune responses. The probiotic strains which outnumber the pathogenic strains in both water and animal gut had many advantages such as nutrition and disease resistance as reported in previous studies (Nayak, 2010).

The formation and kind of microbial communities in the fish GIT can be influenced by both external and endogenous influences. The interaction of various elements determines the composition of the fish gut microbiome. The adjacent water and microbial communities have a direct impact on fish gut microbiome (Sullam et al., 2012; Roeselers et al., 2011). There is further evidence that the gut microbiota is affected by host genetics, stage of development, immunological condition, and other host-specific factors. Furthermore, nutrition affects the gut flora, which changes in composition based on dietary intake. Thus, as a result, the factors influencing the microbiome of the gut can be classified as follows: (1) ecological and environmental circumstances, (2) host particular, and (3) trophic status and/or feeding behavior. *Bacillus* species are among the most common bacteria and can be found in soil, water, and food.

FACTORS AFFECTING GUT-ASSOCIATED MICROBES

The gut microbiota of aquatic organisms is determined by various factors (Yan et al., 2016; Yukgehnaish et al., 2020), which includes host genetic (Landsman et al., 2019; Perry et al., 2020; Naya-Català et al., 2022), nutrition (Bakke et al., 2013a, 2013b; Ni et al., 2014; Bolnick et al., 2014; Cordero et al., 2015; Xiong et al., 2017; Miyake et al., 2015; Razak et al., 2019; Perry et al., 2020; Ye et al., 2023), feeding habit (Ward et al., 2009; Larsen et al., 2014; Li et al., 2014; Miyake et al., 2015; Ghanbari et al., 2015; Ye et al., 2023), host phylogeny (Larsen et al., 2014; Tzeng et al., 2015; Denison et al., 2020; Huang et al., 2020), trophic level (Clements et al., 2007; Liu et al., 2016), life stage (Tanasomwang and Muroga, 1988; Wang et al., 1993; Hansen and Olafsen, 1999; Moran et al., 2005; Huang et al., 2014; Zhang et al., 2018; Razak et al., 2019), rearing system (Huang et al., 2020; Giatsis et al., 2015; Perry et al., 2020; Romero et al., 2022; Ye et al., 2023), stress (Rajeev et al., 2021), and environmental conditions such as temperature, pH, salinity (MacFarlane et al., 1986; Nayak, 2010), and oxygen availability. Host habitat and environmental factors are the main determinants of gut microbiota in aquatic animals (Sullam et al., 2012; Dehler et al., 2017; Rajeev et al., 2021). Habitat alone accounts for 14.6% of the variation, while health status alone accounts for 7.7% of the variation (Xiong et al., 2015).

Vibrio, Aeromonas, Flavobacterium, Plesiomonas, Pseudomonas, Enterobacteriaceae, *Micrococcus, Acinetobacter, Clostridium, Fusarium,* and *Bacteroides* are all present in the GITs of all fish. Their abundance, on the other hand, is determined by species and environment (Nayak, 2010). The most common bacterial species discovered in all ponds and water sources were *Aeromonas hydrophila, Citrobacter freundii, Escherichia coli, Enterobacter aerogenes, Klebsiella pneumoniae, Pseudomonas* sp., *Bacillus* sp., *Listeria monocytogenes*, and *Staphylococcus* sp. Other bacterial species discovered included *Plesiomonas shigelloides, Serratia liquefaciens, Flavobacterium* sp., *Corynebacterium, Streptococcus* sp., and Micrococcus sp. According to the number of published reports, the most important hydrocarbon-degrading bacteria in both marine and soil settings are *Achromobacter, Acinetobacter, Alcaligenes, Arthrobacter, Bacillus, Flavobacterium, Nocardia, Pseudomonas*, and the coryneforms (Leahy and Colwell 1990).

Different stages of shrimp growth showed changes in microbial composition. Flavobacteriaceae and Rhodobacteraceae *(Alphaproteobacteria)* groups have been reported at all stages of shrimp growth suggesting that these two groups are part of the core gut microbiota (Huang et al., 2014).

Organisms cultured in ponds show a different microbial composition and additional bacterial groups compared to indoor cultures (Huang et al., 2014). Autochthonous bacteria can colonize or be connected with the host's intestine epithelial surface. Allochthonous bacteria, on the other hand, are temporary, coupled with particles of food, or exist in the animal lumen. Autochthonous bacteria also support host physiological processes such as nutrient metabolism, immune modulation, barrier function, and exclusion of competitive pathogens (Ghanbari et al., 2015; Ringø et al., 2016; Diwan et al., 2022; Diwan et al., 2023; Bhatia et al., 2023).The gut microbiota of crustaceans such as shrimp or crab can be dominated by bacteria involved in chitin degradation, while the gut microbiota of mollusks such as oysters or clams can be influenced by factors such as water temperature or salinity (Rungrassamee et al., 2014; Lokmer and Mathias Wegner, 2015).

Giatsis et al. (2015) demonstrated that pH, NO3-N, PO4-P, and conductivity account for 68% of the total variation in bacterial community composition in water. Changes in water physico-chemical properties correlate with differences in the composition and functionality of aquatic bacterial communities between different rearing systems and over time (Giatsis et al., 2015). The gut microbiota of Atlantic cod larvae (*Gadus morhua*) (Bakke et al., 2013a, 2013b), rainbow trout (*Oncorhynchus mykiss*) (Wong et al., 2013), and tilapia larvae (Giatsis et al., 2015) has shown a correlation with the surrounding water microbiota. Abiotic factors in water and soil can indirectly affect the composition and diversity of the gut microbiome by creating a favorable or unfavorable environment for specific microbial species (Razak et al., 2019). Water quality discrepancies between rearing systems are linked to variations in system design and management, both of which influence the composition and functionality of the communities of bacteria in the water source (Yukgehnaish et al., 2020). Understanding the mechanisms that influence fish gut microbiota is critical for designing methods to optimize fish health and improve aquaculture operations, as well as for comprehending the significance of the fish gut microbiota in the functioning of aquatic ecosystems (Kim et al., 2021).

SOIL/SEDIMENT

Sediment microbiota can affect gut microbiota. Some studies have reported that the gut microbiota has a closer relationship with the bacterial community in the sediment (Cornejo-granados et al., 2017). The sediment, water, and shrimp gut microbiomes all contain identical bacterial compositions with varying abundance (Hou et al., 2018). Cluster analysis has shown similar dominant bacterial genera in sediment and shrimp intestine (Huang et al., 2018). Fan et al. (2019) also found that the bacterial composition in shrimp gut and sediment are largely identical, but the relative abundance of bacterial communities is different. The bacterial operational taxonomic unit (OUT) of the gut was very similar to the host habitat, suggesting a transfer of microbiota from the surrounding environment to the gut of aquatic organisms (Sun et al., 2019). The impact of sediment and water microbiota on the composition of the gut microbiome depends on the cultured species, habitat, and their feeding habits (Bi et al., 2021).

The redox potential of sediments is an important environmental factor affecting the development of different bacterial communities. Beta-diversity analysis of fully oxic and oxic–anoxic sediments confirmed the distinct clustering of bacterial communities (Robinson et al., 2016).

WATER PARAMETERS

SALINITY

In aquaculture, water salinity is an important abiotic factor influencing the form and function of both ambient and gut microbiota. It is assumed that salinity stress can affect the stability of the environmental microbiome by increasing potential pathogens and altering environmental microbial populations. For instance, by increasing the water salinity of shrimp ponds, the chance of vibriosis caused by various *Vibrio* bacteria including *V. parahaemolyticus*, *V. harveyi*, *V. campebeli*, and *V. owensi* will be increased which is in part due to the suppression of beneficial bacteria with both endogenous and exogenous origins. Thus, water salinity can affect the composition of bacterial communities in the gut as well as in the surrounding water (Chaiyapechara et al., 2022). The gut bacterial populations of marine fish species have been shown to be enriched with *Vibrio* bacteria adapted to their hosts' saline environment. This implies that bacteria destined to associate with the guts of marine fish species must also be able to endure the salinity of their hosts' environment (Lozupone and Knight 2007; Sullam et al., 2012). The abundance of OTUs in the gut microbiota has been found to depend on water salinity (Zhang et al., 2016a; Liu et al., 2019a, 2019b). The high salinity culture contained more opportunistic pathogen species such as *Vibrio* sp. (MacFariane et al., 1986; Hou et al., 2020).

The differences in the microbial composition of the fish gut microbiota with respect to salinity can be explained by host environmental adaptation. Greater microbial diversity and species richness have been observed in freshwater fish than in marine fish (Zhao et al., 2020; Kim et al., 2021). The gut bacterial community of freshwater fish and marine fish showed distinct clustering, indicating that salinity is a major factor in shaping the gut microbiome of fish (Kim et al., 2021). The abundance of bacterial species in the gut microbiota differed between freshwater and marine aquatic organisms, suggesting the role of salinity in shaping the gut microbiota of Asian sea bass (Morshed et al., 2023).

Striped bass gut microbiota from estuary and marine habitats is dominated by opportunistic fish infections. Higher salinity in coastal waters favored *Vibrio* species in the gut of striped bass. Total gut bacterial counts in striped bass from estuarine environments were consistently higher than those in marine environments. Estuarine environments may provide more favorable conditions for the growth and abundance of gut bacteria in striped bass (MacFariane et al., 1986).

Salinity can affect the distribution of certain classes of *Proteobacteria* in the gut microbiota of pike fry (Dulski et al., 2020). Specifically, freshwater fish harbor more vitamin B12-synthesizing bacteria, such as *Cetobacterium somerae*, to meet their dietary needs, while seawater fish have a greater abundance of bacteria involved in sulfur metabolism (Kim et al., 2021). The pathogenic bacterium *Vibrio harveyi* is mentioned as a biotic factor that significantly affects the microbiota of the shrimp gut. It has been suggested that gut microbial diversity is reduced when the host is infected with pathogens (Deris et al., 2022).

The proportion of bacteria considered opportunistic increased, while the proportion of bacteria considered commensal or beneficial decreased when the host was exposed to hyposaline or hypersaline stress (Zhang et al., 2016a). Under high salinity culture conditions, opportunistic pathogen species, such as *Vibrio* sp., accumulate in the shrimp intestine. The transition to high salinity affects the abundance and presence of certain bacterial species in the shrimp gut and causes differences in gut bacterial signatures (Hou et al., 2020).

Schmidt et al. (2015) observed a complete turnover of dominant bacterial taxa within the host microbiome after salinity acclimation. Even a small increase in salinity led to changes in the microbiome of the fish. Competitive interactions and niche appropriation play an important role in shaping the fish microbiome rather than stochastic colonization by surrounding tank water communities.

pH

Acidic exposure to Tambaqui fish induced dysbiosis in the microbiota of cutaneous mucus, feces, and environmental water. The gut microbiota of Tambaqui showed strong resilience due to the ability of fish to migrate between neutral and acidic pH water for feeding during the rainy season (Sylvain et al., 2016). The drop in water pH was found to be related to the occurrence and severity of white feces disease (WFD) in shrimp (Alfiansah et al., 2020).

DISSOLVED OXYGEN

A strong association between gut microbiome β-diversity and water dissolved oxygen (DO) levels was observed in Mexican tetra, *Astyanax mexicanus*, where microbiome composition differed significantly between riverine and cave-adapted populations. Distinct clustering of gut microbiota was observed at low and high DO levels (Ornelas-García et al., 2018).

Under normoxic conditions, the relative abundance of microbiota associated with various beneficial metabolic pathways increased in *Bostrichthys sinensis*. Hypoxic conditions disrupted the homeostasis of the gut microbiota in *B. sinensis*, leading to a reduction in the diversity and richness of the gut microbiota and changing its composition. However, no disruption of the antioxidant system and major organs was observed in hypoxic environments (Fan et al., 2020).

Hypoxic stress has been found to have significant effects on the intestinal microflora of largemouth bass. Hypoxia-induced oxidative stress in the gut can result in elevated amounts of reactive oxygen species (ROS), which might further disrupt the microbiota. Hypoxia also affects the number of species, species richness, and community composition of microflora, leading to disruption and imbalance of gut microflora (Song et al., 2022).

TEMPERATURE

Specific and dynamic seasonal variations in bacterial population structure have been observed in the freshwater aquaculture system of *Litopenaeus vannamei* (Tang et al., 2014). Water with higher temperatures showed a significantly higher diversity and abundance of bacterial composition compared to water with lower temperatures (Tang et al., 2014).

The amount of bacteria in water and sediment samples varies with temperature fluctuations. The abundance of certain bacterial genera *Pseudomonas* and *Aeromonas* showed a positive correlation with water temperature, while *Flavobacterium* and *Bdellovibrio* showed a negative correlation (Zhang et al., 2016b). Temperatures and ammonium levels were shown to be strongly linked with microbial populations in diseased culture ponds, whereas the pH level and DO were found to be closely associated with the bacterial communities in healthy culture ponds (Sun et al., 2019). In the yellowtail kingfish, non-optimal elevated temperatures of the water (particularly 26°C) resulted in a decrease in the overall number of OTUs associated with Alteromonadaceae in the structure of the microbes of the skin mucosa, intestinal mucosa, and digestive system, leading to poor health (Horlick et al., 2020).

BIOLOGICAL OXYGEN DEMAND (BOD) AND CHEMICAL OXYGEN DEMAND (COD)

The harmful bacterial species *Pseudomonas* and *Aeromonas* were inversely connected with COD, but the harmful bacterial genus *Flavobacterium* and the beneficial bacterial genus *Bdellovibrio* were positively correlated with COD. The COD values correlated negatively with particular bacterial populations in the sediment (Zhang et al., 2016b). Low levels of BOD and CO in the culture tanks affected the common carp (*Cyprinus carpio*) gut microbiome by increasing *Fusobacteria* and *Proteobacteria* and decreasing the overall number of *Cetobacterium somerae* (Mamuad et al., 2021).

AMMONIA, NITRITE, AND NITRATE

Ammoniacal nitrogen is a key environmental element for microbial populations in *Penaeus vannamei* culture. Ammoniacal nitrogen shows a negative link with the pathogenic bacterial species *Pseudomonas* and *Aeromonas*, and a beneficial relationship with the probiotic bacterial genera *Bacillus* (Zhang et al., 2016b).

Nitrite and nitrate levels in aquaculture systems were found to be significantly connected to the prevalence of potentially harmful bacteria such as *Vibrio, Tenacibaculum*, and *Photobacterium*. Variations in ammonia and nitrite levels have also been reported to damage aquatic animals' immunological function and make them more susceptible to disease (Sun et al., 2019).

NUTRIENT AVAILABILITY

Total nitrogen in water correlates positively with *Flavobacterium*, and negatively with *Rhodopseudomonas* (Zhang et al., 2016b). The amounts of nitrogen and phosphorus in the aquatic environment were discovered to have a considerable impact on the gut microbiota of the sea bass, *Lateolabrax japonicus*. Higher nitrogen concentrations have shown a negative correlation with gut bacterial communities by reducing the proportion of probiotics and the abundance and diversity of beneficial bacteria in the gut of *L. japonicus* (Zhu et al., 2023).

MICROPLASTICS

A positive correlation between environmental and gut microplastic abundance was observed, leading to changes in the surrounding water, sediment, and gut microbiome in aquaculture (Yan et al., 2021; Liu et al., 2023). Chronic exposure to microplastics alters the abundance and composition of the gut microbiota and leads to dysbiosis, which adversely affects host health (Fackelmann and Sommer, 2019; Feng et al., 2021).

POLLUTANTS

Pollutants such as pesticides, polychlorinated biphenyls, PCBs, polybrominated diphenyl ethers (PBDEs), heavy metals, nanoparticles, pharmaceuticals, microplastics, and endocrine-disrupting chemicals can induce qualitative and functional changes in the gut microbiota of aquatic organisms. Exposure to pollutants can be beneficial for microbiota that have higher resistance to pollutants (Evariste et al., 2019).

FEED-BIOTECHNOLOGY

Biotechnology applies biology to develop technologies to improve the advancement in various fields. The application of biotechnology has improved the health and well-being of animals and aquatic organisms. The major applications of biotechnology in aquaculture includes the improvement of nutritional value or digestibility of feed through fermentation technology, maintaining water quality by the application of probiotics, development of genetically improved varieties, development of certain feed additives like immunoceuticals and probiotics. With the rapid development of aquaculture, application of biotechnology finds its way to improve feed quality and fish nutrition. Insights into digestibility and nutrient utilization promote the efficient feed utilization. Studies on nutrient–gene interactions, functional roles of nutrients, and metabolic programming of fish will help to achieve the goal of maximum utilization of feed nutrients in fish.

APPLICATION OF BIOTECHNOLOGY IN NUTRITION METHODOLOGIES

The current trend of replacing conventional feed ingredients or improving the nutritional value of plant-based ingredients applies the biotechnological tools like solid-state fermentation (SSF), addition of exogenous enzymes, or bioconversion of feed ingredients or feed. The fermentation of feed ingredients using various microorganisms like *Bacillus, Saccharomyces, Aspergillus*, or *Lactobacillus* improves the digestibility, increases the nutritional value, and maintains a healthy microbial load in the feed. This can support the beneficial microflora and fauna in the gut wall and avoid the establishment of pathogenic microbes such as coliforms (Gatesoupe, 1999, 2005). The microbial fermentation of ingredients results in the availability of prebiotics and metabolites for the fish which can support the digestion, metabolism, and gut health of the animals. Microbial fermentation results in reduced anti-nutritional factors and toxins in fish feed (Shamna et al., 2015). Similarly, the availability of microbial enzymes like cellulase, xylanase, and phytase in the fermented feed ingredients can improve the digestibility and mineral availability in fish (Kaushik et al., 1995).

BACILLUS AS A BIOTECHNOLOGY TOOL IN AQUACULTURE

Bacillus species can produce heat and cold-resistant spores which will help them to be dormant for years (Nicholson, 2004). *Bacillus subtilis* is also FDA approved as non-toxic to humans and animals and is hence commonly employed in biotechnological procedures. *B. subtilis* can degrade proteins, polysaccharides, and complex lipids, as well as utilize carbon and nitrogen sources efficiently (Ochoa-Solano and Olmos-Soto, 2006; Olmos et al., 2011). *B. subtilis* also secrete enzymes which are helpful in the purification of the same and its application in feed (Gu et al., 2018) and also secrete many peptide antibiotics, secondary metabolites, proteins, etc. (Stein 2005; Ongena and Jacques 2008; Olmos-Soto and Contreras-Flores, 2003; Harwood et al., 2018). Furthermore, *L. vannamei* fed with plant components and *B. subtilis* showed improved growth, survival, and resistance to ammonia and oxygen levels (Olmos et al., 2011). *B. subtilis* produces enzymes that effectively break down plant components and improve digestion, making it one of the best probiotic strains (Arellano and Olmos 2002; Ochoa-Solano and Olmos-Soto, 2006; Akhter et al. 2015; Nawaz et al., 2018). The microbial inhibitory property of *B. subtilis* can be attributed to the release of antibacterial substances, alteration of pH of the growing medium, and efficiency in nutrient utilization by outcompeting the pathogenic strains in addition to the tolerance to low pH (Iwashita, et al., 2015). There are evidence of extracellular mechanisms of the species which produce cellulase, lipase, and protease which can enhance the growth parameters and nutrient utilization as evidenced in Nile tilapia and sea cucumber (Liu et al., 2016; Zhao et al., 2012).

Bacillus cereus, B. subtilis, B. coagulans, and *B. licheniformis* are some of the important species that find several applications in the fields of biotechnology, especially as probiotics for human, aquaculture, and veterinary uses (Muras et al., 2021). *B. licheniformis* is shown to have beneficial effects on alleviating several disease conditions in humans such as liver (Shobharani et al., 2019), cardiovascular, and neurological diseases (Ramirez-Olea et al., 2022). *Bacilli* are naturally capable of producing a wide range of useful extracellular enzymes, including amylase, glucoamylase, protease, pectinase, and cellulase, which are used as functional food supplements in a variety of commercial and pharmaceutical applications (Ghani et al., 2013). Besides, they also produce several proteins, antimicrobial compounds, vitamins, pigments, and carotenoids, which add to their importance in the food chain (Elshaghabee et al., 2017).

Combination use of *Lactobacillus* sp. and *Bacillus* sp. constitutes an important component of the intestinal microbiota. However, these were regarded as transient gut microorganisms for a considerable period of time owing to their soil origin. Of late, several studies have reported the presence of actively colonizing *Bacillus* sp. in the GIT of humans and animals (Ilinskaya et al., 2017). Vegetative cells of *Bacillus* harbor the rhizosphere of decomposing plants, and, in soil, they occur in the form of spores that germinate in the gut of humans and animals (Hong et al., 2009). Compared to other anaerobic bacteria, facultatively aerobic bacilli constitute a lesser proportion of the intestinal microflora, still influencing the gut microbial community in a big way (Ilinskaya et al., 2017).

Traditional fermented products are a rich source of bacteria belonging to *Bacillus* sp., which carry beneficial bioactive properties with anticancer, antioxidant, cardioprotective, and neurological effects, to mention a few among many others. These bacteria are commonly used as major components of probiotic preparations and commercially available for various purposes. Probiotics refer to live microorganisms fed to the host in required amounts in order to improve the host health (FAO/WHO, 2002). These compounds find widespread application as animal feeds, especially in the poultry and aquaculture sectors, which in addition to alleviating gastrointestinal orders, promote the growth and other desirable physiological characteristics of cultured animals. The use of probiotics is boosted by the development of antimicrobial resistance as a result of unregulated use of commercial antibiotics. Probiotics are used as a safe and cost-effective alternative to synthetic antibiotics, where the natural gut microbiome of the animals remains undisturbed, resulting in health enhancement of the animals (Hong et al., 2005).

Microbial organisms play several roles other than probiotic functions in the GIT which has industrial applications. One such important characteristic is the production of exogenous enzymes that can be judiciously used for the SSF of plant-based ingredients (Gosh et al., 2018). SSF is a technique widely used for detoxifying and reducing anti-nutritional factors in aquafeed ingredients (Rajesh and Raj, 2010; Shamna et al., 2015, 2017). A few reports showed that SSF by bacilli isolated from fish gut reduced the anti-nutritional factors in it mainly because of phytase, cellulose, or xylanase-like exogenous enzymes which was produced by bacilli (Bairagi et al., 2004; Ramachandran and Ray 2007; Khan and Ghosh, 2013a, 2013b). Ramachandran and Ray (2007) reported that *Bacillus* species could reduce the antinutrients and fiber content when black gram seed meal was fermented. *Bacillus cereus* (Sn-1 & Sn-3) when used as a fermenting agent for aquafeed increased the protein level and suggested that this method can enhance nutritional value. Sesame oil cake bio-fermented with *B. subtilis subsp. Subtilis* reduced not only the fiber but also tannin, trypsin inhibitor, and phytic acid (Das and Ghosh, 2015). One of the dominant cellulolytic bacteria in grass carp is *Bacillus* (Li et al., 2016) and Banerjee et al. (2016) identified five different species of *Bacillus* as potential NSP-degrading bacteria in Indian Major Carps and Exotic Carps. Bacteria isolated from rohu gut (*B. circulans, B. cereus*, and *B. pumilus)* from *Labeo bata* gut (*B. licheniformis and B. Subtilis*) and from tilapia gut (*Bacillus circulans* and *Bacillus megaterium*) showed cellulose degrading property during SSF (Ghosh et al., 2002; Saha et al., 2006; Mondal et al., 2010). *Bacillus* species like *B. subtilis* and *B. velesensis* isolated from Pacu gut (Peixoto et al., 2011) and carp gut (*B. altitudinis, B. pumilus, B. tequilensis, B. megaterium*, and *B. pumilus*) (Banerjee et al., 2016) exhibited the capability of xylan degrading property. *B. pumillus* SE5 isolated from the gut of *Epinephelus coioides* reduced the anti-nutrients in soybean meal during SSF and supported the replacement of fishmeal up to 40% with fermented soybean meal in Japanese seabass diet (Rahimnejad et al., 2019). Mukherjee et al. (2016) isolated autochthonus *Bacillus* species (*Bacillus stratosphericus* CM1FG7, *Bacillus aerophilus* CM1HG5, and *Bacillus licheniformis* CM3FG19) from mrigal gut which exhibited xylanase-producing activity. SSF using *Bacillus tequilensis* (isolated from grass carp intestine) and *Bacillus pumilus* (isolated from rohu intestine) exhibited xylan-degrading activity (Banerjee et al., 2016). Cotton seed meal when fermented with *Bacillus coagulants*, the gossypol content reduced from 923.80 to 167.90 mg/kg (Zhang et al., 2022). Ray et al. (2007) reported that *Bacillus* isolated from common carp gut, *Bacillus subtilis* CY5 and *B. circulans* TP3, is aiding in cellulose production and improving the digestibility of plant-based feeds. The protease-producing capacity of *B. cereaus* isolated from mullet gut is assessed by Esakkiraj et al. (2009) and observed that at 3% NaCl, it gave an optimum yield of protease. *Lemna polyrhiza* leaf meal biotreated with fish gut-isolated *Bacillus* species reduced the fiber content and antinutrients and enhanced the free fatty acids and amino acid contents when incorporated in the diet up to 30% (Bairagi et al., 2002a, 2002b). Another research by Bairagi et al. (2004) reported that *Leucaena leucocephala* leaf meal inoculated with Bacilli isolated from fish intestine decreased crude fiber, cellulose, and anti-nutrients like tannin and mimosine. Grass pea seed meal fermented with gut-isolated *Bacillus* sp. decreased anti-nutritional factors and increased apparent digestibility (Ramachandran et al., 2005). Feed conversion and protein efficiency were improved when black gram pre-treated with *Bacillus* from common carp gut (Ramachandran and Ray, 2007). A feeding trial with *Bacillus* sp.-fermented Eichhornia leaf meal showed a higher growth rate and feed conversion efficiency in fish at 30% inclusion levels (Saha and Ray, 2011a, 2011b).

The microbial digestion of fibers with the support of exogenous enzymes is very well studied in aquafeed and observed that such digestion results in the production of several metabolites like vitamin B12, SCFAs, and polysaccharides (Rowland et al., 2018; Jisha et al., 2023). Moreover, the removal of anti-nutrients not only enhances the digestibility of feed but also enhances the content of other nutrients (Ghosh et al., 2018). There were a few attempts to improve the nutritional values and bioactivity of soybean meal using *Bacillus* strain. Zhang et al. (2018) studied the effects of fermentation of soybean meal with *Bacillus subtilis natto* on nutritional composition and anti-nutrient removal and reported that total protein and total essential amino acids were increased, while total fiber content was decreased after fermentation. *Bacillus* isolates, TP-5 and TP-7, enhanced the nutritional profile and digestibility of soybean meal (Nualkul et al., 2022). Furthermore, Das and Ghosh (2015) reported that treating sesame oil cake with *B. subtilis subsp. Subtilis* enhanced the fatty acids, minerals, and amino acid content. However, Banerjee et al. (2016) observed a decrease in lysine and leucine levels and an increase in other amino acids when fermentation of soybean meal is done with *Bacillus*. The essential amino acid concentration was increased when soy pulp fermented with *Bacillus cereus* after the eighth day (Imran and Wei, 2019a, 2019b). Roy et al. (2016) reported that phytase-producing bacilli (*B. licheniformis* LF1 and *B. licheniformis* LH1) isolated from rohu reduce phytate content of sesame seed meal. Mukherjee et al. (2016) isolated *Bacillus aerophilus* CM1HG5 from mrigal gut-exhibited phytase secretion capacity. Serra et al. (2019) observed that the marine bacilli isolated from the intestine of European seabass hydrolyzed non-starch polysaccharides and the isolates showed a high or medium carbohydrate hydrolyzing potential. One study showed that *B. coagulans* used as a fermenting agent for different aquafeed ingredients like soybean meal, mixed ingredients, and wheat bran revealed that essential amino acids, crude fat, and protein were increased with an increase in fermenting time till 96 h (Imelda et al., 2008).

Adelina et al. (2020) observed the utilization of feather meal fermented using *B. subtilis* in the diet of silver pompano, *Trachinotus blochii*. The microbial hydrolysis approach in fermentation is a keratin breakdown process used to boost the protein digestibility of chicken feathers (Brandelli et al., 2015). *Bacillus subtilis* is one of the keratins-degrading bacteria which produce an enzyme that can hydrolyze a variety of soluble and insoluble proteins. In the study, the bacteria produced keratinase, protease, lipase, and amylase enzymes, which aid in the digestion of complex compounds in fish feed. Another report showed that fermentation of feather meal with *B. subtilis* enhanced the protein digestion rate from 39.09 to 48.75% (Adelina and Lukistyowati, 2017). Adelina and Lukistyowati (2017) also observed that fish fed a diet containing 10% fermented feather meal exhibited the best mean specific growth rate of *Lates calcarifer* and had the highest feed digestibility of 65.16%. Similarly, the digestibility of dry chicken feather treated with *B. subtilis* increased to 54.20% (Zerdani et al., 2004). From these studies, it can be well understood that the autochthonous exogenous enzyme produced by bacilli improves the digestibility of plant-based ingredients and reduces anti-nutrients in it.

Wang et al. (2022) attempted to study the effect of *B. subtilis* HGcc-1 isolated from the intestine of tilapia as an agent for SSF of rice bran and soybean meal. In the study, the fermented product was added in a high-fat diet and fed to zebrafish to understand the gut health, microbial growth, fat metabolism, and liver health. The study supported the effect of fermented products by *Bacillus subtilis* HGcc-1 at 0.1% upregulated expression of immune and fat hydrolyzing genes.

Bacillus species has been found to be abundant in the gut microbiome of several freshwater fishes of India as it has high adhesion capacity and colony-forming ability (Ghosh and Setlow, 2010; Dutta and Ghosh, 2015; Mukherjee et al., 2017). Similarly, *Bacillus* sp. is a recognized probiotic component in shrimp feeds due to its high shelf life. However, the application of *Bacillus* in feed biotechnology and immune-physiological effects in fish needs more investigations (Gosh et al., 2018).

METHODS OF DELIVERY

The mode of delivery is critical in ensuring a probiotic's optimal efficacy. It is critical to investigate the best method for administering or delivering probiotics in order to keep the antibiotics alive and functional. The administration of probiotics is primarily determined by parameters such as the kind and mode of supplementation, dosage level, and duration of treatment. *Bacillus* probiotics are most commonly used in aquaculture via feed and straight addition to water.

ADDITION IN DIET (ORAL)

Because the fish gut contains enzymes, the probiotic provided with the diet must be firm with the ability to colonize the GIT. The addition of *Bacillus* species to the meal not only decreases infection but also stimulates development and immunity.

THROUGH LIVE FEED (ORAL)

Live feed organisms can be encapsulated and administered in which the organisms act as strong vectors for transferring probiotic strains into the fish. The method works incredibly well for injecting probiotics into the intestinal tracts of young fishes. This process is costly and received less attention than other methods of examination.

WATER APPLICATION (BATH)

Direct application of microbial strains into the water promotes the health of the fish by increasing water quality, modifying the composition of microbes present in the water and sediments, and creating an environment that is more favorable to aquatic organism welfare.

CONCLUSION

Aquaculture has recently transitioned from conventional to scientific farming. The importance of microbiome studies and interplay in farming is emphasized by the "one health approach" in sustainable aquaculture production. For decades, many *Bacillus* species have been employed in aquaculture as probiotics and bioremediation. "Omic" technology can be used to learn more about *Bacillus*' method of action on aquatic creatures and ecosystems. Furthermore, there has been little research on how *Bacillus* and other interacting microbes work in the culture system. The latest concept of post-biotics production through SSF stresses the importance of bacterial fermentation, especially a versatile organism like *Bacillus* in feed biotechnology. The increasing urbanization of aquaculture has raised our reliance on closed systems like RAS, biofloc, and aquaponic systems, necessitating additional research into the extent and potential of *Bacillus* application in these systems. *Bacillus* has a good impact on aquaculture through a variety of processes because of its probiotic effect which is determined by various methods of actions. Because of its enzymatic breakdown capacity, ability to utilize organic materials as an energy and carbon source, and heterotrophic nitrification, it is bioremediating in nature.

REFERENCES

Adelina A., Feliatra F., Siregar Y. I. and Suharman I., 2020. Utilization of feather meal fermented Bacillus subtilis to replace fish meal in the diet of silver pompano, *Trachinotus blochii* (*Lacepede*, 1801). AACL Bioflux *13*(1): pp.100–108.

Adelina, A. N. and Lukistyowati I., 2017. [Utilization of chicken feathers as a protein substitute for fish meal in white snapper diet (*Lates calcarifer*, Bloch)]. Research Report, Research and Community Service Institution, Riau University, 58 pp. [in Indonesian)]

Akhter, N., Wu, B., Memon, A. M. and Mohsin, M., 2015. Probiotics and prebiotics associated with aquaculture: A review. *Fish & Shellfish Immunology*, *45*(2), pp.733–741.

Alfiansah, Y. R., Peters, S., Harder, J., Hassenrück, C. and Gärdes, A., 2020. Structure and co-occurrence patterns of bacterial communities associated with white faeces disease outbreaks in Pacific white-leg shrimp *Penaeus vannamei* aquaculture. *Scientific Reports*, *10*(1), p.11980.

Almand, E. A., Moore, M. D., Outlaw, J. and Jaykus, L. A., 2017. Human norovirus binding to select bacteria representative of the human gut microbiota. *PLoS ONE* 12, p.e0173124. doi: 10.1371/journal.pone.0173124

Angthong, P., Uengwetwanit, T., Arayamethakorn, S., Chaitongsakul, P., Karoonuthaisiri, N. and Rungrassamee, W., 2020. Bacterial analysis in the early developmental stages of the black tiger shrimp (*Penaeus monodon*). *Scientific Reports*, *10*(1), p.4896.

Arellano, C. F. and Olmos, S. J., 2002. Thermostable a-1,4- and a-1–6-glucosidase enzymes from Bacillus spp. isolated from a marine environment. *World Journal of Microbiology and Biotechnology*, *18*, pp.791–795.

Bairagi, A., Sarkar Ghosh, K., Sen, S. K. and Ray, A. K., 2002a. Enzyme producing bacterial flora isolated from fish digestive tracts. *Aquaculture International*, *10*, pp.109–121.

Bairagi, A., Sarkar Ghosh, K., Sen, S. K. and Ray, A. K., 2002b. Duck weed (*Lemna polyrhiza*) leaf meal as a source of feedstuff in formulated diets for rohu (*Labeo rohita* Ham.) fingerlings after fermentation with a fish intestinal bacterium. *Bioresource Technology 85*, pp.17–24.

Bairagi, A., Sarkar Ghosh, K., Sen, S. K. and Ray, A. K., 2004. Evaluation of nutritive value of *Leucaena leucocephala* leaf meal inoculated with fish intestinal bacteria *Bacillus subtilis* and *Bacillus circulans* in formulated diets for rohu, *Labeo rohita* (Hamilton) fingerlings. *Aquaculture Research 35*, pp.436–446.

Bakke, I., Skjermo, J., Vo, T. A. and Vadstein, O., 2013a. Live feed is not a major determinant of the microbiota associated with cod larvae (G adus morhua). *Environmental Microbiology Reports*, *5*(4), pp.537–548.

Bakke, T., Klungsøyr, J. and Sanni, S., 2013b. Environmental impacts of produced water and drilling waste discharges from the Norwegian offshore petroleum industry. *Marine Environmental Research*, *92*, pp.154–169.

Banerjee, S., Mukherjee, A., Dutta, D. and Ghosh, K., 2016. Non-starch polysaccharide degrading gut bacteria in Indian major carps and exotic carps. *Jordan Journal of Biological Sciences 9*(1), pp.69–78.

Bates, J. M., Mittge, E., Kuhlman, J., Baden, K. N., Cheesman, S. E. and Guillemin, K., 2006. Distinct signals from the microbiota promote different aspects of zebrafish gut differentiation. *Developmental Biology*, *297*(2), pp.374–386.

Bhatia, K., Thakur, K., Sharma, D., Mahajan, D., Sharma, A. K., Brar, B., Kumari, H., Pankaj, P. P. and Kumar, R., 2023. Gut Microbiota of *Salmo trutta fario* and *Oncorhynchus mykiss*: Implications for fish health and aquaculture management. *International Journal of Oceanography & Aquaculture*, *7*(2), p.000242.

Bi, S., Lai, H., Guo, D., Liu, X., Wang, G., Chen, X., Liu, S., Yi, H., Su, Y. and Li, G., 2021. The characteristics of intestinal bacterial community in three omnivorous fishes and their interaction with microbiota from habitats. *Microorganisms*, *9*(10), p.2125.

Bolnick, D. I., Snowberg, L. K., Hirsch, P. E., Lauber, C. L., Org, E., Parks, B., Lusis, A. J., Knight, R., Caporaso, J. G. and Svanbäck, R., 2014. Individual diet has sex-dependent effects on vertebrate gut microbiota. *Nature Communications*, *5*(1), p.4500.

Brandelli, A., Sala, L. and Kalil, S. J., 2015. Microbial enzymes for bioconversion of poultry waste into added-value products. *Food Research International*, *73*, pp.3–12.

Chaiyapechara, S., Uengwetwanit, T., Arayamethakorn, S., Bunphimpapha, P., Phromson, M., Jangsutthivorawat, W., Tala, S., Karoonuthaisiri, N. and Rungrassamee, W., 2022. Understanding the host-microbe-environment interactions: Intestinal microbiota and transcriptomes of black tiger shrimp *Penaeus monodon* at different salinity levels. *Aquaculture*, *546*, p.737371.

Clements, K. D., Pasch, I. B., Moran, D. and Turner, S. J., 2007. Clostridia dominate 16S rRNA gene libraries prepared from the hindgut of temperate marine herbivorous fishes. *Marine Biology*, *150*, pp.1431–1440.

Cordero, H., Guardiola, F. A., Tapia-Paniagua, S. T., Cuesta, A., Meseguer, J., Balebona, M. C., Moriñigo, M.Á. and Esteban, M.Á., 2015. Modulation of immunity and gut microbiota after dietary administration of alginate encapsulated *Shewanella putrefaciens* Pdp11 to gilthead seabream (Sparus aurata L.). *Fish & Shellfish Immunology*, *45*(2), pp.608–618.

Cornejo-Granados, F., Lopez-Zavala, A. A., Gallardo-Becerra, L., Mendoza-Vargas, A., Sánchez, F., Vichido, R., Brieba, L. G., Viana, M. T., Sotelo-Mundo, R. R. and Ochoa-Leyva, A., 2017. Microbiome of Pacific White leg shrimp reveals differential bacterial community composition between Wild, Aquacultured and AHPND/EMS outbreak conditions. *Scientific Reports*, *7*(1), p.11783.

Das, P. and Ghosh, K. (2015) Improvement of nutritive value of sesame oil cake in formulated diets for rohu, *Labeo rohita* (Hamilton) after bio-processing through solid state fermentation by a phytase-producing fish gut bacterium. *International Journal of Aquatic Biology* 3, pp.89–101.

Dehler, C. E., Secombes, C. J. and Martin, S. A., 2017. Environmental and physiological factors shape the gut microbiota of Atlantic salmon parr (*Salmo salar L.*). *Aquaculture*, *467*, pp.149–157.

Denison, E. R., Rhodes, R. G., McLellan, W. A., Pabst, D. and Erwin, P. M., 2020. Host phylogeny and life history stage shape the gut microbiome in dwarf (*Kogia sima*) and pygmy (*Kogia breviceps*) sperm whales. *Scientific Reports*, *10*(1), pp.1–13.

Deris, Z. M., Iehata, S., Gan, H. M., Ikhwanuddin, M., Najiah, M., Asaduzzaman, M., Wang, M., Liang, Y., Danish-Daniel, M., Sung, Y. Y. and Wong, L. L., 2022. Understanding the effects of salinity and *Vibrio harveyi* on the gut microbiota profiles of *Litopenaeus vannamei*. *Frontiers in Marine Science*, *9*, p.974217.

Dimitroglou, A., Merrifield, D. L., Carnevali, O., Picchietti, S., Avella, M., Daniels, C., Güroy, D. and Davies, S. J., 2011. Microbial manipulations to improve fish health and production–a Mediterranean perspective. *Fish & Shellfish Immunology*, *30*(1), pp.1–16.

Diwan, A. D., Harke, S. N., Gopalkrishna and Panche, A. N., 2022. Aquaculture industry prospective from gut microbiome of fish and shellfish: An overview. *Journal of Animal Physiology and Animal Nutrition*, *106*(2), pp.441–469.

Diwan, A.D., Harke, S. N. and Panche, A. N., 2023. Host-microbiome interaction in fish and shellfish: An overview. *Fish and Shellfish Immunology Reports*, p.100091.

Dulski, T., Kujawa, R., Godzieba, M. and Ciesielski, S., 2020. Effect of salinity on the gut microbiome of pike fry (Esox lucius). *Applied Sciences*, *10*(7), p.2506.

Dutta, D. and Ghosh, K., 2015. Screening of extracellular enzyme-producing and pathogen inhibitory gut bacteria as putative probiotics in mrigal, *Cirrhinus mrigala* (Hamilton, 1822). *International Journal of Fisheries and Aquatic Studies*, *2*(4), pp.310–318.

Ehling-Schulz, M., Lereclus, D. and Koehler, T. M., 2019. The Bacillus cereus group: *Bacillus* species with pathogenic potential. *Microbiology Spectrum*, *7*(3), pp.10–1128.

Elshaghabee, F. M. F., Rokana, N., Gulhane, R. D., Sharma, C. and Panwar, H., 2017. Bacillus as potential probiotics: Status, concerns, and future perspectives. *Frontiers in Microbiology*, 8, p.1490. doi: 10.3389/fmicb.2017.01490

Esakkiraj, P., Immanuel, G., Sowmya, S. M., Iyapparaj, P. and Palavesam, A., 2009 Evaluation of protease-producing ability of fish gut isolate Bacillus cereus for aqua feed. *Food and Bioprocess Technology*, 2, pp.383–390.

Evariste, L., Barret, M., Mottier, A., Mouchet, F., Gauthier, L. and Pinelli, E., 2019. Gut microbiota of aquatic organisms: A key endpoint for ecotoxicological studies. *Environmental Pollution*, *248*, pp.989–999.

Fackelmann, G. and Sommer, S., 2019. Microplastics and the gut microbiome: How chronically exposed species may suffer from gut dysbiosis. *Marine Pollution Bulletin*, *143*, pp.193–203.

Fan, J., Chen, L., Mai, G., Zhang, H., Yang, J., Deng, D. and Ma, Y., 2019. Dynamics of the gut microbiota in developmental stages of *Litopenaeus vannamei* reveal its association with body weight. *Scientific Reports*, *9*(1), pp.1–10.

Fan, S., Li, H. and Zhao, R., 2020. Effects of normoxic and hypoxic conditions on the immune response and gut microbiota of *Bostrichthys sinensis*. *Aquaculture*, *525*, p.735336.

FAO/WHO, 2002. *Guidelines for the Evaluation of Probiotics in Food. Food and Agriculture Organization of the United Nations and World Health Organization Working Group Report*. Rome: Food and Agriculture Organization.

Feng, S., Zeng, Y., Cai, Z., Wu, J., Chan, L. L., Zhu, J. and Zhou, J., 2021. Polystyrene microplastics alter the intestinal microbiota function and the hepatic metabolism status in marine medaka (*Oryzias melastigma*). *Science of the Total Environment*, *759*, p.143558.

Fidopiastis, P.M., Bezdek, D.J., Horn, M.H. and Kandel, J.S., 2006. Characterizing the resident, fermentative microbial consortium in the hindgut of the temperate-zone herbivorous fish, *Hermosilla azurea* (Teleostei: Kyphosidae). *Marine Biology*, *148*, pp.631–642.

Gatesoupe, F.J., 1999. The use of probiotics in aquaculture. *Aquaculture*, *180*(1–2), pp.147–165.

Gatesoupe, F.J., 2007. Live yeasts in the gut: Natural occurrence, dietary introduction, and their effects on fish health and development. *Aquaculture*, *267*(1–4), pp.20–30.

Gatesoupe, J., 2005. Probiotics and prebiotics for fish culture, at the parting of the ways. *Aqua Feeds: Formulation & Beyond*, *2*(3), pp.3–5.

Ghanbari, M., Kneifel, W. and Domig, K.J., 2015. A new view of the fish gut microbiome: Advances from next-generation sequencing. *Aquaculture*, *448*, pp.464–475.

Ghani, M., Ansari, A., Aman, A., Zohra, R. R., Siddiqui, N. N. and Qader, S. A. U. 2013. Isolation and characterization of different strains of *Bacillus licheniformis* for the production of commercially significant enzymes. *Pakistan Journal of Pharmaceutical Sciences*, *26*, pp.691–697.

Ghosh, K., Sen, S. K., Ray A. K. (2002) Characterization of bacilli isolated from gut of rohu, *Labeo rohita*, fingerlings and its significance in digestion. *Journal of Applied Aquaculture*, *12*, pp.33–42.

Ghosh, S. and Setlow, P., 2010. The preparation, germination properties and stability of super dormant spores of Bacillus cereus. *Journal of Applied Microbiology*, *108*(2), pp.582–590.

Giatsis, C., Sipkema, D., Smidt, H., Heilig, H., Benvenuti, G., Verreth, J. and Verdegem, M., 2015. The impact of rearing environment on the development of gut microbiota in tilapia larvae. *Scientific Reports*, *5*(1), pp.1–15.

Gómez, G.D. and Balcázar, J.L., 2008. A review on the interactions between gut microbiota and innate immunity of fish. *FEMS Immunology & Medical Microbiology*, *52*(2), pp.145–154.

Gu, Y., Xu, X., Wu, Y., Niu, T., Liu, Y., Li, J., Du, G. and Liu, L., 2018. Advances and prospects of Bacillus subtilis cellular factories: From rational design to industrial applications. *Metabolic Engineering*, *50*, pp.109–121.

Hansen, G.H. and Olafsen, J.A., 1999. Bacterial interactions in early life stages of marine cold water fish. *Microbial Ecology*, *38*, pp.1–26.

Harwood, C.R., Mouillon, J.M., Pohl, S. and Arnau, J., 2018. Secondary metabolite production and the safety of industrially important members of the Bacillus subtilis group. *FEMS Microbiology Reviews*, *42*(6), pp.721–738.

Henderson, R.J. and Millar, R.M., 1998. Characterization of lipolytic activity associated with a Vibrio species of bacterium isolated from fish intestines. *Journal of Marine Biotechnology*, *6*(3), pp.168–173.

Hong, H. A., Duc le, H. and Cutting, S. M., 2005. The use of bacterial spore formers as probiotics. *FEMS Microbiology Reviews*, *29*, pp.813–835. doi: 10.1016/j.femsre. 2004.12.001

Hong, H. A., To, E., Fakhry, S., Baccigalupi, L., Ricca, E. and Cutting, S. M., 2009. Defining the natural habitat of *Bacillus* spore-formers. *Research in Microbiology*, *160*, pp.375–379. doi: 10.1016/j.resmic.2009.06.006

Horlick, J., Booth, M.A. and Tetu, S.G., 2020. Alternative dietary protein and water temperature influence the skin and gut microbial communities of yellowtail kingfish (*Seriola lalandi*). *PeerJ*, *8*, p.e8705.

Hou, D., Huang, Z., Zeng, S., Liu, J., Weng, S. and He, J., 2018. Comparative analysis of the bacterial community compositions of the shrimp intestine, surrounding water and sediment. *Journal of Applied Microbiology*, *125*(3), pp.792–799.

Hou, D., Zhou, R., Zeng, S., Wei, D., Deng, X., Xing, C., Yu, L., Deng, Z., Wang, H., Weng, S. and He, J., 2020. Intestine bacterial community composition of shrimp varies under low-and high-salinity culture conditions. *Frontiers in Microbiology*, *11*, p.589164. https://doi.org/10.1016/B978-0-12-804024-9.00011-2.

Huang, F., Pan, L., Song, M., Tian, C. and Gao, S., 2018. Microbiota assemblages of water, sediment, and intestine and their associations with environmental factors and shrimp physiological health. *Applied Microbiology and Biotechnology*, *102*, pp.8585–8598.

Huang, L., Guo, H., Chen, C., Huang, X., Chen, W., Bao, F., Liu, W., Wang, S. and Zhang, D., 2020. The bacteria from large-sized bioflocs are more associated with the shrimp gut microbiota in culture system. *Aquaculture*, *523*, p.735159.

Huang, Z., Li, X., Wang, L. and Shao, Z., 2014. Changes in the intestinal bacterial community during the growth of white shrimp, *Litopenaeus vannamei*. *Aquaculture Research*, *47*(6), pp.1737–1746.

Ilinskaya, O. N., Ulyanova, V. V., Yarullina, D. R. and Gataullin, I. G., 2017. Secretome of Intestinal Bacilli: A natural guard against pathologies. *Frontiers in Microbiology*, *8*, p.1666. https://doi.org/10.3389/fmicb.2017.01666

Imelda, J., Paulraj, R. and Bhatnagar, D, 2008. Effect of solid state fermentation on nutrient composition of selected feed ingredients. *Indian Journal of Fisheries*, *55*(4), pp.327–332.

Imran, S. Z. and Wei, L. S., 2019. Selected essential amino acid enhancement by *Bacillus cereus* from Solid State Fermentation of Soy Pulp Through Carbon Concentration and Fermentation Period. *Journal of Aquaculture Research & Development*, *10*(566), p.2.

Itoi, S., Okamura, T., Koyama, Y. and Sugita, H., 2006. Chitinolytic bacteria in the intestinal tract of Japanese coastal fishes. *Canadian Journal of Microbiology*, *52*(12), pp.1158–1163.

Iwashita, M.K.P., Nakandakare, I.B., Terhune, J.S., Wood, T. and Ranzani-Paiva, M.J.T., 2015. Dietary supplementation with Bacillus subtilis, *Saccharomyces cerevisiae* and *Aspergillus oryzae* enhance immunity and disease resistance against *Aeromonas hydrophila* and *Streptococcus iniae* infection in juvenile tilapia *Oreochromis niloticus*. *Fish & Shellfish Immunology*, *43*(1), pp.60–66.

Jami, M., Ghanbari, M., Kneifel, W. and Domig, K.J., 2015. Phylogenetic diversity and biological activity of culturable *Actinobacteria* isolated from freshwater fish gut microbiota. *Microbiological Research*, *175*, pp.6–15.

Jisha, K., Gayathri, G., Gopikrishnan, V., Song, J.J., Soytong, K. and Prabha, T.R., 2023. Fish gut microbiota: A source of novel metabolites–A review.

Kaushik, S.J., Cravedi, J.P., Lalles, J.P., Sumpter, J., Fauconneau, B. and Laroche, M., 1995. Partial or total replacement of fish meal by soybean protein on growth, protein utilization, potential estrogenic or antigenic effects, cholesterolemia and flesh quality in rainbow trout, *Oncorhynchus mykiss. Aquaculture, 133*(3–4), pp.257–274.

Khan, A. and Ghosh, K., 2013a. Phytic acid-induced inhibition of digestive protease and a-amylase in three Indian major carps: An *in vitro* study. *Journal of the World Aquaculture Society, 44*, pp.853–859.

Khan, A. and Ghosh, K., 2013b. Evaluation of phytase production by fish gut bacterium, *Bacillus subtilis* for processing of *Ipomea aquatica* leaves as probable aquafeed ingredient. *Journal of Aquatic Food Product Technology, 22*, pp.508–519.

Kim, P.S., Shin, N.R., Lee, J.B., Kim, M.S., Whon, T.W., Hyun, D.W., Yun, J.H., Jung, M.J., Kim, J.Y. and Bae, J.W., 2021. Host habitat is the major determinant of the gut microbiome of fish. *Microbiome, 9*(1), p.166.

Kim, S. K., Guevarra, R. B., Kim, Y. T., Kwon, J., Kim, H., Cho, J. H., et al. (2019). Role of probiotics in human gut microbiome-associated diseases. *Journal of Microbiology and Biotechnology, 29*, pp.1335–1340. doi: 10.4014/jmb.1906.06064

Landsman, A., St-Pierre, B., Rosales-Leija, M., Brown, M. and Gibbons, W., 2019. Impact of aquaculture practices on intestinal bacterial profiles of Pacific white leg shrimp *Litopenaeus vannamei. Microorganisms, 7*(4), p.93.

Larsen, A.M., Mohammed, H.H. and Arias, C.R., 2014. Characterization of the gut microbiota of three commercially valuable warm water fish species. *Journal of applied microbiology, 116*(6), pp.1396–1404.

Lauzon, H.L., Magnadottir, B., Gudmundsdottir, B.K., Steinarsson, A., Arnason, I.O. and Gudmundsdottir, S., 2010. Application of prospective probionts at early stages of Atlantic cod (*Gadus morhua* L.) rearing. *Aquaculture Research, 41*(10), pp.e576–e586.

Leahy, J.G. and Colwell, R.R., 1990. Microbial degradation of hydrocarbons in the environment. *Microbiological Reviews*, 54(3), pp.305–315.

Li, H., Wu, S., Wirth, S., Hao, Y., Wang, W., Zou, H. et al., 2016. Diversity and activity of cellulolytic bacteria, isolated from the gut the gut contents of grass carp (*Ctenopharyngodon idellus*) (Valenciennes) fed on Sudan grass (Sorghum sudanese) or artificial feedstuffs. *Aquaculture Research, 47*, pp.153–164.

Liu, X., Li, D.Q., Zhao, W., Yu, D., Cheng, J.G., Luo, Y., Wang, Y., Yang, Z.X., Yao, X.P., Wu, S.S. and Wang, W.Y., 2019b, June. Sequencing and analysis of gut microbiota in forest musk deer from Qinghai and Sichuan. In *BIBE 2019; The Third International Conference on Biological Information and Biomedical Engineering* (pp.1–7). VDE.

Li, X.M., Zhu, Y.J., Yan, Q.Y., Ringø, E. and Yang, D.G., 2014. Do the intestinal microbiotas differ between paddlefish (*Polyodon spathala*) and bighead carp (*Aristichthys nobilis*) reared in the same pond? *Journal of Applied Microbiology, 117*(5), pp.1245–1252.

Liu, B., Guo, H., Zhu, K., Guo, L., Liu, B., Zhang, N., Jiang, S. and Zhang, D., 2019a. Salinity effect on intestinal microbiota in golden pompano *Trachinotus ovatus* (Linnaeus, 1758). *Israeli Journal of Aquaculture – Bamidgeh, 71*, p.1538.

Liu, H., Guo, X., Gooneratne, R., Lai, R., Zeng, C., Zhan, F. and Wang, W., 2016. The gut microbiome and degradation enzyme activity of wild freshwater fishes influenced by their trophic levels. *Scientific Reports, 6*(1), pp.1–12.

Liu, M.J., Guo, H. Y., Gao, J., Zhu, K. C., Guo, L., Liu, B. S., Zhang, N., Jiang, S. G. and Zhang, D. C., 2023. Characteristics of microplastic pollution in golden pompano (*Trachinotus ovatus*) aquaculture areas and the relationship between colonized-microbiota on microplastics and intestinal microflora. *Science of the Total Environment, 856*, p.159180.

Lokmer, A. and Mathias Wegner, K., 2015. Hemolymph microbiome of Pacific oysters in response to temperature, temperature stress and infection. *The ISME Journal, 9*(3), pp.670–682.

Lozupone, C.A. and Knight, R., 2007. Global patterns in bacterial diversity. *Proceedings of the National Academy of Sciences, 104*(27), pp.11436–11440.

MacDonald, N. L., Stark, J. R. and Austin, B. (1986) Bacterial microflora in the gastrointestinal tract of Dover sole (*Solea solea* L.), with emphasis on the possible role of bacteria in the nutrition of the host. *FEMS Microbiology Letters, 35*, pp.107–111.

MacFariane, R.D., McLaughlin, J.J. and Bullock, G.L., 1986. Quantitative and qualitative studies of gut flora in striped bass from estuarine and coastal marine environments. *Journal of Wildlife Diseases, 22*(3), pp.344–348.

Mamuad, L., Lee, S.H., Jeong, C.D., Ramos, S., Miguel, M., Son, A.R., Kim, S.H., Cho, Y.I. and Lee, S.S., 2021. Ornamental fish, *Cyprinus carpio*, fed with fishmeal replacement *Ptecticus tenebrifer* and *Tenebrio molitor. Aquaculture Research, 52*(3), pp.980–990.

Merrifield, D.L., Dimitroglou, A., Foey, A., Davies, S.J., Baker, R.T., Bøgwald, J., Castex, M. and Ringø, E., 2010. The current status and future focus of probiotic and prebiotic applications for salmonids. *Aquaculture*, *302*(1–2), pp.1–18.

Michl, S.C., Ratten, J.M., Beyer, M., Hasler, M., LaRoche, J. and Schulz, C., 2017. The malleable gut microbiome of juvenile rainbow trout (*Oncorhynchus mykiss*): Diet-dependent shifts of bacterial community structures. *PloS one*, *12*(5), p.e0177735.

Miyake, S., Ngugi, D. K. and Stingl, U., 2015. Diet strongly influences the gut microbiota of sturgeon fishes. *Molecular Ecology*, *24*(3), pp.656–672.

Mondal, S., Roy, T., Ray, A. K. (2010) Characterization and identification of enzyme-producing bacteria isolated from the digestive tract of bata, *Labeo bata*. *Journal of World Aquaculture Society*, *41*, pp.369–376.

Moran, D., Turner, S. J. and Clements, K. D., 2005. Ontogenetic development of the gastrointestinal microbiota in the marine herbivorous fish *Kyphosus sydneyanus*. *Microbial Ecology*, *49*, pp.590–597.

Morshed, S.M., Chen, Y.Y., Lin, C.H., Chen, Y.P. and Lee, T.H., 2023. Freshwater transfer affected intestinal microbiota with correlation to cytokine gene expression in Asian sea bass. *Frontiers in Microbiology*, *14*, p.1097954.

Mountfort, D.O., Campbell, J. and Clements, K.D., 2002. Hindgut fermentation in three species of marine herbivorous fish. *Applied and Environmental Microbiology*, *68*(3), pp.1374–1380.

Mukherjee, A., Dutta, D., Banerjee, S., Ringø, E., Breines, E.M., Hareide, E., Chandra, G. and Ghosh, K., 2016. Potential probiotics from Indian major carp, *Cirrhinus mrigala*. Characterization, pathogen inhibitory activity, partial characterization of bacteriocin and production of exo-enzymes. *Research in Veterinary Science*, 108, pp.76–84.

Mukherjee, S., Sahu, P. and Halder, G., 2017. Microbial remediation of fluoride-contaminated water via a novel bacterium *Providencia vermicola* (KX926492). *Journal of Environmental Management*, *204*, pp.413–423.

Muras A., Romero, M., Mayer, C. and Otero, A. 2021. Biotechnological applications of *Bacillus licheniformis*. *Critical Reviews in Biotechnology*, *41*, pp.609–627. 10.1080/07388551.2021.1873239

Nawaz, A., Bakhsh, J., Irshad, S., Hoseinifar SH, Xiong, H., 2018. The functionality of prebiotics as immunostimulant: Evidences from trials on terrestrial and aquatic animals. *Fish and Shellfish Immunology*, *76*, pp.272–278. doi.org/10.1016/j.fsi.2018.03.004

Naya-Català, F., Piazzon, M.C., Torrecillas, S., Toxqui-Rodríguez, S., Calduch-Giner, J.À., Fontanillas, R., Sitjà-Bobadilla, A., Montero, D. and Pérez-Sánchez, J., 2022. Genetics and nutrition drive the gut microbiota succession and host-transcriptome interactions through the gilthead sea bream (*Sparus aurata*) Production Cycle. *Biology*, *11*(12), p.1744.

Nayak, S.K., 2010. Role of gastrointestinal microbiota in fish. *Aquaculture Research*, *41*(11), pp.1553–1573.

Ni, J., Yan, Q., Yu, Y. and Zhang, T., 2014. Factors influencing the grass carp gut microbiome and its effect on metabolism. *FEMS Microbiology Ecology*, *87*(3), pp.704–714.

Nicholson, W. L. (2004) Ubiquity, longevity, and ecological roles of Bacillus spores. In: Ricca E, Henriques, A. O., Cutting, S. M. (eds) *Bacterial spore formers: Probiotics and emerging applications*. Horizons Bioscience, Norfolk, UK, pp.1–15

Nualkul, M., Yuangsoi, B., Hongoh, Y., Yamada, A., and Deevong, P. 2022. Improving the nutritional value and bioactivity of soybean meal in solid-state fermentation using *Bacillus* strains newly isolated from the gut of the termite *Termes propinquus*. *FEMS Microbiology Letters*, *369*(1), p.fnac044

Ochoa-Solano, J.L. and Olmos-Soto, J., 2006. The functional property of Bacillus for shrimp feeds. *Food Microbiology*, *23*(6), pp.519–525.

Olmos, J., Acosta, M., Mendoza, G. and Pitones, V., 2020. Bacillus subtilis, an ideal probiotic bacterium to shrimp and fish aquaculture that increase feed digestibility, prevent microbial diseases, and avoid water pollution. *Archives of Microbiology*, *202*, pp.427–435.

Olmos-Soto, J. and Contreras-Flores, R., 2003. Genetic system constructed to overproduce and secrete proinsulin in Bacillus subtilis. *Applied Microbiology and Biotechnology*, *62*, pp.369–373.

Ongena, M. and Jacques, P., 2008. *Bacillus lipopeptides*: Versatile weapons for plant disease biocontrol. *Trends in Microbiology*, 16(3), pp.115–125.

Ornelas-García, P., Pajares, S., Sosa-Jiménez, V.M., Rétaux, S. and Miranda-Gamboa, R.A., 2018. Microbiome differences between river-dwelling and cave-adapted populations of the fish Astyanax mexicanus (De Filippi, 1853). *PeerJ*, *6*, p.e5906.

Peixoto S. B., Cladera-Olivera F., Daroit D. J. and Brandelli, A. 2011. Cellulase-producing *Bacillus* strains isolated from the intestine of Amazon basin fish. *Aquaculture Research*, *42*, pp.887–891.

Pérez, T., Balcázar, J. L., Ruiz-Zarzuela, I., Halaihel, N., Vendrell, D., De Blas, I. and Múzquiz, J. L., 2010. Host–microbiota interactions within the fish intestinal ecosystem. *Mucosal Immunology*, *3*(4), pp.355–360.

Perry, W.B., Lindsay, E., Payne, C.J., Brodie, C. and Kazlauskaite, R., 2020. The role of the gut microbiome in sustainable teleost aquaculture. *Proceedings of the Royal Society B*, *287*(1926), p.20200184.

Qazi, J. I., Nadir, S. and Shakir, H. A. 2012. Solid state fermentation of fish feed with amylase producing bacteria. *Punjab University Journal of Zoology*, *27*(1), pp.1–7.

Rahimnejad, S., Lu, K., Wang, L., Song, K., Mai, K., Davis, D. A. and Zhang, C., 2019. Replacement of fish meal with *Bacillus pumillus* SE5 and Pseudozyma aphidis ZR1 fermented soybean meal in diets for Japanese seabass (*Lateolabrax japonicus*). *Fish & Shellfish Immunology*, *84*, pp.987–997.

Rajeev, R., Adithya, K. K., Kiran, G.S. and Selvin, J., 2021. Healthy microbiome: A key to successful and sustainable shrimp aquaculture. *Reviews in Aquaculture*, *13*(1), pp.238–258.

Rajesh, N. and Raj, R.P., 2010. Value addition of vegetable wastes by solid-state fermentation using *Aspergillus niger* for use in aquafeed industry. *Waste Management*, *30*(11), pp.2223–2227.

Ramachandran, S., Bairagi, A. and Ray, A. K., 2005. Improvement of nutritive value of Grass pea (Lathyrus sativus) seed meal in the formulated diets for rohu, *Labeo rohita* (Ham.) fingerlings after fermentation with a fish gut bacterium. *Bioresource Technology*, *96*, pp.1465–1472.

Ramachandran, S., and Ray, A. K., 2007. Nutritional evaluation of fermented black gram seed meal for rohu *Labeo rohita* (Hamilton) fingerlings. *Journal of Applied Ichthyology*, *23*, pp.74–79.

Ramachandran, S. and Ray, A. K., 2008. Effect of different processing techniques on the nutritive value of grass pea, *Lathyrus sativus* L., seed meal in compound diets for Indian major carp rohu, *Labeo rohita* (Hamilton), fingerlings. *Archives of Polish Fisheries*, *16*, pp.189–202.

Ramirez-Olea, H., Reyes-Ballesteros, B. and Chavez-Santoscoy, R. A., 2022. Potential application of the probiotic *Bacillus licheniformis* as an adjuvant in the treatment of diseases in humans and animals: A systematic review. *Frontiers in Microbiology*, *13*, p.993451. doi: 10.3389/fmicb.2022.993451.

Rao, K. P., Chennappa, G., Suraj, U., Nagaraja, H., Raj, A. C. and Sreenivasa, M. Y., 2015. Probiotic potential of Lactobacillus strains isolated from sorghum- based traditional fermented food. *Probiotics and Antimicrobial Proteins*, *7*, pp.146–156. doi: 10.1007/s12602-015-9186-6

Ray, A. K., Bairagi, A., Ghosh, K. S., and Sen, S. K., 2007. Optimization of fermentation conditions for cellulase production by B*acillus subtilis* cy5 and B*acillus circulans* tp3 isolated from fish gut. *Acta Ichthyologica Et Piscatoria*, *37*(1), pp.47–53.

Ray, A. K., Ghosh, K. and Ringø, E. J. A. N., 2012. Enzyme-producing bacteria isolated from fish gut: A review. *Aquaculture Nutrition*, *18*(5), pp.465–492.

Ray, B. and Bhunia, A., 2007. *Fundamental food microbiology*. CRC Press.

Razak, S. A., Griffin, M. J., Mischke, C.C., Bosworth, B. G., Waldbieser, G. C., Wise, D. J., Marsh, T. L. and Scribner, K. T., 2019. Biotic and abiotic factors influencing channel catfish egg and gut microbiome dynamics during early life stages. *Aquaculture*, *498*, pp.556–567.

Ringø, E. Z. Z. V., Zhou, Z., Vecino, J. G., Wadsworth, S., Romero, J., Krogdahl, Å., Olsen, R. E., Dimitroglou, A., Foey, A., Davies, S. and Owen, M., 2016. Effect of dietary components on the gut microbiota of aquatic animals. A never-ending story? *Aquaculture Nutrition*, *22*(2), pp.219–282.

Robinson, G., Caldwell, G. S., Wade, M. J., Free, A., Jones, C. L. and Stead, S. M., 2016. Profiling bacterial communities associated with sediment-based aquaculture bioremediation systems under contrasting redox regimes. *Scientific Reports*, *6*(1), pp.1–13.

Roeselers, G., Mittge, E. K., Stephens, W. Z., Parichy, D. M., Cavanaugh, C. M., Guillemin, K. and Rawls, J. F., 2011. Evidence for a core gut microbiota in the zebrafish. *The ISME Journal*, 5(10), pp.1595–1608.

Romero, J., Díaz, O., Miranda, C.D. and Rojas, R., 2022. Red cusk-eel (*Genypterus chilensis*) gut microbiota description of wild and aquaculture specimens. *Microorganisms*, *10*(1), p.105.

Romero, J. and Navarrete, P., 2006. 16S rDNA-based analysis of dominant bacterial populations associated with early life stages of coho salmon (*Oncorhynchus kisutch*). *Microbial Ecology*, *51*, pp.422–430.

Rowland, I., Gibson, G., Heinken, A., Scott, K., Swann, J., Thiele, I. and Tuohy, K., 2018. Gut microbiota functions: Metabolism of nutrients and other food components. *European Journal of Nutrition*, *57*, pp.1–24.

Roy, T., Dan, S. K., Banerjee, G., Nandi, A., Ghosh, P. and Ray, A. K., 2016. Comparative efficacy of phytase from fish gut bacteria and a commercially available phytase in improving the nutritive value of sesame oilseed meal in formulate diets for fingerlings of rohu, *Labeo rohita* (Actinopterygii: Cypriniformes: Cyprinidae). *Acta Ichthyologica et Piscatoria*, *46*, pp.9–23.

Rungrassamee, W., Klanchui, A., Maibunkaew, S., Chaiyapechara, S., Jiravanichpaisal, P. and Karoonuthaisiri, N., 2014. Characterization of intestinal bacteria in wild and domesticated adult black tiger shrimp (*Penaeus monodon*). *PloS one*, *9*(3), p.e91853.

Saha, S. and Ray, A. K., 2011. Evaluation of nutritive value of water hyacinth (*Eichhornia crassipes*) leaf meal in compound diets for rohu, *Labeo rohita* (Hamilton, 1822) fingerlings after fermentation with two bacterial strains isolated from fish gut. *Turkish Journal of Fisheries and Aquatic Sciences*, *11*, pp.199–209

Saha S., Roy R. N., Sen S. K., Ray A. K. (2006) Characterization of cellulase- producing bacteria from the digestive tract of tilapia, *Oreochromis mossambica* (Peters) and grass carp, *Ctenopharyngodon idella* (Valenciennes). *Aquaculture Research*, *37*, pp.380–388.

Schmidt, V. T., Smith, K. F., Melvin, D. W. and Amaral-Zettler, L. A., 2015. Community assembly of a euryhaline fish microbiome during salinity acclimation. *Molecular Ecology*, *24*(10), pp.2537–2550.

Schultz, M., Burton, J. P. and Chanyi, R. M., 2017. Chapter 11 - Use of *Bacillus* in *Human Intestinal Probiotic Applications*, Editor(s): Martin H. Floch, Yehuda Ringel, W. Allan Walker, The Microbiota in Gastrointestinal Pathophysiology, Academic Press.

Serra, C. R., Almeida, E. M., Guerreiro, I., Santos, R., Merrifield, D.L., Tavares, F., Oliva-Teles, A. and Enes, P., 2019. Selection of carbohydrate-active probiotics from the gut of carnivorous fish fed plant-based diets. *Scientific Reports*, *9*(1), p.6384.

Shamna, N., Sardar, P., Sahu, N. P., Pal, A. K., Jain, K. K. and Phulia, V., 2015. Nutritional evaluation of fermented jatropha protein concentrate in *Labeo rohita* fingerlings. *Aquaculture Nutrition*, *21*(1), pp.33–42.

Shamna, N., Sardar, P., Sahu, N. P., Phulia, V., Rajesh, M., Fawole, F. J., Pal, A.K. and Angel, G., 2017. Heamato-immunological and physiological responses of *Labeo rohita* fingerlings to dietary fermented *Jatropha curcas* protein concentrate. *Animal Feed Science and Technology*, *232*, pp.198–206.

Shobharani P., Muthukumar S. P., Kizhakayil D. and Halami P. M., 2019. Strain-specific quantification of native probiotic *Bacillus* spp. and their effect on liver function and gut microflora of experimental rats. *Probiotics and Antimicrobial Proteins*, *11*, pp.478–492. 10.1007/s12602-018-9391-1

Silva-Brito, F., Alexandrino, D. A., Jia, Z., Mo, Y., Kijjoa, A., Abreu, H., Carvalho, M. F., Ozório, R. and Magnoni, L., 2021. Fish performance, intestinal bacterial community, digestive function and skin and fillet attributes during cold storage of gilthead seabream (*Sparus aurata*) fed diets supplemented with Gracilaria by-products. *Aquaculture*, *541*, p.736808.

Smriga, S., Sandin, S. A. and Azam, F., 2010. Abundance, diversity, and activity of microbial assemblages associated with coral reef fish guts and feces. *FEMS Microbiology Ecology*, *73*(1), pp.31–42.

Soltani, M., Ghosh, K., Hoseinifar, S. H., Kumar, V., Lymbery, A. J., Roy, S. and Ringø, E., 2019. Genus Bacillus, promising probiotics in aquaculture: Aquatic animal origin, bio-active components, bioremediation and efficacy in fish and shellfish. *Reviews in Fisheries Science & Aquaculture*, *27*(3), pp.331–379.

Song, Z., Ye, W., Tao, Y., Zheng, T., Qiang, J., Li, Y., Liu, W. and Xu, P., 2022. Transcriptome and 16S rRNA analyses reveal that hypoxic stress affects the antioxidant capacity of largemouth bass (*Micropterus salmoides*), resulting in intestinal tissue damage and structural changes in microflora. *Antioxidants*, *12*(1), p.1.

Stein, T., 2005. Bacillus subtilis antibiotics: Structures, syntheses and specific functions. *Molecular Microbiology*, *56*(4), pp.845–857.

Stubbendieck, R. M., and Straight, P. D., 2016. Multifaceted interfaces of bacterial competition. *Journal of Bacteriology*, *198*, pp.2145–2155. doi: 10.1128/JB.00275-16

Sugita, H., Kawasaki, J. and Deguchi, Y., 1997. Production of amylase by the intestinal microflora in cultured freshwater fish. *Letters in Applied Microbiology*, *24*(2), pp.105–108.

Sullam, K. E., Essinger, S. D., Lozupone, C. A., O'connor, M. P., Rosen, G. L., Knight, R. O. B., Kilham, S. S. and Russell, J. A., 2012. Environmental and ecological factors that shape the gut bacterial communities of fish: A meta-analysis. *Molecular Ecology*, *21*(13), pp.3363–3378.

Sun, F., Wang, Y., Wang, C., Zhang, L., Tu, K. and Zheng, Z., 2019. Insights into the intestinal microbiota of several aquatic organisms and association with the surrounding environment. *Aquaculture*, *507*, pp.196–202.

Sylvain, F. É., Cheaib, B., Llewellyn, M., Gabriel Correia, T., Barros Fagundes, D., Luis Val, A. and Derome, N., 2016. pH drop impacts differentially skin and gut microbiota of the Amazonian fish tambaqui (*Colossoma macropomum*). *Scientific Reports*, *6*(1), pp.1–10.

Tamang, J. P., Shin, D. H., Jung, S. J., and Chae, S. W., 2016. Functional properties of microorganisms in fermented foods. *Frontiers in Microbiology*, *7*, p.578. doi: 10.3389/fmicb.2016.00578

Tanasomwang, V. and Muroga, K., 1988. Intestinal microflora of larval and juvenile stages in Japanese flounder (*Paralichthys olivaceus*). *Fish Pathology*, *23*(2), pp.77–83.

Tang, Y., Tao, P., Tan, J., Mu, H., Peng, L., Yang, D., Tong, S. and Chen, L., 2014. Identification of bacterial community composition in freshwater aquaculture system farming of *Litopenaeus vannamei* reveals distinct temperature-driven patterns. *International Journal of Molecular Sciences*, *15*(8), pp.13663–13680.

Tzeng, T. D., Pao, Y. Y., Chen, P.C., Weng, F. C. H., Jean, W. D. and Wang, D., 2015. Effects of host phylogeny and habitats on gut microbiomes of oriental river prawn (*Macrobrachium nipponense*). *PloS one*, *10*(7), p.e0132860.

Wang, A., Meng, D., Hao, Q., Xia, R., Zhang, Q., Ran, C., Yang, Y., Li, D., Liu, W., Zhang, Z. and Zhou, Z.. 2022. Effect of supplementation of solid-state fermentation product of Bacillus subtilis HGcc-1 to high-fat diet on growth, hepatic lipid metabolism, epidermal mucus, gut and liver health and gut microbiota of zebrafish *Aquaculture*, 560, p.738542.

Wang, H., Liu, P., Hu, T. and Chen X., 1993. Study on the intestinal microflora of carp in freshwater culture ponds. *Acta Hydrobiologica Sinica*, *18*, pp.354–359.

Wang, J., Tang, H., Zhang, C., Zhao, Y., Derrien, M., Rocher, E., et al. (2015). Modulation of gut microbiota during probiotic-mediated attenuation of metabolic syndrome in high fat diet-fed mice. *The ISME Journal*, *9*, pp.1–15. doi: 10.1038/ ismej.2014.99

Ward, N. L., Steven, B., Penn, K., Methé, B.A. and Detrich, W. H., 2009. Characterization of the intestinal microbiota of two Antarctic notothenioid fish species. *Extremophiles*, *13*, pp.679–685.

Willmott, M. E., Clements, K. D. and Wells, R. M., 2005. The influence of diet and gastrointestinal fermentation on key enzymes of substrate utilization in marine teleost fishes. *Journal of Experimental Marine Biology and Ecology*, *317*(1), pp.97–108.

Wong, S., Waldrop, T., Summerfelt, S., Davidson, J., Barrows, F., Kenney, P. B., Welch, T., Wiens, G. D., Snekvik, K., Rawls, J. F. and Good, C., 2013. Aquacultured rainbow trout (*Oncorhynchus mykiss*) possess a large core intestinal microbiota that is resistant to variation in diet and rearing density. *Applied and Environmental Microbiology*, *79*(16), pp.4974–4984.

Xiong, J., Wang, K., Wu, J., Qiuqian, L., Yang, K., Qian, Y. and Zhang, D., 2015. Changes in intestinal bacterial communities are closely associated with shrimp disease severity. *Applied Microbiology and Biotechnology*, *99*, pp.6911–6919.

Xiong, J., Zhu, J., Dai, W., Dong, C., Qiu, Q. and Li, C., 2017. Integrating gut microbiota immaturity and disease-discriminatory taxa to diagnose the initiation and severity of shrimp disease. *Environmental Microbiology*, *19*(4), pp.1490–1501.

Yan, M., Li, W., Chen, X., He, Y., Zhang, X. and Gong, H., 2021. A preliminary study of the association between colonization of microorganism on microplastics and intestinal microbiota in shrimp under natural conditions. *Journal of Hazardous Materials*, *408*, p.124882.

Yan, Q., Li, J., Yu, Y., Wang, J., He, Z., Van Nostrand, J. D., Kempher, M.L., Wu, L., Wang, Y., Liao, L. and Li, X., 2016. Environmental filtering decreases with fish development for the assembly of gut microbiota. *Environmental Microbiology*, *18*(12), pp.4739–4754.

Ye, C., Geng, S., Zhang, Y., Qiu, H., Zhou, J., Zeng, Q., Zhao, Y., Wu, D., Yu, G., Gong, H. and Hu, B., 2023. The impact of culture systems on the gut microbiota and gut metabolome of bighead carp (*Hypophthalmichthys nobilis*). *Animal Microbiome*, *5*(1), pp.1–17.

Yukgehnaish, K., Kumar, P., Sivachandran, P., Marimuthu, K., Arshad, A., Paray, B. A. and Arockiaraj, J., 2020. Gut microbiota metagenomics in aquaculture: Factors influencing gut microbiome and its physiological role in fish. *Reviews in Aquaculture*, *12*(3), pp.1903–1927.

Zerdani, I., Faid, M. and Malki, A., 2004. Feather wastes digestion by new isolated strains Bacillus sp. in Morocco. *African Journal of Biotechnology* *3*(1), pp.67–70.

Zhang, H., Sun, Z., Liu, B., Xuan, Y., Jiang, M., Pan, Y., Zhang, Y., Gong, Y., Lu, X., Yu, D. and Kumar, D., 2016b. Dynamic changes of microbial communities in *Litopenaeus vannamei* cultures and the effects of environmental factors. *Aquaculture*, *455*, pp.97–108.

Zhang, J., Zhang, C., Zhu, Y., Li, J. and Li, X., 2018. Biodegradation of seven phthalate esters by *Bacillus mojavensis* B1811. *International Biodeterioration & Biodegradation*, *132*, pp.200–207.

Zhang, M., Sun, Y., Liu, Y., Qiao, F., Chen, L., Liu, W. T., Du, Z. and Li, E., 2016a. Response of gut microbiota to salinity change in two euryhaline aquatic animals with reverse salinity preference. *Aquaculture*, *454*, pp.72–80.

Zhang, Z., Yang, D., Liu, L., Chang, Z. and Peng, N., 2022. Effective gossypol removal from cottonseed meal through optimized solid-state fermentation by Bacillus coagulans. *Microbial Cell Factories*, *21*(1), p.252.

Zhao, R., Symonds, J. E., Walker, S. P., Steiner, K., Carter, C.G., Bowman, J. P. and Nowak, B.F., 2020. Salinity and fish age affect the gut microbiota of farmed Chinook salmon (*Oncorhynchus tshawytscha*). *Aquaculture*, *528*, p.735539.

Zhao, Y., Zhang, W., Xu, W., Mai, K., Zhang, Y. and Liufu, Z., 2012. Effects of potential probiotic Bacillus subtilis T13 on growth, immunity and disease resistance against *Vibrio splendidus* infection in juvenile sea cucumber *Apostichopus japonicus*. *Fish & Shellfish Immunology*, *32*(5), pp.750–755.

Zheng, Y., Yu, M., Liu, J., Qiao, Y., Wang, L., Li, Z., Zhang, X. H. and Yu, M., 2017. Bacterial community associated with healthy and diseased Pacific white shrimp (*Litopenaeus vannamei*) larvae and rearing water across different growth stages. *Frontiers in Microbiology*, *8*, p.1362.

Zhu, Z., Xu, Y. M., Liang, J. H., Huang, W., Chen, J. D., Wu, S. T., Huang, X. H., Huang, Y. H., Zhang, X. Y., Sun, H. Y. and Qin, Q.W., 2023. Relationship of environmental factors in pond water and dynamic changes of gut microbes of sea bass *Lateolabrax japonicus*. *Frontiers in Microbiology*, *14*, p.1086471.

14 Safety of *Bacillus* Probiotics in Aquaculture

Environmental and Ecological Considerations

Mehdi Soltani
University of Tehran, Tehran, Iran
Murdoch University, Perth, Australia

Einar Ringø
UiT The Arctic University of Norway, Tromsø, Norway

Preetham Elumalai
Cochin University of Science and Technology, Kochi, Kerala, India

14.1 INTRODUCTION

With our developed knowledge in the ecology of gastrointestinal microbial and the mechanisms of action of micro-organisms as probiotics, the number of probiotic products for use in nutrition and/or in the rearing water of aquaculture species is increasing. Application of live micro-organisms including *Bacillus* spp. as probiotics is becoming a routine alternative to growth promoters and immune enhancers in aquaculture production. Several *Bacillus* probiotics have been demonstrated as effective tools in improving finfish and shellfish growth performance, increasing disease resistance, and the spread of the enteric pathogens in gastrointestinal tract.

However, despite such attractive and high demand for the use of *Bacillus* probiotics as a potentially viable solution to address the issues of growth, survival, and increasing antibiotic resistance, further studies on the effect, mechanism of action, and safety of probiotics are required to obtain a consistent efficacy and a feasible economic benefit. The effectiveness and potency of a *Bacillus* probiotic in a fish/shellfish is the outcome of the interaction between the host and the probiotic. However, host–probiotic–environment interactions are some issues requiring serious attention. While generally considered safe, there is little indication that probiotics are absolutely safe and it has been agreed that "zero risk does not exist" (Marteau, 2001). The uncertainty would, therefore, always exist about the safety of *Bacillus* probiotics. Researches on the minimum required dose of a particular *Bacillus* probiotic to attain envisioned benefits and maximum dose rate that could be used without any negative side effects on the target fish/shellfish support to ensure the benefits and minimize the risk of the probiotics. Further studies are also required to define whether a *Bacillus* probiotic used in aquatic animal nutrition enters the human food chain and what are its effect on human health status. In addition, data about specific protections regarding handling by susceptible people, for example, immunocompromised humans, could be helpful for a further reduction of risks. The severities of the regulations on the use of spore-forming probiotics (*Bacillus* spp.) in aquaculture vary, even in developed countries, and, therefore, regulation of probiotics must be based on the assessment by scientific and exerted people studying the safety and other aspects such

DOI: 10.1201/9781003503811-14

as efficacy and potency of a particular probiotic micro-organism such as *Bacillus* probiotics. This chapter addresses the pathogenic *Bacillus* bacteria for the aquaculture species as well as the issues concerning the safety of *Bacillus* probiotics in aquaculture.

14.2 PATHOGENIC *BACILLUS*

From the available literature, there are a few reports describing the pathogenesis of *Bacillus* bacteria naturally affecting fish and shellfish. Furthermore, no report has been published on the experimental pathogenicity of these *Bacillus* spore formers in aquatic animals so far. Goodwin et al. (1994) reported *B. mycoides* as the causative agent of a superficial epizootic disease in commercial channel catfish in Alabama, USA. The affected fish had pale areas or ulcers on the dorsal area and focal necrosis of epaxial muscle with chains of Gram-positive bacilli identified as *B. mycoides*. When healthy catfish were subjected to the isolated bacterium either intramuscularly or subcutaneously at 1.6×10^4 cfu/fish, the development of lesions resembled those in natural epizootic was seen. A possible explanation for the congestion and lack of bleeding seen in the affected fish could be due to toxin products by these bacilli bacteria, for example, *B. cereus* that are able to cause a disseminated intravascular coagulation.

Wang et al. (2000) reported a new bacterial white spot syndrome caused by *B. subtilis* in cultured tiger shrimp *P. monodon* in Malaysian shrimp farming. The affected shrimp showed white spots similar to symptoms caused by white spot viral disease (WSVD), but the affected shrimp was active and grew normally with no significant morbidity and mortality. The appeared white spots were lichen like with the puncture centers unlike the melanized dots in the white spots caused by WSVD. Microscopically, a degeneration and discoloration of the cuticle of the epicuticle and underlying cuticular layers were detected. It was suggested that such disease may be associated with the regular use of *B. subtilis* probiotic in shrimp ponds. *B. subtilis* can excrete enzymes of protease, amylase, glucanase, and lipase (Shady, 1997) suggesting that the bacterium's ability to lyse the shrimp cuticle composition, that is, chitin, calcium carbonate, and lipid (Branson, 1993).

In an experimental study by Sineva et al. (2009), the expression of *B. cereus* hemolysin II in *B. subtilis* rendered the bacterium being pathogenic for the crustacean *Daphnia magna* when challenged with the expressed *B. subtilis* at 0^4–10^6 cfu L^{-1} at water temperature 20°C ± 5°C. The lethal concentrations 50% (LC_{50}) on the fifth day of the experiment for the expressed *B. subtilis* and *B. cereus* were 5.4×10^5 and 4.5×10^5 cfu mL^{-1}, respectively.

More recently, a new bacterial white patch disease caused by *B. cereus* has been reported from different *L. vannamei* aquaculture farms industry in India in 2015 (Velmurugan et al., 2015). The disease caused continuous morbidity and mortality, and the affected shrimps showed white opaque patches in the carapace, necrosis, whitish blue coloration, loss of appetite, and pale white muscles. A mortality of up to 70% was observed within 3–5 days at the acute stage of the disease outbreak. The isolated *B. cereus* strains represented high virulence factors including hemolytic and lipase activities, and mortality of up to 100% occurred in *L. vannamei* and *Artemia franciscana* after challenging them with the bacterium intraperitoneally at 10^6 cfu/shrimp or via bath at 10^8 cfu/mL in both shrimp and *Artemia* at unknown water temperature. *B. cereus* is capable of proliferating in a wide range of environments including soils, clays, sediment, dust, mineral water, and processed foods and is able to secrete protease, amylase, glucanase, and lipase (Hendriksen et al., 2006), thus, it may be able to invade the suppressed aquatic animals under an adverse environmental condition, which is suitable for expressing the virulence genes of the bacterium (Andreeva et al., 2007).

14.3 SAFETY OF *BACILLUS*

The possibility of *Bacillus* probiotics used in aquatic animal feed or the rearing water entering the human food chain should be considered an important public health issue. However, there are no data available related to the risk of human food contamination with *Bacillus* probiotics used in aquatic animals. Transfer of an antibiotic resistance via the presence of transmissible antibiotic resistance genes in some probiotic bacteria is one of the critical risk issues. Also, infections from the probiotics and existence of enterotoxins and emetic toxins in probiotic bacteria are other important risks that require serious attention in the aquaculture industry. However, almost all published data relating to *Bacillus* probiotics in aquaculture are associated with their efficacy rather than the safety. It is also worth saying that safety evaluation and data on a specific *Bacillus* strain probiotic must not be considered common to similar probiotic strains because the safety and risk assessment of each probiotic should be considered based on a case-by-case basis. The severity of a negative effect by a particular probiotic is associated with the level of susceptibility to immune-physiological conditions of the target aquatic animal, for example, early hatched fish larvae. Therefore, it is possible that a *Bacillus* probiotic strain is considered as a safe under particular conditions but it is unsafe under other conditions. Like antibiotics, it is feasible to say that there is no specific probiotic to be regarded as 100% safe. Also, the existence of unwanted/contaminated bacteria or their toxic substances in the form of a probiotic can be also considered as another important safety and quality issue. Sometimes such contaminants may be a more significant issue than the specific quality of the probiotics. At the present time, probiotics such as *Bacillus* species used in aquatic animal feed or in their water cultures are generally considered safe. However, some of the bacterial probiotic species potentiate some risks of transmission of antibiotic resistance to some of pathogenic bacteria or production of some their enterotoxins (Anadón et al., 2006).

14.3.1 Common Risks Associated with the Use of *Bacillus* Asprobiotics in Aquaculture

Despite the safety of *Bacillus* probiotics in water or in the feed of aquatic animals, it is worth to say that protection of human, animals, and the environment from potentially unsafe probiotic bacteria is an essential issue.. Generally, the risks associated with the use of *Bacillus* probiotics in aquaculture sector can be summarized as follows (Marteau, 2001; FAO/WHO, 2002; Doron & Snydman, 2015): (a) transfer of antibiotic resistance from the probiotics to other pathogenic bacteria; (b) gastrointestinal or systemic infection of the target animal fed with the probiotic; (c) gastrointestinal or systemic infection of the consumers of the animal products produced by animals fed with the probiotics; (d) gastrointestinal or systemic infection of the handlers of animal or aquatic animal feed; (e) release of infectious bacteria or their toxic substances to the environment from the animal production system; (f) sensitization of the external tissues such as skin, eye, and mucus membrane in the handlers of the probiotics; (g) toxic effects in the host due to the production of toxins by the bacteria contaminated in the probiotics; and (h) hypersensitivity of the immune system reactions in susceptible hosts. In other words, any particular *Bacillus* species planned to be used in the aquatic animals feed or as the bioremediation of their rearing water quality must be carefully tested according to the guidelines described by QPS status (EFSA, 2007). It is crucially important that the *Bacillus* spp. that does not meet the requirements of QPS status must be investigated for detailed safety assessment criteria prior to be considered as a market product (EFSA, 2007). The safety evaluation of a *Bacillus* species to detect QPS status is generally a basis for a *Bacillus* probiotic, and, at the moment, the following QPS measures are the most recommended criteria (EFSA, 2007): (i) harboring transferable antibiotic resistance gene; (ii) identification to the strain level; (iii) evidence of infectious diseases; (iv) possible cause of excessive stimulation of the immune system; (v) possible negative effect on the growth performance of the target animal; (vi) possible suppressing of the target animal immune status; and (vii) possible co-infection with other potentially pathogenic micro-organisms, for example, toxin production. Among a list of more than 100 species of micro-organisms of Gram-positive non-sporulating bacteria, (i) *Bacillus* species, (ii) yeasts, and (iii) filamentous fungi, 13 of them belonged to the *Bacillus* spp. including: *B. amyloliquefaciens, B. atrophaeus, B, clausii B. coagulans, B. fusiformis, B. lentus, B. Licheniformis, B. megaterium, B. mojavensis, B. pumilus, B. Subtilis, B. Vallismortis*, and *Geobacillus stearothermophillus* (EFSA, 2007).

14.3.2 The Possible Risk Assessment

Assessment of *Bacillus* probiotics in aquatic animal diets or in their rearing water against the potential risks is a significant issue. The *Bacillus* used as probiotics need to be recognized to strain level with any infection in humans and aquatic animals. Also, such probiotics must not be the carrier of transferable antibiotic resistance genes. *Bacillus* probiotics that are able to produce toxins or cause hypersensitivity reactions in the target host are not suitable for probiotics. Since 2007, the European Food Safety Authority (EFSA) has been using a concept [Qualified Presumption of Safety, QPS), European approach for the assessment of the safety of probiotics] as a generic risk assessment tool to assess the safety of a micro-organism such as *Bacillus* bacteria (EFSA, 2007). As these spore-forming bacteria are becoming popular as promising probiotics for use in aquatic animal feed or in their rearing water because of their tolerance to fluctuations in the water temperatures that make them easier to handle during the manufacture, storage, and transportation of feed. For instance, a number of 13 *Bacillus* species including *B. subtilis, B. amyloliquefaciens, B. licheniformis, B. coagulans*, and *B. megaterium* have been recognized by the EFSA to be used as probiotics for animal feed including fish and shellfish (EFSA, 2013).

14.3.3 Safety of Commonly Used *Bacillus* as Probiotics

It is notable that the application of *Bacillus* probiotics as the spore-forming bacteria is not risk-free, as some *Bacillus* species including *B. anthracis, B. cereus*, and *B. thuringiensis* are pathogenic bacteria in animals and humans (Damgaard et al., 1997; Hernandez et al., 1998; Little & Ivins, 1999; Kotiranta et al., 2000; Raymond et al., 2010). Despite well-known data about the pathogenicity of some of these *Bacillus* bacteria such as *B. anthracis* and *B. cereus*, no evidence for the pathogenesis of other *Bacillus* spp. in terrestrial animals including human beings.

The production of emetic toxin cereulide and enterotoxins hemolysin BL and non-hemolytic enterotoxin and cytotoxin K by *B. cereus* causes a serious illness in humans (Schoeni & Wong, 2005). From 333 strains from various species of *Bacillus tested by* From et al. (2005), eight *Bacillus* strains belonging to *B. subtilis, B. mojavensis, B. pumilus*, and *B. fusiformis* were able to produce cytotoxic and emetic toxins. Also, some *Bacillus* species, such as *B. cereus*, have been reported to cause mastitis in cattle (Parkinson et al., 1999) and *B. licheniformis* was associated with abortion in cattle (Agerholm et al., 1997).

14.3.4 Antibiotic Resistance in *Bacillus*

The transfer of genes associated with antibiotic resistance to some pathogenic bacteria is now one of the crucial issues associated with probiotics. This is because many bacterial species currently used as probiotics are capable of harboring transferable antibiotic resistance genes. Hence, it is very important to pay attention to the quality assurance criteria when some microbes such as *Bacillus* are used as probiotics; and only bacteria with proven absence of transferable antibiotic resistance genes be considered as the potential probiotic. Like *Lactobacillus*, *Bacillus* bacteria present greater risk of transfer antibiotic resistance genes, as many species of these genera have transferable antibiotic resistance genes. However, the status of antibiotic resistance genes in many microbial strains such as *Bacillus* spp. used as probiotics has not been determined. The presence of antibiotic resistance genes may not be a serious issue if such genes are intrinsic in chromosomes and not transferable (Bajagai et al., 2016). However, a precaution should be taken to avoid *Bacillus* sp. with acquired genes being used as probiotics.

This is because the antibiotic resistance has frequently been demonstrated in some *Bacillus* species. For instance, *B. subtilis*, a frequently used probiotic in aquaculture, can harbor conjugative transposons (e.g., Tn5397), which can transfer resistance to tetracycline encoded by the *tet (M)* gene (Mullany et al., 1990; Roberts et al., 1999). Phelan et al. (2011) reported another transferable tetracycline resistance gene, *tet (L)*, in a *Bacillus* sp. encoded by a plasmid. Macrolides are a very important class of antibiotics widely used to control human and animal infections, and Monod et al. (1986) demonstrated that the *B. subtilis* contains macrolide-lincosamide-streptogramin B (MLS) resistance determinants on the plasmid. These authors illustrated that MLS determinant is homologous to the *erm(C)* gene, one of the 19 analogous *erm* resistance genes. Gryczan et al. (1984) found that the most prevalent antibiotic resistance gene is *erm(D)* that encodes the determinants for the resistance to MLS, although the transferability of the determinants encoded by this gene has not been confirmed so (EFSA, 2007).

14.3.5 *Bacillus* Food Poisoning

Minimum attention has been paid to the food poisoning of *Bacillus* spp. via aquatic animals' carcasses so far. However, some *Bacillus* such as *B. cereus* can cause some distinct types of food poisoning in humans via consumption of contaminated aquatic animal products. The signs of diarrhea and abdominal pain can occur 8–16 hours after consumption of the contaminated fish and shellfish meats (Turnbull, 1996). The symptoms of food poisoning caused by other *Bacillus* species such as *B. subtilis* and *B. licheniformis* are less well defined in humans, although symptoms of diarrhea and nausea have been reported 1–14 hours after consumption of the contaminated fish. The food poisoning episode by the *Bacillus* spp. usually occurs due to spore's survival during uncompleted cooking or pasteurization which can result in germination and multiplication when the food is inadequately refrigerated. The symptoms of *B. cereus* food poisoning are caused by a toxin or toxins produced in the food during this multiplication (Turnbull, 1996).

14.3.6 Clinical Manifestations in the Contaminated Consumers

Apart the *B. anthrax* as the best-known *Bacillus* disease, in recent years, the role of other *Bacillus* species has been increasingly implicated in a wide range of human infections such as abscesses, bacteremia, septicemia, wound and burn infections, ear infections, endocarditis, meningitis, ophthalmitis, osteomyelitis, peritonitis, and respiratory and urinary tract infections (Turnbull, 1996). Most of these occur as secondary or mixed infections or immune-deficient or otherwise immunocompromised hosts, for example, alcoholic and diabetic humans. However, a remarkable proportion is primary infections in healthy individuals. Some of these infections are severe or lethal. The most frequently *Bacillus* species infecting the consumers are: *B cereus*, followed by *B licheniformis* and *B subtilis*. Also, there are some occasional reports of infecting of the human consumers by *B. alvei, B. brevis, B. circulans, B. coagulans, B. macerans, B. pumilus, B. sphaericus*, and *B thuringiensis* (Turnbull, 1996). As the secondary invaders, *Bacillus* species may intensify the previous infections by producing either tissue-damaging toxins or metabolites such as penicillinase that interfere with treatment (Turnbull, 1996).

14.4 CONCLUSION

Based on our knowledge and the available data, the application of probiotics including *Bacillus* spp. is not without risk. The use of *Bacillus* probiotics could be responsible for a range of risks in aquatic animal health, the environment, and human health, varying from mild to serious reactions and life-threatening infections. Also, the safety data of one particular strain of a *Bacillus* probiotic is not applicable to other closely related one; and the existing data are insufficient to assert a 100% safety for a group strains of *Bacillus* species as a probiotic. The hazard evaluation of a particular *Bacillus* sp. Is, therefore, highly recommended on a case-by-case basis. There are minimum information concerning the pathogenicity of *Bacillus* bacteria particularly those used as probiotics in aquaculture species. The available data also required further studies as there is a lack of data regarding some pathogenesis aspects in the target fish/shellfish. Also, besides existing reports on some *Bacillus* such as *B. subtilis, B. mycoides*, and *B. cereus* as the pathogen in either fish or shrimp, no data about other species of *Bacillus* are available so far.

Although the safety *Bacillus* species is based on the detection of the absence of enterotoxins and emetic toxins, it is very important to note that some *Bacillus* bacteria such as *B. anthracis, B. cereus*, and *B. thuringiensis* are serious pathogens in humans and animals (Damgaard et al., 1997; Hernandez et al., 1998; Little & Ivins, 1999; Kotiranta et al., 2000.; Raymond et al., 2010). Despite good information available about the pathogenesis of *B. anthracis* and *B. cereus*, no evidence of pathogenic effects for other endospore-forming bacteria is available. The emetic toxin (cereulide), enterotoxins hemolysin, non-hemolytic enterotoxin, and cytotoxin products by *B. cereus* are well-known toxins that can affect humans (Granum & Lund, 1997; Schoeni & Wong, 2005). Also, some strains of *B. subtilis, B. mojavensis, B. pumilus, B. licheniformis*, and *B. fusiformis* are able to produce emetic toxins and cytotoxins even causing some disorders in terrestrial animals, but no data available regarding aquatic animals (Agerholm et al., 1997; Parkinson et al., 1999; From et al. 2005). Additionally, the reports of antibiotic resistance in some *Bacillus* species such as *B. subtilis* (Mullany et al., 1990; Roberts et al., 1999; Phelan et al., 2011) have raised the safety and risk issues concerning the application of *Bacillus* probiotics in aquaculture, particularly if they are applied as the bioremediation in the rearing waters where the risks of bacterial resistance can considerably increase. Additionally, the reports of antibiotic resistance in some *Bacillus* species such as B. subtilis show that this bacterium carriers conjugative transposons which is able to transfer resistance to some antibiotics encoded by some genes (Mullany et al., 1990; Roberts et al., 1999; Phelan et al., 2011). Therefore, establishing identity, existing knowledge, possible pathogenicity, and intended end-use are some required QPS for the assessment of a *Bacillus* sp. as a probiotic.

BIBLIOGRAPHY

Agerholm J, Willadsen C, Nielsen TK, Giese SB, Holm E, Jensen L, Agger J (1997) Diagnostic studies of abortion in Danish dairy herds. *Journal of Veterinary Medicine*, 44, pp.551–558.

Anadón A, Martínez-Larrañaga MR, Martínez MA (2006) Probiotics for animal nutrition in the European Union. Regulation and safety assessment. *Regulatory Toxicology and Pharmacology*, 45, pp.91–95.

Andreeva ZI, Nesterenko VF, Fomkina MG, Ternovsky VI, Suzina NE, Bakulina AY, Solonin AS, Sineva EV (2007) The properties of Bacillus cereus hemolysin II pores depend on environmental conditions. *Biochemistry and Biophysics Acta* 1768, pp.253–263.

Bajagai YS, Klieve AV, Dart PJ, Bryden WL (2016) Probiotics in Animal Nutrition:production, impact and regulation. In: Harinder P.S. Makkar (editor), FAO Animal Production and Health Paper No. 179. Rome, p.92.

Branson E (1993) Basic anatomy and physiology. In: Brown L (ed.) *Aquaculture for veterinarians, fish husbandry and medicine*. Pergamon Press, Oxford, pp.1–30.

Damgaard PH, Granum PE, Bresciani J, Torregrossa MV, Eilenberg J, Valentino L (1997). Characterization of *Bacillus thuringiensis* isolated from infections in burn wounds. *FEMS Immunological Medical Microbiology*, 18(1): pp.47–53.

Doron S, Snydman DR (2015) Risk and safety of probiotics. *Clinical Infectious Diseases*, 60 (suppl. 2): pp.S129–S134.

EFSA [European Food Safety Authority]. (2007) Opinion of the scientific committee on a request from EFSA on the introduction of a qualified presumption of safety (QPS) approach for assessment of selected micro-organisms referred to EFSA. *The EFSA Journal*, 587, pp.1–16.

EFSA (2013) Scientific opinion on the maintenance of the list of QPS biological agents intentionally added to food and feed (2013 update). Safety, QPS, bacteria, yeast, fungi, virus. First published in the EFSA Journal: 14 November 2013, Last Updated: 12 September 2017. DOI:10.2903/j.efsa.2013.3449

FAO/WHO (2002) Guidelines for the evaluation of probiotics in food. http://www.fda.gov/ohrms/dockets/dockets/95s0316/95s-0316-rpt0282-tab-03-ref-19-joint-faowho-vol219.pdf

From C, Pukall R, Schumann P, Hormazábal V, Granum PE (2005) Toxin-producing ability among *Bacillus* spp. outside the *Bacillus cereus* group. *Applied and Environmental Microbiology*, 71(3): pp.1178–1183.

Goodwin AE, Spencer, Roy J Jr, Grizzle JM, Goldsby MTJ (1994) *Bacillus mycoides*: a bacterial pathogen of channel catfish. *Diseases of Aquatic Organisms*, 18, pp.173–179.

Granum PE, Lund T (1997) *Bacillus cereus* and its food poisoning toxins. *FEMS Microbiology Letters*, 157(2): pp.223–228.

Gryczan T, Israeli-Reches M, Del Bue M, Dubnau D (1984) DNA sequence and regulation of ermD, a macrolide-lincosamide-streptogramin B resistance element from *Bacillus licheniformis*. *Molecular and General Genetics*, 194(3): pp.349–356.

Hendriksen, NB, Hansen, BM, Johansen, JE (2006) Occurrence and pathogenic potential of *Bacillus cereus* group bacteria in a sandy loam. *Antonie Van Leeuwenhoek*, 89, pp.239–249.

Hernandez E, Ramisse F, Ducoureau JP, Cruel T, Cavallo J-D (1998) *Bacillus thuringiensis* subsp. *konkukian* (serotype H34) superinfection: case report and experimental evidence of pathogenicity in immunosuppressed mice. *Journal of Clinical Microbiology*, 36(7): pp.2138–2139.

Kotiranta A, Lounatmaa K, Haapasalo M (2000). Epidemiology and pathogenesis of *Bacillus cereus* infections. *Microbial Infection*, 2(2): pp.189–198.

Little SF, Ivins BE (1999) Molecular pathogenesis of *Bacillus anthracis* infection. *Microbial Infection*, 1(2): pp.131–139.

Marteau P (2001) Safety aspects of probiotic products. *Scandinavian Journal of Nutrition*, 45, pp.22–24.

Monod M, DeNoya C, Dubnau D (1986) Sequence and properties of pIM13, a macrolide-lincosamide-streptogramin B resistance plasmid from *Bacillus subtilis*. *Journal of Bacteriology*, 167(1): pp.138–147.

Mullany P, Wilks M, Lamb I, Clayton C, Wren B, Tabaqchali S (1990) Genetic analysis of a tetracycline resistance element from *Clostridium difficile* and its conjugal transfer to and from *Bacillus subtilis*. *Journal of General Microbiology*, 136(7): pp.1343–1349.

Raymond B, Johnston PR, Nielsen-LeRoux C, Lereclus D, Crickmore N (2010) *Bacillus thuringiensis*: an impotent pathogen? *Trends in Microbiology*, 18(5): pp.189–194.

Roberts AP, Pratten J, Wilson M, Mullany P (1999) Transfer of a conjugative transposon, Tn5397 in a model oral biofilm. *FEMS Microbiology Letters*, 177(1): pp.63–66.

Parkinson TJ, Merrall M, Fenwick SG (1999) A case of bovine mastitis caused by *Bacillus cereus*. *New Zealand Veterinary Journal*, 47(4): pp.151–152.

Phelan RW, Clarke C, Morrissey JP, Dobson AD, O'Gara F, Barbosa TM (2011) Tetracycline resistance-encoding plasmid from *Bacillus* sp. strain# 24, isolated from the marine sponge *Haliclona simulans*. *Applied and Environmental Microbiology*, 77(1): pp.327–329.

Schoeni JL, Lee Wong AC (2005) *Bacillus cereus* food poisoning and its toxins. *Journal of Food Protection*, 68(3): pp.636–648.

Sineva EV, Andreeva-Kovalevskaya ZI, Shadrin AM, Gerasimov YL, Ternovsky VI, Teplova VV et al. (2009) Expression of *Bacillus cereus* hemolysin II in *Bacillus subtilis* renders the bacteria pathogenic for the crustacean *Daphnia magna*. *FEMS Microbiology Letters*, 299(1): pp.110–119. DOI:10.1111/j.1574-6968.2009.01742.x

Turnbull PCB (1996) Chapter 15 *Bacillus*, In: Baron S (editor) *Medical Microbiology*. 4th edition, Galveston (TX): University of Texas Medical Branch at Galveston.

Velmurugan S, Palanikumar P, Velayuthani P, Donio MBS, Michael Babu M, Lelin C, et al. (2015) Bacterial white patch disease caused by *Bacillus cereus*, a new emerging disease in semi-intensive culture of *Litopenaeus vannamei*. *Aquaculture* 444, pp.49–54.

Wang YG, Lee KL, Najiah M, Shariff M, Hassan MD (2000) A new bacterial white spot syndrome (BWSS) in cultured tiger shrimp *Penaeus monodon* and its comparison with white spot syndrome (WSS) caused by virus. *Diseases of Aquatic Organisms* 41, pp.9–18.

Index

Pages in **bold** refer to tables.

A

B

C

D

E

M

N

O

P

9781032822730